AF389449

Effects of Plants' Ingredients on Dough and Final Product

Effects of Plants' Ingredients on Dough and Final Product

Editor

Silvia Mironeasa

MDPI • Basel • Beijing • Wuhan • Barcelona • Belgrade • Manchester • Tokyo • Cluj • Tianjin

Editor
Silvia Mironeasa
Ştefan cel Mare University of
Suceava
Romania

Editorial Office
MDPI
St. Alban-Anlage 66
4052 Basel, Switzerland

This is a reprint of articles from the Special Issue published online in the open access journal *Applied Sciences* (ISSN 2076-3417) (available at: https://www.mdpi.com/journal/applsci/special_issues/Plants_Ingredients).

For citation purposes, cite each article independently as indicated on the article page online and as indicated below:

LastName, A.A.; LastName, B.B.; LastName, C.C. Article Title. *Journal Name* **Year**, *Volume Number*, Page Range.

ISBN 978-3-0365-3733-7 (Hbk)
ISBN 978-3-0365-3734-4 (PDF)

Contents

About the Editor

Silvia Mironeasa (Professor Habilitate, PhD. Eng.) joined the Faculty of Food Engineering, "Ştefan cel Mare" University of Suceava, Romania, in 2008, where she teaches different courses and applications in the food product engineering field. She holds a PhD in Materials Science and Engineering from 2004 and became a PhD supervisor in Food Engineering in 2017, when she sustained her habilitation thesis. Silvia Mironeasa has expertise in food engineering, food processing, food quality analysis, in the design of experiments, and data analysis. Her research activities focus on enhancing raw materials' characteristics by applying various physical treatments, such as grinding, heat treatment, etc., and by valorizing vegetable byproducts to improve the nutritional profile and the quality attributes of various foods, such as baked goods, pasta, etc. She has been involved in 31 interdisciplinary research projects, published more than 130 scientific papers, and is author of 29 patents under evaluation and 6 published patents.

applied sciences

Editorial

Current Approaches in Using Plant Ingredients to Diversify Range of Bakery and Pasta Products

Silvia Mironeasa

Faculty of Food Engineering, Ștefan cel Mare University of Suceava, 13 Universitatii Street, 720229 Suceava, Romania; silviam@fia.usv.ro

Citation: Mironeasa, S. Current Approaches in Using Plant Ingredients to Diversify Range of Bakery and Pasta Products. *Appl. Sci.* **2022**, *12*, 2794. https://doi.org/10.3390/app12062794

Received: 22 February 2022
Accepted: 7 March 2022
Published: 9 March 2022

Publisher's Note: MDPI stays neutral with regard to jurisdictional claims in published maps and institutional affiliations.

There is a growing interest in the industry to manufacture food products containing health-promoting nutrients and to prevent nutrition-related disorders. The addition of plant ingredients represents a valuable strategy to improve food nutritional properties, contributing to reducing nutrient deficiencies. The use of natural ingredients, composite and gluten-free flours or applying physical treatments to raw materials in the manufacture of bakery and pasta products represent a common practice. Plant ingredients encompassing polysaccharides and fibers, prebiotics and probiotics, lipids, vitamins and minerals, and antioxidants added in food recipes are of interest for both researchers and those working in the food industry. Dough, the intermediate product of baked goods and pasta, is typically made from wheat. Wheat dough presents unique and particular rheological properties that enable the bulk to be stretched and deform without breaking; it is elastic and tenacious, and able to maintain its assigned shape even when subjected to physical stress. Wheat flour is a versatile raw material due to the variable quality of wheat and the small amount of improvements needed to be added in order to obtain the desired bread quality. In this Special Issue, Vartolomei and Turtoi [1] considered the use of rosehip powder as a natural alternative to synthetic ascorbic acid in breadmaking. As stated by these authors, wheat bread with rosehip powder presented an enhanced physicochemical profile, with the volume and porosity being higher than wheat bread, while the elasticity decreased slightly. Some researchers have approached the enrichment of wheat dough bread with plant-based ingredients of high nutritional profile, such as legume flours, which contain valuable proteins related to amino acid composition, complementary to that of cereals. As an example, soybeans and lentils have been subjected to convenient processing techniques such as germination to enhance the nutritional profile of the legumes, as well their sensory profile, as investigated in this Special Issue by Ungureanu-Iuga et al. [2]. As observed in this work, the simultaneous incorporation of these germinated legume flours in wheat flour affected dough behavior during mixing, extension, and fermentation. Baked goods with high nutritive value and minimum impairment of quality can be accomplished due to the germination process.

The enrichment of bakery and pasta products with bioactive compounds and associated fibers, along with the increase in starch resistance to digestion, have gained noticeable importance, and can be considered a valuable approach regarding the enhancement of human health through the diet, knowing that starch possess a high glycemic index. One way to reduce the glycemic index of starch-based food consists of modifying flour by hydrothermal processing, which determines the increase in slowly digestible starch and resistant starch content. Another way to reduce the digestibility of starch is the supplementation of food with phenolic compounds from fruit by-products. In this Special Issue, Iuga et al. [3] provide an in-depth assessment of starch digestibility as affected simultaneously by the parameters of hydrothermal treatment and by the addition of grape peels. Polyphenolic compounds' associated fibers from grape peels and hydrothermal treatment exhibited a synergistic effect on wheat dough and pasta characteristics, reducing resistant starch content and rising fiber content, total polyphenols and antioxidant activity. At the same time, a clean label product was generated by using food industry waste and applying an eco-friendly treatment.

The use of whole flour from cereals or pseudocereals in bread making has become a concern in the bakery industry. One of the reasons for their use is related to their nutritional profile which encompasses fiber components, especially beta-glucans and arabinoxylans, as well as bioactive substances accompanying dietary fiber. Due to the fact that most types of whole flours have unfavorable baking and sensory characteristics, many research groups have focused on various milling technologies in order to eliminate this inconvenience. In this sense, Skřivan et al. [4] presented a special disintegration equipment and evaluated the impact of novel grinding technology on microstructure and properties of the flours. According to them, this technology is suitable for achieving fine granulation of bran particles and to does not result in very substantial damage of starch granules, especially in rye and wheat.

Currently, it is a challenge to produce flour of different particle sizes with high concentrations of components such as proteins, fibers, minerals, etc. Moreover, the milling fraction varies in physical, chemical and functional–technological properties, and offers the possibility to produce flour with specific characteristics for diverse uses in food products. One the other hand, pseudocereals have gained interest in the bakery industry due to their nutritional value, being richer in fiber, lipids, minerals and vitamins than wheat. In addition, due to their high protein quality, the supplementation of wheat flour with pseudocereals solves the wheat protein deficiency in lysine, threonine and methionine. However, the replacement of wheat flour with pseudocereals impacts dough and bread quality, since these grains do not contain gluten. This aspect is approached by Coțovanu and Mironeasa [5] in an extensive study related to the substitution of wheat flour with different milling fractions of buckwheat seed flour. The results showed that the dough's rheological properties were affected in various ways, depending on the amount and particle size of buckwheat flour, thus influencing the quality of bread.

A current trend regarding bakery products includes the diversification of the assortment range to satisfy consumer demands. Some consumers want baked goods without sodium or with a low sodium content, while others want gluten-free products. In this Special Issue, these aspects were approached. The review written by Codină et al. [6] presents sodium chloride effects on bread making from technological and sensory points of view, revealing different options for salt reduction in foods, focusing on bakery products. Moreover, the physiological role of sodium chloride, its effect on the human body and legislative recommendations on its consumption, were highlighted. Various gluten-free raw materials can be used to replace wheat flour in the production of bakery products, satisfying consumers needs for a gluten-free diet. The performance of different grains in gluten-free bread applications was investigated by Banu and Aprodu [7]. Their work presented a comparative analysis of quinoa, sorghum, millet and rice flours and breads in terms of proximate composition, resistant starch, antioxidant activity and total phenolic content. As observed in the work, the quinoa, sorghum and millet may represent reasonable alternatives for developing functional gluten-free bread due their huge potential. The use of gluten-free raw material in bread making represents a challenge, since in the absence of gluten, dough presents poor viscoelastic properties, being unable to develop a protein network, a fact that impacts gluten-free bread quality. By including some alternative ingredients such as hydrocolloids, enzymes, emulsifier, and fibers or fruit and vegetable by-products in the formulation of gluten-free bread, the viscoelastic properties can be improved, as well the nutritional and sensory profile. In this regard, Djeghim et al. [8] found that different types of fruits and vegetable by-products could be used as economical and inexpensive methods to improve gluten-free bread, enhancing the quality characteristics up to a certain addition level, depending on the by-product type.

Baked goods may have a low protein content due to the reduced amino acid profile possessed by cereals. The use of ingredients from oil seeds has been approached in many studies over recent years due to the fact they have higher protein content than cereals.

Moreover, they are rich in omega-6 and omega-9 essential fatty acids, fiber, and natural antioxidant compounds, comprising tocopherol, beta-carotene, chlorogenic acid, caffeic acid and flavonoids. The oil seeds can be added as raw or processed material in the food formulation products. As an example, the use of roasted flaxseed flour in biscuits with partial substitution of wheat flour has been reported to result in promising healthy and nutritious alternative to consumers, as observed by Man et al. [9]. According to them, the replacement of wheat flour with 25% roasted flaxseed flour has improved the manufactured biscuits from nutritional, textural and sensorial points of view, without affecting their aftertaste. In another study published in this Special Issue, Szydłowska-Czerniak et al. [10] investigated the possibility of manufacturing biscuits with new and attractive characteristics for the consumer, with improved antioxidant potential. According to their results, by the incorporation of rapeseed press cake flour and the replacement of margarine by rapeseed oil in the formula of wheat flour-based biscuits, the nutritional quality of biscuits was enhanced in terms of antioxidant activity. Nevertheless, these biscuits may result in a critical loss of consumers' acceptance and purchase intent, with the transferal of this ingredient incorporation to the industry level and the launching of these biscuits to the market being unfeasible.

The works included in this Special Issue underline the current tendencies in the use of plants' ingredients to improve the technological, nutritional, and sensory properties of bakery and pasta products. Moreover, the employment of some physical treatments, the optimization of the amount added, and the characteristics of such plant ingredients used to diversify the range of bakery and pasta products were highlighted.

Funding: This research received no external funding.

Acknowledgments: I want to thank all the authors and peer reviewers for their valuable contributions to this Special Issue.

Conflicts of Interest: The authors declare no conflict of interest.

References

1. Vartolomei, N.; Turtoi, M. The Influence of the Addition of Rosehip Powder to Wheat Flour on the Dough Farinographic Properties and Bread Physico-Chemical Characteristics. *Appl. Sci.* **2021**, *11*, 12035. [CrossRef]
2. Ungureanu-Iuga, M.; Atudorei, D.; Codină, G.G.; Mironeasa, S. Rheological Approaches of Wheat Flour Dough Enriched with Germinated Soybean and Lentil. *Appl. Sci.* **2021**, *11*, 11706. [CrossRef]
3. Iuga, M.; Batariuc, A.; Mironeasa, S. Synergistic Effects of Heat-Moisture Treatment Regime and Grape Peels Addition on Wheat Dough and Pasta Features. *Appl. Sci.* **2021**, *11*, 5403. [CrossRef]
4. Skřivan, P.; Sluková, M.; Jurkaninová, L.; Švec, I. Preliminary Investigations on the Use of a New Milling Technology for Obtaining Wholemeal Flours. *Appl. Sci.* **2021**, *11*, 6138. [CrossRef]
5. Coţovanu, I.; Mironeasa, S. Buckwheat Seeds: Impact of Milling Fractions and Addition Level on Wheat Bread Dough Rheology. *Appl. Sci.* **2021**, *11*, 1731. [CrossRef]
6. Codină, G.G.; Voinea, A.; Dabija, A. Strategies for Reducing Sodium Intake in Bakery Products, a Review. *Appl. Sci.* **2021**, *11*, 3093. [CrossRef]
7. Banu, I.; Aprodu, I. Assessing the Performance of Different Grains in Gluten-Free Bread Applications. *Appl. Sci.* **2020**, *10*, 8772. [CrossRef]
8. Djeghim, F.; Bourekoua, H.; Różyło, R.; Bieńczak, A.; Tanaś, W.; Zidoune, M.N. Effect of By-Products from Selected Fruits and Vegetables on Gluten-Free Dough Rheology and Bread Properties. *Appl. Sci.* **2021**, *11*, 4605. [CrossRef]
9. Man, S.M.; Stan, L.; Păucean, A.; Chiş, M.S.; Mureşan, V.; Socaci, S.A.; Pop, A.; Muste, S. Nutritional, Sensory, Texture Properties and Volatile Compounds Profile of Biscuits with Roasted Flaxseed Flour Partially Substituting for Wheat Flour. *Appl. Sci.* **2021**, *11*, 4791. [CrossRef]
10. Szydłowska-Czerniak, A.; Poliński, S.; Momot, M. Optimization of Ingredients for Biscuits Enriched with Rapeseed Press Cake—Changes in Their Antioxidant and Sensory Properties. *Appl. Sci.* **2021**, *11*, 1558. [CrossRef]

applied
sciences

Article

The Influence of the Addition of Rosehip Powder to Wheat Flour on the Dough Farinographic Properties and Bread Physico-Chemical Characteristics

Nicoleta Vartolomei and Maria Turtoi *

Cross-Border Faculty, Dunarea de Jos University of Galati, 47 Domneasca Street, 800008 Galati, Romania;
nicoleta.vartolomei@ugal.ro
* Correspondence: maria.turtoi@ugal.ro

Featured Application: Rosehip fruits have a high vitamin C content and can be used in bread-making as a natural alternative to synthetic ascorbic acid. The studied form was rosehip powder added to wheat flour and used to obtain bread with properties similar to that prepared from wheat flour with ascorbic acid as an improver.

Abstract: An in-depth analysis of wheat flour (WF) substituted with 0.5–2.5% rosehip powder (Rp) concerning the proximate composition, dough farinographic properties, and bread physico-chemical characteristics was performed. The purpose of this study was to investigate whether the use of Rp as a natural alternative for synthetic ascorbic acid in breadmaking was appropriate. A sample of wheat flour with an ascorbic acid addition of 2 mg/100 g was also used. Rp showed higher ash, carbohydrates, and fibre content, as well as lower moisture and protein content compared to wheat flour, and a vitamin C content of 420 ± 16.09 mg/100 g. A proximate composition analysis revealed a decrease in moisture, protein, and wet gluten, and an increase in ash, carbohydrates, and fibres for the flour mixtures compared with WF. Farinographic properties were positively influenced by the Rp addition and the high fibre content in the flour mixtures. Water absorption increased from 58.20% (WF) to 61.90% (2.5% Rp). Dough stability increased for the 0.5–1.0% Rp addition, then slightly decreased. The physico-chemical properties of bread prepared from flour mixtures showed a significant increase in height: 100.10 ± 0.14 mm (WF)–115.50 ± 0.14 mm (1.5% Rp), specific volume: 142.82 cm^3/100 g (WF)–174.46 cm^3/100 g (1.5% Rp), moisture: $41.81 \pm 0.40\%$ (WF)–$43.92 \pm 0.15\%$ (2.0% Rp), and porosity: $87.75 \pm 1.06\%$ (WF)–$89.40 \pm 0.57\%$ (2.5% Rp). The results indicated that the Rp used in breadmaking to replace synthetic ascorbic acid could be suitable.

Keywords: ascorbic acid; bread; dough; farinograph; rosehip powder; wheat flour

Citation: Vartolomei, N.; Turtoi, M. The Influence of the Addition of Rosehip Powder to Wheat Flour on the Dough Farinographic Properties and Bread Physico-Chemical Characteristics. *Appl. Sci.* **2021**, *11*, 12035. https://doi.org/10.3390/app112412035

Academic Editor: Silvia Mironeasa

Received: 9 November 2021
Accepted: 10 December 2021
Published: 17 December 2021

Publisher's Note: MDPI stays neutral with regard to jurisdictional claims in published maps and institutional affiliations.

1. Introduction

As one of humanity's most important staple foods over time, bread has evolved into various forms, with distinctive and different characteristics. Bakers have made, and continue to make, traditional varieties based on accumulated knowledge, adapting existing methods and developing new ones, aiming at the best use of the available raw materials to obtain bread with the desired quality [1].

Wheat flour is the primary raw material for bread production in most of the world's regions and is a versatile raw material, accepting many additions in specific proportions [2,3]. Because the quality of wheat and flour varies greatly, improvers are added to wheat flour and dough in small amounts to obtain the desired and consistent bread quality [4]. The improver selection is made according to what needs to be improved and the flour's main technological properties, especially the power of flour and its ability to form gases to obtain the best possible result [5].

The most-used improvers in the bakery are amylolytic enzymes such as α-amylase from cereal malt [6,7]), fungal α-amylase from *Bacillus subtilis* [8] or *Lactobacillus plantarum* [9], β-amylase [10], and fungal amyloglucosidase from *Aspergillus niger* [11]; as well as proteolytic enzymes such as malt proteases [12], proteases from *A. oryzae* [13] and vegetable proteases (papain, bromelain); pentosanases such as exogenous or fungal xylanases [14–16]; fungal lipases [17]; lipoxygenases from wheat or soy flour [12]; transglutaminase produced by the bacterium *Streptoverticillium* [12]; cellulases [10]; natural emulsifiers such as lecithin from oil demucilagination [18,19]; synthetic emulsifiers such as polyalcohol esterified with paraffin chains [20]; oxidizing agents such as L-ascorbic acid (E300) [21], KBr (E924) [22,23], KI (E917) [24]; and reducing agents such as L-cysteine [25].

Ascorbic acid (AA) has been used as an improver in breadmaking since 1935 [26]. AA is added either into flour or directly into the dough. The role of AA in baking is to mediate the oxidation reactions that stabilise the dough for preserving its elastic and viscous properties so that the dough can retain gases and go through the stages of the breadmaking process (stretching, shaping, etc.) [26,27]. In addition to improving the ability of gluten to retain gases, ascorbic acid also contributes to faster proofing, as well as obtaining a piece of bread with a larger volume, a more delicate crumb, smaller and more pores, and to reduce the thickness of the bread crust. These changes also result in a softer crumb, making the bread look fresh for longer [28].

Although AA, the reduced form of vitamin C, is naturally present in many fruits and vegetables, most of the AA that supplies the bakery industry is obtained through chemical reactions that use glucose as a raw material [29]. Therefore, AA is classified as the chemical compound, E300 [30].

Sahi [31] presented some research results performed at Campden BRI, UK, where synthetic AA was replaced with AA-rich plant materials. Acerola cherry (*Malpighia emarginata*) extract was used in the research. Disclosed experimental data showed that the addition of acerola cherry extract leads to a dough with good farinographic properties. Thus, the dough development time was equal to that obtained for the control sample with synthetic AA. The softening degree in kneading tests in the farinograph was reduced to 64 Brabender Units (BU) but was close to dough with synthetic AA (70 BU) and lower than the control (120 BU). The bread volume, crumb structure, and texture were similar to bread obtained with chemical AA [31]. This research paved the way for more extensive possibilities of plant material selections to replace synthetic AA, especially when a clean label is desired [31].

Fruits of rosehip (*Rosa canina* L.) are recognised as a vegetal source rich in vitamin C, with the content in fresh rosehip varying between 100 and 1.400 mg/100 g (0.1–1.4%), with average values of 400–800 mg/100 g [32,33].

Rosehip fruits have been used, as a powder or an extract, in various formulas in baking to enhance the bread's nutritional value [34–36]. However, no research or trial has been developed to investigate the replacement of synthetic AA with rosehip fruits. Therefore, the research presented in this work is original and aims to study the influence of the rosehip powder (Rp) addition as a natural substitute for synthetic AA on the dough farinographic properties and bread quality.

2. Materials and Methods

2.1. Materials

The white wheat flour (WF, type 550) was purchased from Dizing SRL, Brusturi, Neamt County, a local milling and breadmaking company. It was not treated with ascorbic acid or other improvers.

The rosehip (*Rosa canina* L.) fruits were collected from the Dofteana area, Bacau County, at the maturity stage to benefit from the higher content of bioactive compounds. After discarding the achenes (the real fruits) containing the seeds, the dried sepals, and the remainder of anthers and filament, the hypanthium (fleshy shell or pulp) was dried in the dark at atmospheric conditions to avoid the loss of vitamins. The dried pulp of rosehip fruits was ground by an ultra-centrifugal laboratory mill, ZM 200 Retsch (18.000 rpm,

4 min) to obtain the Rp, then was sieved to select the particles around 180 μm in size, similar to wheat flour granularity. The obtained Rp was stored in airtight brown glass jars in the dark and was kept cold until use.

The Rp additions were selected based on the regular supplementation of wheat flour with ascorbic acid and the vitamin C content of Rp. Usually, the wheat flour is supplemented with 2 to 10 g ascorbic acid per 100 kg flour (10 mg/100 g), where the higher quantities in the interval are used for flour with lower quality indicators. Calculations based on the vitamin C content of Rp, just before the addition to flour (420 ± 16.09 mg/100 g), served to select the Rp addition levels used in the research, which were 0.5–2.5%.

Other ingredients used for breadmaking, i.e., salt and compressed yeast, were purchased from the local market (Bacau, Romania). Pakmaya compressed yeast was produced by Rompak SRL, Pascani, Iasi County, Romania.

Ascorbic acid was purchased from Enzymes & Derivates, Costișa, Neamt County, which is a local supplier of ingredients for the food industry.

Chemicals and reagents, such as ethanol (C_2H_5OH, with a minimum of 95% *w/w*), concentrated sulphuric acid (98%), copper sulphate ($CuSO_4 \cdot 5H_2O$), boric acid (H_3BO_3), clorhydric acid, and borax ($Na_2B_4O_7 \cdot 10H_2O$) of an analytical grade were purchased from Merck (Darmstadt, Germany). Calcium chloride, natrium hydroxide, natrium chloride, and phenolphthalein of an analytical grade were purchased from Sigma-Aldrich (Steinheim, Germany).

2.2. Proximate Compositions

The proximate compositions of WF, Rp, and mixtures of wheat flour with 0.5%, 1.0%, 1.5%, 2.0%, and 2.5% of Rp were determined according to the official methods of analysis from the Association of Official Analytical Chemists [37] and the American Association of Cereal Chemists International [38], and in force standards, as follows: moisture content (SR 90:2007 [39]), ash content (SR EN ISO 2171:2010 [39]), protein content (SR EN ISO 20483:2014 [39]) through the Kjeldahl method using a nitrogen-to-protein conversion factor of 6.25 for Rp and 5.7 for wheat flour and flour mixtures (AACC method 46-11.02 [38]), lipids through an organic solvent extraction (Soxhlet method) (AOAC method 983.23 [37]), carbohydrates through the Luff–Schoorl iodometric method (AOAC method 920.183 [37]), wet gluten (SR 90:2007) [39], dry gluten (SR EN ISO 21415-3:2007) [39], the gluten deformation index, acidity, the sedimentation index, granularity (SR 90:2007) [39], and ascorbic acid (vitamin C) through the indophenol method (AOAC method 967.21 [37]).

2.3. Farinographic Measurements

The Brabender farinograph (Brabender Model E, Duisburg, Germany), according to SR EN ISO 5530-1:2015 [39] and the AACC method 54-21 [38], was used to determine the dough farinographic properties, namely: water absorption (WA in %), dough development time (DT in min), dough stability (DS in min), the softening degree (SfD in BU), and the Farinograph quality number (QN).

2.4. The Breadmaking Procedure

The doughs were prepared with the direct (one-stage) method (SR 91:2007 [39]) using the following recipe per 100 g of wheat flour: 1.5% salt, 1.8% compressed baker's yeast, 0–2.5% Rp, and water (58–65 mL), according to the water absorption of each mixture of WF and Rp. The quantities were calculated for 3 kg of wheat flour to obtain around 5 kg of dough. Mixing and kneading were performed in a Diosna dough mixer with a removable bowl, a 12 kg maximum dough capacity, and 0.44/0.9 kW of motor power. After 10 min of kneading, the dough was fermented for 150 min at 28–30 °C in the mixing bowl, divided into pieces of 380 g, moulded, and introduced into baking trays. After the additional 50–60 min of leavening at 30 °C in a continuous proofer, the samples were baked for 20 min at 230–240 °C. Proofing and baking were accomplished on the technological flow of the Dizing breadmaking company.

2.5. Bread Characterization

The bread samples were stored for 120 min at room temperature before analysis.

The physico-chemical analysis of bread samples included the determination of the bread height, moisture, porosity, elasticity, and acidity of the crumb (SR 91:2007 [39]), as well as measuring the volume through the rapeseed displacement method (SR 91:2007 [39]).

2.6. Statistical Analysis

The results are reported as the average values of two or three replicates, along with their standard deviations (three replicates, especially for Rp and sensitive analyses). A one-way analysis of variance (ANOVA) was carried out with the Microsoft Excel programme, Microsoft Office 2010, to detect significant differences among results. Fisher's least significant difference (LSD) test at a 95.0% confidence level, used to determine differences between values, was applied using the Statgraphics Centurion XVI.I software (Statgraphics Technologies, Inc., The Plains, Virginia, USA).

3. Results and Discussion

3.1. The Proximate Composition of the Flours

The proximate compositions of Rp, WF, and the flour mixtures (WF–Rp) obtained with the addition of 0.5–2.5% are shown in Table 1.

Table 1. Proximate composition of rosehip powder, wheat flour, and flour mixtures.

Samples	Moisture, %	Ash, %	Proteins, %	Wet Gluten, %	Carbohydrates, %	Fibres, %	Vitamin C, mg/100 g
Rp	13.40 ± 0.15 a	6.50 ± 0.07 b	4.89 ± 0.11 a	–	73.66 ± 0.19 b	8.63 ± 0.12 b	420 ± 16.09 b
WF	14.15 ± 0.06 b	0.55 ± 0.01 a	13.45 ± 0.03 b	34.10 ± 0.07 c	70.68 ± 0.29 a	0.10 ± 0.07 a	–
WF–Rp 0.5%	14.14 ± 0.01 b	0.59 ± 0.04 a	13.42 ± 0.05 b	33.93 ± 0.16 bc	70.74 ± 0.51 a	0.12 ± 0.05 a	2.0 ± 0.20 a
WF–Rp 1.0%	14.14 ± 0.01 b	0.61 ± 0.04 a	13.36 ± 0.04 b	33.76 ± 0.09 abc	70.83 ± 0.32 a	0.14 ± 0.11 a	4.0 ± 0.36 a
WF–Rp 1.5%	14.14 ± 0.01 b	0.64 ± 0.06 a	13.32 ± 0.07 b	33.59 ± 0.15 abc	70.85 ± 0.15 a	0.16 ± 0.07 a	6.0 ± 0.30 a
WF–Rp 2.0%	14.13 ± 0.02 b	0.67 ± 0.06 a	13.27 ± 0.05 b	33.42 ± 0.19 ab	71.00 ± 0.17 a	0.18 ± 0.09 a	8.0 ± 0.17 a
WF–Rp 2.5%	14.13 ± 0.01 b	0.70 ± 0.07 a	13.21 ± 0.04 b	33.25 ± 0.07 a	71.04 ± 0.23 a	0.20 ± 0.06 a	10 ± 0.26 a
WF–AA	14.15 ± 0.05 b	0.55 ± 0.02 a	13.45 ± 0.04 b	34.10 ± 0.09 c	70.68 ± 0.14 a	0.10 ± 0.06 a	2.0 ± 0.00 a

WF—wheat flour, Rp—rosehip powder, WF–Rp—mixtures of wheat flour with rosehip powder, WF–AA—wheat flour with ascorbic acid addition of 2 mg/100 g. Different letters within the column indicate the values are statistically different at a 95.0% confidence level.

Rp showed a vitamin C content of 420 ± 16.09 mg/100 g, which belonged to the interval of 100–1.400 mg/100 g [32,33,40–43]. Moreover, Rp contains a high ash content, indicating a high mineral salts content, low protein content, and high levels of fibres and carbohydrates (Table 1).

Because Rp has a moisture content ($13.40 \pm 0.15\%$) lower than wheat flour ($14.15 \pm 0.06\%$), the addition of Rp in wheat flour resulted in a slight decrease in the moisture of the flour mixtures as the added amount of Rp increased. Thus, the lowest values of moisture (14.13%) were obtained with the addition of 2.0 and 2.5% of Rp into the wheat flour. According to the ANOVA, the differences between the moisture values of the samples were not significant ($p > 0.05$). This finding is due to the small amounts of the Rp additions. The results are consistent with other literature [44,45].

The results presented in Table 1 indicated that due to the increase of the Rp addition, the ash content increased from $0.55 \pm 0.01\%$ in the control to $0.70 \pm 0.07\%$ in the sample with 2.5%. The ash content of the Rp used was $6.5 \pm 0.07\%$ which explains the ash increase in the flour mixtures. The results are similar to other data presented in the literature. Koletta et al. [46] reported an ash content of 0.63% in wheat flour and 1.26% in barley flour, and Cvetković et al. [47] obtained an ash content of 3.71% in dried rosehips.

The protein content (Table 1) of the flours decreased from $13.45 \pm 0.03\%$ in the control to $13.21 \pm 0.04\%$ in the sample with 2.5% Rp, i.e., a reduction of 1.78% compared to the control sample. This decrease is due to Rp, which has a lower protein content ($4.89 \pm 0.11\%$) than wheat flour.

Van Hung et al. [48] studied the quality of dough and bread made from whole wheat flour, intending to delay the ageing of bread. The wholemeal flour with which the studies were carried out had a significantly higher protein content than commercially available white wheat flour (13.5% compared to 12.6%).

Dall'Asta et al. [49] reported a study on the influence of a chestnut flour addition to white wheat flour on the bread's physicochemical properties and volatile components. The mixtures made were 80/20 and 50/50 white soft wheat flour/chestnut flour, with white flour alone as the control sample. As chestnut flour has a lower protein content than wheat flour, the obtained mixtures also had a lower protein content than the control sample. Thus, the protein content ranged from 13.2% for wheat flour to 5.8% for the 50/50 wheat flour/chestnut flour mixture.

According to the ANOVA, the differences between the protein content of WF and WF–Rp were not significant ($p > 0.05$).

The lipid content of Rp was 0.76 ± 0.05%, quite close to that of WF (0.83 ± 0.07%). The flour mixtures showed lipid contents situated between these values, almost without any variation, as the ANOVA analysis confirmed ($p > 0.05$) (data not shown).

A study by Sun et al. [50], regarding the addition of wheat germ flour (3%, 6%, 9%, and 12%) to white flour to obtain Chinese steamed bread, does not mention the lipid content of the resulted mixtures, only that of white flour (1.3%).

The different evolutions of lipid content could be achieved when the ingredients used have higher values. Pınarlı et al. [51] obtained pasta (macaroni) from durum wheat flour (semolina) mixed with 15% wheat germ, obtaining a significant improvement in nutritional value. Other authors, e.g., Arshad et al. [52] mentioned that they replaced up to 25% of wheat flour with defatted wheat germ and obtained improved functional and nutritional properties for prepared cookies.

The carbohydrate content for the control sample and flour mixtures had a very slight growth tendency with increased amounts of Rp added (Table 1). Wheat flour has a carbohydrate content of 70.68 ± 0.29% and Rp has a carbohydrate content of 73.66 ± 0.19%. The highest value of carbohydrate content, 71.04%, was obtained with the addition of 2.5% Rp, representing an increase of only 0.51% compared to the carbohydrate content of the control sample. According to the ANOVA analysis, this variation was not significant ($p > 0.05$) compared to the carbohydrate content of the control sample because the Rp was added in tiny proportions.

The fibre content of WF–Rp varied with Rp addition values, increasing by about 0.02% for every 0.5% of Rp added, from 0.1 ± 0.07% in white flour to 0.2 ± 0.06% in the mixture with an addition of 2.5% Rp. According to the ANOVA single-factor statistical analysis, the differences were not significant ($p > 0.05$).

The total fibre content of whole wheat flour obtained by Van Hung et al. [48] was 15.3%, divided into 11.2% insoluble and 4.1% soluble fibres. The white wheat flour used in their research contained 3.4% total fibre with 1.6% insoluble and 1.8% soluble fibres because most of the fibre, both soluble and insoluble, was removed with bran and germs. The authors [48] believed that wholemeal wheat flour with a high fibre content, especially insoluble fibre, is a suitable raw material for high-fibre foods. However, these authors found that a high fibre content dilutes gluten proteins during dough kneading, leading to a soft and inelastic dough. Thus, bread made from whole wheat flour has a significantly smaller specific volume and larger pores than bread made from white flour [48].

The wet gluten in flour mixtures with Rp varies inversely with the addition of Rp, decreasing from 34.10 ± 0.07% in the control sample to 33.25 ± 0.07% with a 2.5% addition of Rp due to the absence of gluten-forming proteins in Rp. According to the single-factor ANOVA statistical analysis, the variation was not significant ($p > 0.05$), with the decrease of gluten content representing about 0.17% at each addition stage.

Banu et al. [44] found a reduction in wet gluten from 25.0 ± 0.15% for the control sample–white wheat flour to 24.4 ± 0.19%, 24.2 ± 0.19%, 24.1 ± 0.17%, 22.9 ± 0.16%, 21.0 ± 0.15%, 19.9 ± 0.15%, and 18.4 ± 0.14% for mixtures of white wheat flour with wheat flour bran (3%, 5%, 10%, 15%, 20%, 25%, and 30%, respectively), the variations being significant ($p < 0.05$). The reduction of wet gluten is greater in this study due to the higher additions of wheat flour bran.

The sedimentation index, a parameter that reflects the quantity and quality of proteins in flour (wheat), increased from 25.0 for wheat flour to 29.0 for flour mixtures with 1.5% and 2.5% Rp, respectively, the variations being significant ($p < 0.05$). These values are in the range of 20–39, corresponding to a good quality flour for baking [4]. The additions of Rp in white wheat flour had tiny values (0.5–2.5%), and the protein content and wet gluten varied very little. Therefore, it could be stated that the addition of Rp did not affect the quality of proteins in WF. However, the sedimentation index increased almost linearly for the first mixtures, namely, the samples with a 0.5–1.0% Rp addition (27.0 and 28.0) and stagnated for the subsequent three samples (29.0), namely, those with a 1.5%, 2.0%, and 2.5% Rp addition (data not shown in Table 1).

The vitamin C content in flour mixtures (Table 1) reflected the addition of 0.5–2.5% of Rp in wheat flour to obtain the mixture flours. The differences were significant ($p < 0.05$). It is equal to (0.5% Rp), or higher than, the usual AA addition to dough in breadmaking when WF has a good quality. AA is a pure and more stable compound than vitamin C in Rp, which can be degraded easily during storage time.

3.2. Dough Farinographic Properties

The farinographic properties of the dough, namely, water absorption, the dough development time, stability, softening degree, and the farinograph quality number are presented in Table 2.

Table 2. Farinographic properties of doughs.

Samples	Water Absorption (500 BU), %	Dough Development Time, min	Stability, min	Softening Degree (12 min after Maximum), BU	Farinograph Quality Number
WF	58.20 ± 0.00 a	6.70 ± 0.00 a	11.00 ± 0.00 a	58.00 ± 0.00 a	116.00 ± 0.00 a
WF–Rp 0.5%	59.70 ± 0.14 b	6.75 ± 0.49 a	11.50 ± 0.42 a	58.50 ± 0.71 a	125.50 ± 6.36 a
WF–Rp 1.0%	60.35 ± 0.07 c	6.60 ± 0.14 a	11.30 ± 0.71 a	65.00 ± 2.83 a	122.50 ± 0.71 a
WF–Rp 1.5%	60.80 ± 0.00 d	6.70 ± 0.42 a	11.20 ± 1.13 a	78.00 ± 1.41 b	118.00 ± 0.00 a
WF–Rp 2.0%	61.25 ± 0.07 e	7.10 ± 0.14 a	10.55 ± 0.07 a	87.00 ± 1.41 c	116.50 ± 0.71 a
WF–Rp 2.5%	61.90 ± 0.00 f	7.20 ± 0.14 a	9.70 ± 0.14 a	91.00 ± 1.41 c	115.50 ± 4.95 a
WF–AA	58.20 ± 0.00 a	6.60 ± 0.14 a	11.20 ± 0.00 a	60.00 ± 1.41 a	121.00 ± 1.41 a

WF—wheat flour, Rp—rosehip powder, WF–Rp—mixtures of wheat flour with rosehip powder, WF–AA—wheat flour with ascorbic acid addition of 2 mg/100 g. Different letters within the column indicate the values are statistically different at a 95.0% confidence level.

Bread dough is a viscoelastic material that shrinks when sheared, combining Hooke's solid and a non-Newtonian viscous liquid. It has a non-linear rheological behaviour but, at very low stresses, it has a linear behaviour. The stress at which the dough has a linear behaviour depends on the type of dough, the mixing method, and the testing method [53]. Information on the rheological properties of the dough helps predict potential applications of wheat flour and the quality of the finished product [54].

The farinograph, a physical dough testing device, was used to assess how adding Rp to wheat flour affected the rheological properties of the dough. It allowed for the evaluation of the performance of flour mixtures and the breadmaking potential. The two z-shaped blades of the farinograph mixer rotate at constant speeds and subject the dough to mixing at a constant temperature [54].

3.2.1. Water Absorption

Flour hydration is essential because it affects the functional properties and the quality of bread [55,56].

The water absorption of the wheat flour and flour mixtures, corrected for 500 BU (Table 2), showed a linear ($R^2 = 0.94$) increase with the increase of the Rp addition to the flour, from 58.20 ± 0.00% for the control to 61.90 ± 0.00% for the 2.5% Rp addition. According to the ANOVA single-factor statistical analysis, the differences were significant

($p < 0.05$). The control sample with AA at 10 g/100 kg showed a water absorption identical to that of the control sample.

The increase in water absorption is widely discussed in the literature. Thus, numerous papers mention the positive influence of protein quantity and quality on water absorption [55,57–59]. Moreover, Hefnawy et al. [60] and Mohammed et al. [61] reported increased water absorption with increased protein content with chickpeas in wheat flour.

Since the addition of Rp to wheat flour decreased the protein content of the flour mixtures, the reason for the increase in water absorption must be sought somewhere else, for example, in the influence of fibre content. Gómez et al. [62] added purified fibres of various origins (oranges, peas, cocoa, coffee, wheat, and microcrystalline cellulose) to wheat flour bread. They obtained an increase in water absorption with an increase in the percentage of added fibres. For example, with the addition of 2% and 5% orange fibre, the water absorption was 63.00% and 65.40%, respectively, compared to 58.70% in the control sample (white wheat flour). The water absorption capacity increased with the addition of fibres due to the hydroxyl groups in the chemical structure of the fibres that allows the binding of water through hydrogen bonds [62]. Many other researchers have come to the same conclusion. Thus, introducing high-fibre wheat bran to white wheat flour increased the water absorption from $57.10 \pm 0.20\%$ in the control sample to $68.20 \pm 0.60\%$ for an addition of 30% wheat bran [44]. Similarly, Lauková et al. [63] obtained an increase in water absorption from $58.00 \pm 0.70\%$ for the control sample to $75.30 \pm 0.70\%$ for an addition of 15% hydrated apple powder. Similar results were reported by Kohajdová et al. [64] with the addition of carrot powder (water absorption increased from $60.67 \pm 0.35\%$ for the control sample to $72.01 \pm 0.25\%$ for a 15% addition of carrot powder) and Ajila et al. [65] with the addition of mango peel powder (water absorption increased from 60.00% for the control sample to 68.00% for a 10% addition of mango peel powder).

3.2.2. Dough Development Time

The dough development time (DT) is measured by adding water to the flour until the dough reaches its maximum consistency without breaking. Water hydrates the flour components during the mixing phase, and the dough develops [63].

The DT for the control sample was 6.70 ± 0.00 min and varied between 6.60 ± 0.14 min (1.0% Rp) and 7.20 ± 0.14 min (2.5% Rp) for the flour mixtures. According to the ANOVA, the differences were not significant ($p > 0.05$), although close to the limit ($p = 0.054$), so the Rp addition had a relatively small influence on DT (Table 2). The wheat flour control sample with the AA addition had a DT of 6.60 ± 0.14 min, lower than the WF control, reflecting that the AA addition to the flour reduces the DT.

A similar effect of DT variations has been reported in several studies. Sudha et al. [66] observed an increase in DT from 1.5 min (control sample) to 3.5 min (an addition of 15% apple pomace with dietary fibre resulting from the manufacture of apple juice), Lauková et al. [63] found an increase in DT from 3.5 min. (control sample) to 11.0 min for an addition of 15% hydrated apple powder in white wheat flour, and Kohajdová et al. [67] observed an increase in DT from 3.43 min (control sample) to 5.53 min (an addition of 15% apple powder).

Kučerová et al. [68] added different Vitacel brand fibres (JRS Company, Rosenberg, Germany) to the wheat flour. They found a slight increase in DT with an increasing fibre addition from 1% to 3%, compared to wheat flour, as follows: from 1.1 min to 1.3 min for wheat fibre (WF 200), 1.5 min to 1.7 min for apple fibre (AF 400), 1.9 min to 2.2 min for potato fibre (PF 200), and from 1.3 min to 1.5 min for bamboo fibre (BF 300).

In the given examples, it is observed that although the DT has increased in direct proportion to the fibre additions, the differences are tiny. Nikolić et al. [69] made a similar statement for flour mixtures with variable buckwheat flour additions. They reported an increase of DT from 1.0 min to 1.2 min for buckwheat flour additions ranging from 1 g/100g to 20 g/100 g flour mixtures and a maximum value of 1.5 min for 30 g/100 g.

The increase in DT was explained by the slowdown in hydration rate and gluten development due to adding ingredients that increased the fibre content in the mixtures obtained [63,66].

3.2.3. Dough Stability

The data relating to the influence of the Rp addition on dough stability (DSt) are shown in Table 2. DSt increased from 11.00 ± 0.00 min (WF control) to 11.50 ± 0.42 min (0.5% Rp), then decreased below the value for the control sample, reaching 9.70 ± 0.14 min for 2.5% Rp. The DSt of the control WF with added AA was 11.20 ± 0.00 min, a little higher than the DSt of the WF sample. According to the ANOVA, the differences were significant ($p < 0.05$).

DSt provides some indications regarding the tolerance of flour to mixing and kneading [70]. Thus, Nassar et al. [71] reported the increase of DSt from 4.9 min for the control sample to 12.4 min and 11.5 min with the addition of 25% flour from orange peels and orange pulp, respectively. The same effect was observed by Kohajdová et al. [72], who reported an increase in DSt after the addition of apple fibre from 6.40 ± 0.15 min for the control sample, to 9.30 ± 0.64 min for the mixture with 15% apple fibre, and from 9.40 min to 10.90 min with the addition of 15% apple pomace fibre [67]. Moreover, Lauková et al. [63] reported a significant increase in DSt from 6.7 min for the control sample to 11.6 min with an addition of 15% hydrated apple powder. These authors explained the rise of DSt through a higher interaction of fibres, water, and proteins in flour [63,67]. On the other hand, the DSt in this research was considerably higher than the values obtained by adding buckwheat flour, in which case the DSt increased from 0.3 ± 0.1 min for the control sample to 4.6 ± 0.3 min with an addition of 30 g of buckwheat flour/100 g of wheat flour [69]. The opposite effect was reported by Sudha et al. [66] after adding apple pomace powder to the wheat flour in different proportions. Thus, the DSt decreased from 4.2 min to 2.1 min with an addition of 15% apple pomace fibre. These results are supported by Rosell et al. [73], who observed a decrease in DSt with increased additions (12.5–25%) of quinoa flour to wheat flour, and Liu et al. [74] reported a reduction in DSt of up to 30% for potato flour additions to wheat flour.

3.2.4. Softening Degree

The softening degree (SfD) in BU is determined from farinograms 10 min after the start of the experiment (data not shown) and 12 min after the curve reached the maximum (Table 2). An ascending variation of SfD at 12 min was obtained for increased Rp additions to the wheat flour. The SfD increased linearly ($R^2 = 0.95$) from 58.00 ± 0.00 (control) to 91.00 ± 1.41 (2.5% Rp). According to the ANOVA, the differences were significant ($p < 0.05$).

There were no findings regarding the degree of softening resulting from the farinographic analysis in the studies mentioned above [44,55,59,62–69].

Instead, other researchers have mentioned a decrease in SfD. For example, Anil [75] observed a reduction in the SfD from 78.0 ± 2.0 BU for the control to 61.0 ± 1.0 BU and 39.0 ± 2.0 BU for the addition of 5.0 and 10.0% crushed and roasted hazelnut powder, respectively. Moreover, Stojceska & Ainsworth [76] noticed an increase in DT and DSt and a decrease in SfD from 25 UB for the control sample to 15, 20, and 5 UB for 10%, 20%, and 30%, respectively, with the addition of brewers spent grains to wheat flour. It should be noted that the proportions of spent grains used in this research were much higher than 0.5–2.5% Rp and could help in increasing the SfD. In another study [77], the authors observed a decrease in the SfD from 60.0 ± 1.0 UB for the control to 20.0 ± 2.0 UB for the samples with additions of fibre concentrates from the pulp of dates, pears, and apples in proportions that ensured a 2% fibre intake.

3.2.5. The Farinograph Quality Number

The values of the farinographic quality number (QN), resulting from Rp additions, are presented in Table 2. It was observed that the QN of samples with added Rp were generally higher than the control sample, except for the sample with an addition of 2.0% Rp, which had the same value (116). The dough obtained from WF with 2 g AA/100 kg had a QN equal to 121, between 1.0% and 1.5% Rp. However, the differences were not significant ($p > 0.05$).

Although no clear correlation can be established between the addition of Rp and the value of the QN, the fact that the QN was higher than the control sample showed that the Rp addition had a positive influence on the QN. This statement is consistent with findings observed by Nikolić et al. [69], who obtained a QN of 52.8 ± 2 for wheat flour (control) and increased values of 67.8 ± 2 for 30% buckwheat flour/100 g flour mixture.

3.3. Bread Characterization

The physico-chemical properties of bread made from WF, WF–Rp, and WF with AA as the improver are shown in Tables 3 and 4.

Table 3. Physical properties of bread (height, weight, and volume).

Samples	Height, mm	Volume, cm³	Weight, g	Specific Volume, cm³/100 g
WF	100.10 ± 0.14 b	486 ± 4.24 a	340.30 ± 1.84 ab	142.82 a
WF–Rp 0.5%	98.75 ± 0.49 a	481 ± 5.66 a	338.10 ± 1.56 ab	142.27 a
WF–Rp 1.0%	113.05 ± 0.07 e	588 ± 4.24 bc	339.00 ± 1.41 ab	173.45 c
WF–Rp 1.5%	115.50 ± 0.14 f	601 ± 5.66 c	344.50 ± 2.12 b	174.46 c
WF–Rp 2.0%	111.15 ± 0.21 d	575 ± 4.95 bc	333.00 ± 1.41 a	172.52 c
WF–Rp 2.5%	108.25 ± 0.07 c	564 ± 5.66 b	337.60 ± 1.56 ab	167.06 b
WF–AA	113.90 ± 0.28 e	588 ± 4.24 bc	336.90 ± 0.85 ab	174.53 c

WF—wheat flour, Rp—rosehip powder, WF–Rp—mixtures of wheat flour with rosehip powder, WF–AA—wheat flour with ascorbic acid addition of 2 mg/100 g. Different letters within the column indicate the values are statistically different at a 95.0% confidence level.

Table 4. Physico-chemical properties of bread crumb.

Samples	Moisture, %	Acidity, Degree	Porosity, %	Elasticity, %
WF	41.81 ± 0.40 a	2.00 ± 0.00 a	87.75 ± 1.06 a	93.30 ± 0.58 d
WF–Rp 0.5%	42.64 ± 0.33 abc	2.10 ± 0.00 a	87.50 ± 0.71 a	91.70 ± 0.58 bcd
WF–Rp 1.0%	42.92 ± 0.22 abc	2.15 ± 0.07 a	90.00 ± 0.71 a	91.50 ± 0.50 bcd
WF–Rp 1.5%	43.31 ± 0.10 bc	2.15 ± 0.07 a	90.70 ± 0.99 a	91.00 ± 0.50 bc
WF–Rp 2.0%	43.92 ± 0.15 c	2.20 ± 0.14 a	89.00 ± 0.71 a	90.30 ± 0.58 ab
WF–Rp 2.5%	42.51 ± 0.34 abc	2.25 ± 0.07 a	89.40 ± 0.57 a	88.50 ± 0.50 a
WF–AA	42.06 ± 0.16 ab	2.15 ± 0.07 a	88.60 ± 1.56 a	92.30 ± 0.58 cd

WF—wheat flour, Rp—rosehip powder, WF–Rp—mixtures of wheat flour with rosehip powder, WF–AA—wheat flour with ascorbic acid addition of 2 mg/100 g. Different letters within the column indicate the values are statistically different at a 95.0% confidence level.

3.3.1. Bread Dimensions

Although the dough was baked in trays, the bread dimensions were measured. It was observed from the beginning that the leavening was differentiated, resulting in pieces of bread of different heights, most likely due to the Rp addition's influence. The values of length (around 235 mm) and width (about 91 mm) of bread did not differ significantly ($p > 0.05$) due to baking in trays. In contrast, the height of the loaves, determined as an average of two duplicates, showed significant differences ($p < 0.05$) (Table 3).

3.3.2. Bread Volume

The volume and specific volume of bread are reported in Table 3.

The bread volume had higher values for samples with added Rp compared to the control (486 ± 4.24 cm^3), except for the addition of 0.5% Rp. The most significant increase in volume was observed for 1.0% and 1.5% Rp, while for the following additions, there was a slight decrease, but the values remain higher than for control. The bread volume with ascorbic acid (WF–AA) had a value between the bread volumes with the addition of 1.0% and 2.0% Rp. According to the ANOVA, the differences were significant ($p < 0.05$). The increase in bread volume could be explained by the Rp addition, which provided the dough with a specific content of AA that contributed to dough stabilization, gluten network strengthening [26,27], and a larger volume of bread [28].

The bread volume is, for most consumers, one of the main criteria for evaluating the bread quality and is implicitly one of the primary decision-making elements of purchasing. Hathorn et al. [78] considered that the importance of bread volume for consumers derives from their desire to consume bread that seems light and not too dense. Thus, even if a large volume is not desirable, consumers associate large volumes with some loaves of bread and small values with others, e.g., flat chapatti, Lebanese bread, and pita bread.

From a technological point of view, the bread volume is an essential external feature as it represents a quantitative measure of the performance of the baking process [79]. The bread volume depends on the quantity and quality of gluten [73] and how the dough is processed. Thus, a well-kneaded and developed dough with a well-formed gluten network will reduce the loss of gas and contribute to an increased bread volume [1]. Moreover, the volume of bread correlates with the moisture of the dough. Thus, Gallagher et al. [80] stated that higher moisture located in the optimal area positively influences bread volume because the bread will have an increased volume.

Osuna et al. [81] studied the composition, sensory properties, and the effects of ascorbic acid and α-tocopherol additions on the oxidative stability of wholemeal bread and vegetable oils. Although they stated that adding AA contributed to increasing the bread volume by improving the dough's stability, the authors did not determine the specific volume or bread volume.

Hallén et al. [82] added chickpea flour obtained from grains, germinated and fermented, in a proportion of 5%, 10%, and 15% to wheat flour. For some additions they obtained increases in the volume of bread loaves compared to the control (1560 mL), for example, an increase up to 1657 mL for the addition of 5%, 1598 mL for 10%, and 1627 mL for 15% chickpea flour; 1667 mL for 5% and 1587 mL for 10% germinated chickpea flour; and 1710 mL for 5% and 1680 mL for 10% fermented chickpea flour. For the rest of the additions, the specific volume was lower than the control. Most of the volume increases were small because the addition of chickpea flour resulted in a stickier dough that was more difficult to process, especially in the case of sprouted chickpeas, due to the reduced content of gluten-forming proteins. The bread obtained was generally more compact, with a denser structure [82].

In a similar study, McWaters et al. [83] added chickpea flour from raw and extruded grains in 15% and 30% proportions. They recorded a decrease in volume for all additions compared to the control. The reduction in volume was higher for the 30% addition compared to 15%, and the reduction in volume was also higher for the flour from extruded chickpea grains due to the decrease in the content of gluten-forming proteins even if the total protein content of the flour mixes increased.

Van Hung et al. [48] stated that the addition of fibre dilutes proteins and interferes with forming the optimal gluten network, confirming the previous statements.

Yamasaengung et al. [84] continued these studies by combining chickpea flour with emulsifiers. If no emulsifier was added to the white bread, the specific volume of the bread was significantly smaller ($p < 0.05$). Therefore, they stated that only emulsifiers significantly affected the specific volume of white bread ($p < 0.05$). The decrease in bread volume caused by chickpea flour can be counteracted by increased water in the dough. In wholemeal bread, the volume increase is influenced by a combined effect of water and emulsifiers [84].

In general, bread volume is correlated with a soft texture and high porosity, while bread density refers to a more compact, harder-textured bread [84].

The inverse of the density, expressed in g/cm^3, is the specific volume in cm^3/g. This parameter results from the calculation that relates the total volume of a loaf of bread to its mass, and it presents the same evolution as the volume of bread. Thus, bread with the addition of Rp had higher values of the specific volume than the control, except for the 0.5% addition of Rp. Sheikholeslami et al. [85] reported a slight increase in the specific volume of bread obtained from wheat flour with the addition of 10% and 20% barley flour with a low proportional coating content for smaller mesh sizes, thus retaining more coating particles with the addition of guar gum. The highest value of the specific volume was obtained for sieve 40 with the addition of 1% guar gum (2.90 ± 0.20 cm^3/g) compared to the control (2.70 ± 0.17 cm^3/g), and was lower for sieve 50 with the addition of 20% barley flour with a low coating content, without the addition of guar gum (2.10 ± 0.17 cm^3/g). Instead, Hathorn et al. [78] obtained a specific volume reduction from 1.7 cm^3/g to 1.4 cm^3/g when adding 50% to 65% sweet potato flour.

In other studies, higher specific bread volumes were reported, for example, 3.3–4.0 cm^3/g for bread prepared with the addition of heat-treated maitake mushroom powder [86], and 3.4–4.4 cm^3/g for wheat bread with added dextrins [87].

3.3.3. Bread Moisture and Acidity

A higher moisture retention in bread is economical and is also required to lengthen shelf life [88].

A higher moisture content was found in bread prepared from WF–Rp mixtures compared to the control (41.81 ± 0.40%). The differences were significant ($p < 0.05$). The increase in bread moisture was probably determined by the increase in the water absorption of WF–Rp mixtures due to the higher fibre content of Rp. However, the bread moisture of samples with an addition of 2.5% Rp and AA as an improver did not follow the increasing tendency, as they were situated between the moistures of the control and the sample with the addition of 0.5% Rp. This might be expected, as several factors not determined in this study could influence the moisture content of bread.

Similar results are reported in the literature. Toasted bread with a high fibre content was obtained by adding 10%, 20%, and 30% of fine or coarse bran that was dark or light in colour, and wheat germ. The bread had a higher moisture content than the bread obtained from unbleached wheat flour (39.29 ± 0.34%) for all types of bran. The increase in moisture was explained by the higher amounts of water used to prepare the dough for samples with bran fractions [88]. Sidhu et al. observed that moisture content was slightly higher for bread with increasing levels of bran addition (10–30%) [89]. Another study reported an increase in moisture with increased levels of polyols [88]. They found a higher moisture content in bread prepared after the incorporation of 4% sorbitol as compared to the control.

Bread acidity (Table 4) values increased for all breads compared to the control (2.0 acidity degrees), reaching 2.25 ± 0.07 acidity degrees for bread prepared with WF–Rp with the 2.5% Rp addition. The differences were not significant ($p > 0.05$) according to the ANOVA one-way statistical analysis. Ascorbic acid is consumed during kneading, and if it somehow remains in excess in the dough, it will be distorted when baked. Other components of Rp, such as the organic acids of rosehip composition, may be responsible for the values obtained for acidity. This issue will be followed in future research.

3.3.4. Bread Crumb Porosity and Elasticity

The breads obtained were characterized by high porosity values (Table 4). Higher porosity was found in bread prepared from WF–Rp mixtures compared to the control (87.75 ± 1.06%), except for the porosity of bread with an addition of 0.5% Rp (87.50 ± 0.71%), which was slightly lower. The porosity values increased by 1.0% and 1.5% with the addition of Rp, then slightly decreased following the same trend as the height and the volume of bread. The differences were significant ($p < 0.05$). Besides dough stabilization and gluten network strengthening [26,27], ascorbic acid used in breadmaking contributed to obtaining bread with a larger amount of smaller and evenly distributed pores, a larger volume [28], and a better porosity. Therefore, the addition of Rp provided the dough with a certain amount of ascorbic acid, which determined the increase in bread height, volume, and porosity.

The elasticity of the bread crumb it is property to return to its original shape after the action of a pressing force, and it depends on the quality and quantity of gluten in the flour and the freshness of the product [5]. It was determined a few hours after cooling and is reported as an average of triplicates $\pm$ SD (Table 4). Bread crumb elasticity showed a slight decrease for all samples compared to the control (93.3 $\pm$ 0.58%), varying from 91.70 $\pm$ 0.58% (0.5% Rp) to 88.50 $\pm$ 0.50% (2.5% Rp) with significant differences ($p < 0.05$). The decrease of bread crumb elasticity with the increase in the Rp addition could be associated with the slight decrease of gluten content in WF–Rp mixtures, as was discussed before (Section 3.1, Table 1). The crumb elasticity of bread with the AA addition was lower (92.30 $\pm$ 0.58%) but close to the control as expected, since its gluten content was equal to that of the control bread.

4. Conclusions

The substitution of wheat flour with rosehip powder influenced the composition of mixture flours by decreasing the moisture, protein, and wet gluten content, increasing the ash, fibre, and carbohydrate content, and introducing vitamin C into mixture flours at levels reflecting the amount of rosehip powder used. Furthermore, it determined changes in dough farinographic properties and bread physico-chemical characteristics. Compared to the control sample and flour where ascorbic acid was used as an improver, the water absorption of all mixture flours increased due to the increased fibre content. The dough development time, dough stability, and softening degree variations were significant, showing a combined influence of vitamin C provided by the rosehip powder, synthetic ascorbic acid, and the high fibre content in the mixture flours. Moreover, the rosehip powder addition positively influenced the farinographic quality number. The bread prepared from wheat flour with the rosehip powder addition showed a positive evolution of physico-chemical properties, such as a significant increase in height, volume, specific volume, moisture, acidity, and porosity, as well as a slight decrease in elasticity as compared to the control bread. These results indicate that rosehip powder could be used in breadmaking to replace synthetic ascorbic acid.

5. Patents

The results of the studies were firstly used in a Romanian patent application no. A/0069 of 19.09.2018: Pâine din făină de grâu cu adaos de pudră de măceșe și procedeu de obținere a acesteia (Bread with rosehip powder addition and process for obtaining the same). The abstract was published in *Buletinul Oficial de Proprietate Industrială*, Secțiunea Brevete de Invenție (*Official Bulletin of Industrial Property*, Patent Section) no. 3/2020, p. 16.

Author Contributions: Conceptualization, M.T. and N.V.; Methodology, N.V. and M.T.; Validation, N.V. and M.T.; Formal analysis, N.V.; Investigation, N.V. and M.T.; Writing—original draft preparation, N.V. and M.T.; Writing—review and editing, M.T.; Visualisation, N.V. and M.T.; Supervision, M.T. All authors have read and agreed to the published version of the manuscript.

Funding: This research received no external funding. Dunarea de Jos University of Galati funded the APC.

Acknowledgments: The authors thank Dizing SRL Brusturi, Neamt County, for technical support. This paper was supported by a grant offered by the Romanian Ministry of Research as Intermediate Body for the Competitiveness Operational Program 2014–2020, call POC/78/1/2/, project number SMIS2014 + 136213, acronym METROFOOD-RO.

Conflicts of Interest: The authors declare no conflict of interest.

References

1. Cauvain, S.P. *The Technology of Breadmaking*, 3rd ed.; Springer International Publishing: Cham, Switzerland, 2015; pp. 1–2, 27–29.
2. Scheuer, P.M.; Francisco, A.; Miranda, M.Z.; Ogilari, P.J.; Torres, G.; Limberger, V.; Montenegro, F.M.; Ruffi, C.R.; Biondi, S. Characterization of Brazilian wheat cultivars for specific technological applications. *Food Sci. Technol. Brazil* **2011**, *31*, 816–826. [CrossRef]
3. Kurek, M.A.; Wyrwisz, J.; Piwińska, M.; Wierzbicka, A. Influence of the wheat flour extraction degree in the quality of bread made with high proportions of β-glucan. *Food Sci. Technol. Brazil* **2015**, *35*, 273–278. [CrossRef]
4. Cauvain, S.P. Improving the texture of bread. In *Texture in Food*; Kilcast, D., Ed.; CRC Press: Boca Raton, FL, USA, 2004; Volume 2, pp. 432–450.
5. Bordei, D. *Tehnologia Modernă a Panificaţiei (Modern Bakery Technology)*; AGIR Publishing House: Bucharest, Romania, 2004; pp. 86–87, 383–384, 517–518.
6. Haghighat-Kharazi, S.; Milani, J.M.; Kasaai, M.R.; Khajeh, K. Use of encapsulated maltogenic amylase in malotodextrins with different formulations in making gluten-free breads. *LWT* **2019**, *110*, 182–189. [CrossRef]
7. Zhang, L.; Li, Z.; Qiao, Y.; Zhang, Y.; Zheng, W.; Zhao, Y.; Huang, Y.; Cui, Z. Improvement of the quality and shelf life of wheat bread by a maltohexaose producing α-amylase. *J. Cereal Sci.* **2019**, *87*, 165–171. [CrossRef]
8. Jones, A.; Lamsa, M.; Frandsen, T.P.; Spendler, T.; Harris, P.; Sloma, A.; Xu, F.; Nielsen, J.B.; Cherry, J.R. Directed evolution of a maltogenic α-amylase from *Bacillus* sp. TS-25. *J. Biotechnol.* **2008**, *134*, 325–333. [CrossRef] [PubMed]
9. Woo, S.; Shin, Y.; Jeong, H.; Kim, J.; Ko, D.; Hong, J.S.; Choi, H.; Shim, J. Effects of maltogenic amylase from *Lactobacillus plantarum* on retrogradation of bread. *J. Cereal Sci.* **2020**, *93*, 102976. [CrossRef]
10. Goesaert, H.; Slade, L.; Levine, H. Amylases and bread firming—An integrated view. *J. Cereal Sci.* **2009**, *50*, 345–352. [CrossRef]
11. Rasiah, I.A.; Sutton, K.H.; Low, F.L.; Lin, H.M.; Gerrard, J.A. Crosslinking wheat dough proteins by glucose oxidase and the resulting effects on bread and croissants. *Food Chem.* **2005**, *89*, 325–332. [CrossRef]
12. Bhat, M.K. Cellulases and related enzymes in biotechnology. *Biotechnol. Adv.* **2000**, *18*, 355–383. [CrossRef]
13. Dahiya, S.; Bajaj, B.K.; Kumar, A.; Tiwari, S.K.; Singh, B. A review on biotechnological potential of multifarious enzymes in bread making. *Process. Biochem.* **2020**, *99*, 290–306. [CrossRef]
14. Al-Widyan, O.; Khataibeh, M.H.; Abu-Alruz, K. The use of xylanases from different microbial origin in bread baking and theie effects on bread qualities. *J. Appl. Sci.* **2008**, *8*, 672–676. [CrossRef]
15. Ahmad, Z.; Butt, M.S.; Ahmed, A.; Riaz, M.; Sabir, S.M.; Farooq, U.; Rehman, F.U. Effect of *Aspergillus niger* xylanase on dough characteristics and bread quality attributes. *J. Food Sci. Technol.* **2012**, *51*, 2445–2453. [CrossRef] [PubMed]
16. Guo, Y.; Gao, Z.; Xu, J.; Chang, S.; Wu, B.; He, B. A family 30 glucurono-xylanase from *Bacillus subtilis* LC9: Expression, characterization and its application in Chinese bread making. *Int. J. Biol. Macromol.* **2018**, *117*, 377–384. [CrossRef] [PubMed]
17. Melis, S.; Meza Morales, W.R.; Delcour, J.A. Lipases in wheat flour bread making: Importance of an appropriate balance between wheat endogenous lipids and their enzymatically released hydrolysis products. *Food Chem.* **2019**, *298*, 125002. [CrossRef] [PubMed]
18. Boutte, T.; Skogerson, L. Stearoyl-2-lactylates and oleoyl lactylates. In *Emulsifiers in Food Technology*, 2nd ed.; Norn, V., Ed.; Wiley-Blackwell: Chichester, UK, 2015; pp. 251–270.
19. Cottrell, T.; van Peij, J. Sorbitan esters and polysorbates. In *Emulsifiers in Food Technology*, 2nd ed.; Norn, V., Ed.; Wiley-Blackwell: Chichester, UK, 2015; pp. 271–275.
20. Gaupp, R.; Adams, W. Diacetyl Tartaric Acids of Monoglycerides (DATEM) and associated emulsifiers in bread making. In *Emulsifiers in Food Technology*, 2nd ed.; Norn, V., Ed.; Wiley-Blackwell: Chichester, UK, 2015; pp. 121–145.
21. Sahi, S.S. Ascorbic acid and redox agents in bakery systems. In *Bakery Products Science and Technology*, 2nd ed.; Zhou, W., Ed.; Wiley-Blackwell: Chichester, UK, 2014; pp. 183–197.
22. IARC Potassium Bromate (Group 2B). In International Agency for Research on Cancer (IARC)—Summaries & Evaluations. 1999, Volume 73, p. 481. Available online: https://inchem.org/documents/iarc/vol73/73-17.html (accessed on 21 November 2020).
23. Shanmugavel, V.; Santhi, K.K.; Kurup, A.H.; Kalakandan, S.; Anandharaj, A.; Rawson, A. Potassium bromate: Effects on bread components, health, environment and method of analysis: A review. *Food Chem.* **2020**, *311*, 125964. [CrossRef] [PubMed]
24. Bakerpedia. L-Cysteine. Available online: https://bakerpedia.com/ingredients/l-cysteine/ (accessed on 17 November 2020).
25. Stoica, A.; Popescu, C.; Barascu, E.; Iordan, M. L-cysteine influence on the physical properties of bread from high extraction flours with normal gluten. *Ann. Food Sci. Technol.* **2010**, *11*, 6–10.
26. Wieser, H. The use of redox agents in breadmaking. In *Breadmaking: Improving Quality*, 2nd ed.; Cauvain, S.P., Ed.; Woodhead Publishing: Oxford, UK, 2012; pp. 447–469.
27. Belitz, H.-D.; Grosch, W.; Schieberle, P. *Food Chemistry*, 4th ed.; Springer: Berlin/Heidelberg, Germany, 2009; pp. 716–718.
28. Campbel, G.M.; Martin, P.J. Bread aeration and dough rheology: An introduction. In *Breadmaking: Improving Quality*, 2nd ed.; Woodhead Publishing: Oxford, UK, 2012; pp. 299–336.
29. Carr, A.C.; Vissers, M.C.M. Synthetic or food-derived vitamin C—Are they equally bioavailable? *Nutrients* **2013**, *5*, 4284–4304. [CrossRef]
30. Verbruggen, R. Food additives in the European Union. In *Food Additives*, 2nd ed.; Branen, A.L., Davidson, P.M., Salminen, S., Thorngate, J.H., III, Eds.; Marcel Dekker: New York, NY, USA, 2002; pp. 109–197.

31. Sahi, S.S. Applications of natural ingredients in baked goods. In *Natural Food Additives, Ingredients and Flavourings*; Baines, D., Seal, R., Eds.; Woodhead Publishing: Oxford, UK, 2012; pp. 318–332.

32. Ziegler, S.J.; Meier, B.; Sticher, O. Fast and Selective Assay of l-Ascorbic Acid in Rose Hips by RP-HPLC Coupled with Electrochemical and/or Spectrophotometric Detection. *Planta Med.* **1986**, *52*, 383–387. [CrossRef] [PubMed]

33. Czyzowska, A.; Klewicka, E.; Pogorzelski, E.; Nowak, A. Polyphenols, vitamin C and antioxidant activity in wines from *Rosa canina* L. and *Rosa rugosa* Thunb. *J. Food Compost. Anal.* **2014**, *39*, 62–68. [CrossRef]

34. Hua, L. Beauty-Maintaining Bread by Utilizing Roses and Chayote and Preparation Method. Thereof. Patent CN 105341087 A, 24 February 2016.

35. Kaiyun, L. Eye-Protecting Anti-Radiation Pineapple Bread and Preparation Method. Thereof. Patent CN 105341087 A, 24 February 2016.

36. Krolevets, A.A. Bread Production Method Comprising Nanostructured Extract of Dry. rosehip. Patent RU 2630250-C1, 6 September 2017.

37. AOAC. *Official Methods of Analysis of the Association of Official Analytical Chemists*, 15th ed.; Helrich, K., Ed.; Methods 920.183, 967.21, 983.23; Association of Official Analytical Chemists: Arlington, VA, USA, 1990.

38. AACC International. *Approved Methods of Analysis*, 11th ed.; Methods 46-11.02, 54-21; American Association of Cereal Chemists International: St. Paul, MN, USA, 2000.

39. ASRO. *Romanian Standards Catalog for Cereal and Milling Products Analysis*; SR 90:2007, SR 91:2007, SR EN ISO 20483:2014, SR EN ISO 21415-1:2007, SR EN ISO 21415-3:2007, SR EN ISO 2171:2010 abd SR EN ISO 5530-1:2015; ASRO: Bucharest, Romania, 2008.

40. Ercisli, S. Chemical composition of fruits in some rose (*Rosa* spp.) species. *Food Chem.* **2007**, *104*, 1379–1384. [CrossRef]

41. Nojavan, S.; Khalilian, F.; Kiaie, F.M.; Rahimi, A.; Arabanian, A.; Chalavi, S. Extraction and quantitative determination of ascorbic acid during different maturity stages of *Rosa canina* L. fruit. *J. Food Compost. Anal.* **2008**, *21*, 300–305. [CrossRef]

42. Roman, I.; Stănilă, A.; Stănilă, S. Bioactive compounds and antioxidant activity of *Rosa canina* L. biotypes from spontaneous flora of Transylvania. *Chem. Cent. J.* **2013**, *7*, 73–82. [CrossRef]

43. Bhave, A.; Schulzova, V.; Chmelarova, H.; Mrnka, L.; Hajslova, J. Assessment of rosehips based on the content of their biologically active compounds. *J. Food Drug Anal.* **2017**, *25*, 681–690. [CrossRef]

44. Banu, I.; Stoenescu, G.; Ionescu, V.S.; Aprodu, I. Effect of the addition of wheat bran stream on dough rheology and bread quality. *Ann. Univ. Dunarea Jos Galati Fascicle VI Food Technol.* **2012**, *36*, 39–52.

45. Al-Attabi, Z.H.; Merghani, T.M.; Ali, A.; Rahman, M.S. Effect of barley flour addition on the physico-chemical properties of dough and structure of bread. *J. Cereal Sci.* **2017**, *75*, 61–68. [CrossRef]

46. Koletta, P.; Irakli, M.; Papageorgiou, M.; Skendi, A. Physicochemical and technological properties of highly enriched wheat breads with wholegrain non wheat flours. *J. Cereal Sci.* **2014**, *60*, 561–568. [CrossRef]

47. Cvetković, B.R.; Filipčev, B.V.; Bodroža-Solarov, M.I.; Bardić, Ž.M.; Sakač, M.B. Chemical composition of dried fruits as a value added ingredient in bakery product. *Food Proc. Qual. Saf.* **2009**, *36*, 15–19.

48. Van Hung, P.; Maeda, T.; Morita, N. Dough and bread qualities of flours with whole waxy wheat flour substitution. *Food Res. Int.* **2007**, *40*, 273–279. [CrossRef]

49. Dall'Asta, C.; Cirlini, M.; Morini, E.; Rinaldi, M.; Ganino, T.; Chiavaro, E. Effect of chestnut flour supplementation on physicochemical properties and volatiles in bread making. *LWT* **2013**, *53*, 233–239. [CrossRef]

50. Sun, R.; Zhang, Z.; Hu, X.; Xing, Q.; Zhuo, W. Effect of wheat germ flour addition on wheat flour, dough and Chinese steamed bread properties. *J. Cereal Sci.* **2015**, *64*, 153–158. [CrossRef]

51. Pınarlı, İ.; İbanoğlu, Ş.; Öner, M.D. Effect of storage on the selected properties of macaroni enriched with wheat germ. *J. Food Eng.* **2004**, *64*, 249–256. [CrossRef]

52. Arshad, M.U.; Anjum, F.M.; Zahoor, T. Nutritional assessment of cookies supplemented with defatted wheat germ. *Food Chem.* **2007**, *102*, 123–128. [CrossRef]

53. Mirsaeedghazi, H.; Emam-Djomeh, Z.; Mousavi, S.M.A. Rheometric measurement of dough rheological characteristics and factors affecting it. *Int. J. Agric. Biol.* **2008**, *10*, 112–119.

54. Mohammed, M.I.O.; Mustafa, A.I.; Osman, G.A.M. Evaluation of wheat breads supplements with Teff (*Eragrostis tef* ZUCC. Trotter) grain flour. *Aust. J. Crop. Sci.* **2009**, *3*, 207–212.

55. Amjid, M.R.; Shehzad, A.; Hussain, S.; Shabbir, M.A.; Khan, M.R.; Shoaib, M. A comprehensive review on wheat flour dough rheology. *Pak. J. Food Sci.* **2013**, *23*, 105–123.

56. Berton, B.; Scher, J.I.; Villieras, F.D.R.; Hardy, J.I. Measurement of hydration capacity of wheat flour: Influence of composition and physical characteristics. *Powder Technol.* **2002**, *128*, 326–331. [CrossRef]

57. Coţovanu, I.; Mironeasa, S. Buckwheat seeds: Impact of milling fractions and addition level on wheat bread dough rheology. *Appl. Sci.* **2021**, *11*, 1731. [CrossRef]

58. Michel, S.; Löschenberger, F.; Ametz, C.; Pachler, B.; Sparry, E.; Bürstmayr, H. Combining grain yield, protein content and protein quality by multi-trait genomic selection in bread wheat. *Theor. Appl. Genet.* **2019**, *132*, 2767–2780. [CrossRef] [PubMed]

59. Paraskevopoulou, A.; Provatidou, E.; Tsotsiou, D.; Kiosseoglou, V. Dough rheology and baking performance of wheat flour-lupin protein isolate blends. *Food Res. Int.* **2010**, *43*, 1009–1016. [CrossRef]

60. Hefnawy, T.M.H.; El-Shourbagy, G.A.; Ramadan, M.F. Impact of adding chickpea (*Cicer arietinum* L.) flour to wheat flour on the rheological properties of toast bread. *Int. Food Res. J.* **2012**, *19*, 521–525.

61. Mohammed, I.; Ahmeda, A.R.; Senge, B. Dough rheology and bread quality of wheat-chickpea flour blends. *Ind. Crops Prod.* **2012**, *36*, 196–202. [CrossRef]

62. Gómez, M.; Ronda, F.; Blanco, C.; Caballero, P.; Apesteguia, A. Effect of dietary fibre on dough rheology and bread quality. *Eur. Food Res. Technol.* **2003**, *216*, 51–56. [CrossRef]
63. Lauková, M.; Kohajdová, Z.; Karovičová, J. Effect of hydrated apple powder on dough rheology and cookies quality. *Potravin. Slovak J. Food Sci.* **2016**, *10*, 506–511. [CrossRef]
64. Kohajdová, Z.; Karovičová, J.; Jurasová, M. Influence of carrot pomace powder on the rheological characteristics of wheat flour dough and on wheat rolls quality. *Acta Sci. Pol. Technol. Aliment.* **2012**, *11*, 381–387.
65. Ajila, C.M.; Leelavathi, K.U.J.S.; Rao, U.P. Improvement of dietary fiber content and antioxidant properties in soft dough biscuits with the incorporation of mango peel powder. *J. Cereal Sci.* **2008**, *48*, 319–326. [CrossRef]
66. Sudha, M.L.; Baskaran, V.; Leelavathi, K. Apple pomace as a source of dietary fiber and polyphenols and its effect on the rheological characteristics and cake making. *Food Chem.* **2007**, *104*, 686–692. [CrossRef]
67. Kohajdová, Z.; Karovičová, J.; Magala, M.; Kuchtová, V. Effect of apple pomace powder addition on farinographic properties of wheat dough and biscuits quality. *Chem. Zvesti* **2014**, *68*, 1059–1065. [CrossRef]
68. Kučerová, J.; Šottníková, V.; Nemodová, Š. Influence of dietary fibre addition on the rheological and sensory properties of dough and bakery products. *Czech. J. Food Sci.* **2013**, *31*, 340–346. [CrossRef]
69. Nikolić, N.; Saka, M.; Mastilovi, J. Effect of buckwheat flour addition to wheat flour on acylglycerols and fatty acids composition and rheology properties. *LWT-Food Sci. Technol.* **2011**, *44*, 650–655. [CrossRef]
70. Fu, L.; Tian, J.-C.; Sun, C.-L.; Li, C. RVA and farinograph properties study on blends of resistant starch and wheat flour. *Agric. Sci. China* **2008**, *7*, 812–822. [CrossRef]
71. Nassar, A.G.; Abdel-Hamied, A.A.; El-Naggar, E.A. Effect of citrus by-products flour incorporation on chemical, rheological and organoleptic characteristics of biscuits. *World J. Agric. Sci.* **2008**, *4*, 612–616.
72. Kohajdová, Z.; Karovičová, J.; Jurasová, M.; Kukurová, K. Effect of the addition of commercial apple fibre powder on the baking and sensory properties of cookies. *Acta Chim. Slov.* **2011**, *4*, 88–97.
73. Rosell, C.M.; Santos, E.; Penella, J.M.S.; Haros, M. Wholemeal wheat bread: A comparison of different breadmaking processes and fungal phytase addition. *J. Cereal Sci.* **2009**, *50*, 272–277. [CrossRef]
74. Liu, X.-L.; Mu, T.-H.; Sun, H.-N.; Zhang, M.; Chen, J.-W. Influence of potato flour on dough rheological properties and quality of steamed bread. *J. Integr. Agric.* **2016**, *15*, 2666–2676. [CrossRef]
75. Anil, M. Using of hazelnut testa as a source of dietary fiber in breadmaking. *J. Food Eng.* **2007**, *80*, 61–67. [CrossRef]
76. Stojceska, V.; Ainsworth, P. The effect of different enzymes on the quality of high-fibre enriched brewer's spent grain breads. *Food Chem.* **2008**, *110*, 865–872. [CrossRef] [PubMed]
77. Bchir, B.; Rabetafika, H.N.; Paquot, M.; Blecker, C. Effect of pear, apple and date fibres from cooked fruit by-products on dough performance and bread quality. *Food Bioproc. Technol.* **2014**, *7*, 1114–1127. [CrossRef]
78. Hathorn, C.S.; Biswas, M.A.; Gichuhi, P.N.; Bovell-Benjamin, A.C. Comparison of chemical, physical, micro-structural, and microbial properties of breads supplemented with sweet potato flour and high-gluten dough enhancers. *LWT* **2008**, *41*, 803–815. [CrossRef]
79. Tronsmo, K.M.; Faergestad, E.M.; Schofield, J.D.; Magnus, S. Wheat protein quality in relation to baking performance evaluated by the Chorleywood bread process and a hearth bread baking test. *J. Cereal Sci.* **2003**, *38*, 205–215. [CrossRef]
80. Gallagher, E.; Kunkel, A.; Gormley, T.R.; Arendt, E.K. The effect of dairy and rice powder addition on loaf and crumb characteristics, and on shelf life (intermediate and long-term) of gluten-free breads stored in a modified atmosphere. *Eur. Food Res. Technol.* **2003**, *218*, 44–48. [CrossRef]
81. Osuna, M.B.; Romero, C.A.; Romero, A.M.; Judis, M.A.; Bertola, N.C. Proximal composition, sensorial properties and effect of ascorbic acid and α-tocopherol on oxidative stability of bread made with whole flours and vegetable oils. *LWT* **2018**, *98*, 54–61. [CrossRef]
82. Hallén, E.; İbanoğlu, S.; Ainsworth, P. Effect of fermented/germinated cowpea flour addition on the rheological and baking properties of wheat flour. *J. Food Eng.* **2004**, *63*, 177–184. [CrossRef]
83. McWaters, K.H.; Phillips, R.D.; Walker, S.L.; McCullough, S.E.; Mensa-Wilmot, Y.; Saalia, F.K.; Hung, Y.-C.; Patterson, S.P. Baking performance and consumer acceptability of raw and extruded cowpea flour breads. *J. Food Qual.* **2004**, *27*, 337–351. [CrossRef]
84. Yamsaengsung, R.; Schoenlechner, R.; Berghofer, E. The effects of chickpea on the functional properties of white and whole wheat bread. *Int. J. Food Sci. Technol.* **2010**, *45*, 610–620. [CrossRef]
85. Sheikholeslami, Z.; Karimi, M.; Komeili, H.R.; Mahfouzi, M. A new mixed bread formula with improved physicochemical properties by using hull-less barley flour at the presence of guar gum and ascorbic acid. *LWT* **2018**, *93*, 628–633. [CrossRef]
86. Seguchi, M.; Morimoto, N.; Abe, M.; Yoshino, Y. Effect of Maitake (*Grifola frondosa*) mushroom powder on bread properties. *J. Food Sci.* **2001**, *66*, 261–264. [CrossRef]
87. Miyazaki, M.; Maeda, T.; Morita, N. Effect of various dextrin substitution for wheat flour on dough properties and bread qualities. *J. Food Res. Int.* **2004**, *37*, 59–65. [CrossRef]
88. Bhise, S.; Kaur, A. Baking quality, sensory properties and shelf life of breads with polyols. *J. Food Sci. Technol.* **2014**, *51*, 2054–2061. [CrossRef]
89. Sidhu, J.S.; Al-Hooti, S.N.; Al-Saqer, J.M. Effect of adding wheat bran and germ fractions on the chemical composition of high-fiber toast bread. *Food Chem.* **1999**, *67*, 365–371. [CrossRef]

Article

Rheological Approaches of Wheat Flour Dough Enriched with Germinated Soybean and Lentil

Mădălina Ungureanu-Iuga [1,2], Denisa Atudorei [1], Georgiana Gabriela Codină [1,*] and Silvia Mironeasa [1]

[1] Faculty of Food Engineering, "Ştefan cel Mare" University of Suceava, 13 Universitatii Street, 720229 Suceava, Romania; madalina.iuga@usm.ro (M.U.-I.); denisa.atudorei@outlook.com (D.A.); silviam@fia.usv.ro (S.M.)
[2] Integrated Center for Research, Development and Innovation in Advanced Materials, Nanotechnologies, and Distributed Systems for Fabrication and Control (MANSiD), "Ştefan cel Mare" University of Suceava, 13th University Street, 720229 Suceava, Romania
* Correspondence: codina@fia.usv.ro

Abstract: Germination is a convenient technique that could be used to enhance the nutritional profile of legumes. Furthermore, consumers' increasing demand for diversification of bakery products represents an opportunity to use such germinated flours in wheat-based products. Thus, this study aimed to underline the effects of soybean germinated flour (SGF) and lentil germinated flour (LGF) on the rheological behavior of dough during different processing stages and to optimize the addition level. For this purpose, flour falling number, dough properties during mixing, extension, fermentation, and dynamic rheological characteristics were evaluated. Response surface methodology (RSM) was used for the optimization of SGF and LGF addition levels in wheat flour, optimal and control samples microstructures being also investigated through epifluorescence light microscopy (EFLM). The results revealed that increased SGF and LGF addition levels led to curve configuration ratio, visco-elastic moduli, and maximum gelatinization temperature rises, while the falling number, water absorption, dough extensibility, and baking strength decreased. The interaction between SGF and LGF significantly influenced ($p < 0.05$) the falling number, dough consistency after 450 s, baking strength, curve configuration ratio, viscous modulus, and maximum gelatinization temperature. The optimal sample was found to contain 5.60% SGF and 3.62% LGF added in wheat flour, with a significantly lower falling number, water absorption, tolerance to kneading, dough consistency, extensibility, and initial gelatinization temperature being observed, while dough tenacity, the maximum height of gaseous production, total CO_2 volume production, the volume of the gas retained in the dough at the end of the test, visco-elastic moduli and maximum gelatinization temperatures were higher compared to the control. These results underlined the effects of SGF and LGF on wheat dough rheological properties and could be helpful for novel bakery products development.

Keywords: germination; lentil; soybean; wheat flour; rheology; microstructure

Citation: Ungureanu-Iuga, M.; Atudorei, D.; Codină, G.G.; Mironeasa, S. Rheological Approaches of Wheat Flour Dough Enriched with Germinated Soybean and Lentil. *Appl. Sci.* **2021**, *11*, 11706. https://doi.org/10.3390/app112411706

Academic Editors: Anabela Raymundo and Stephen Grebby

Received: 30 October 2021
Accepted: 7 December 2021
Published: 9 December 2021

Publisher's Note: MDPI stays neutral with regard to jurisdictional claims in published maps and institutional affiliations.

1. Introduction

Bread is the most consumed food product in the world, with its contribution to the human diet having the greatest importance [1]. However, because most people prefer white bread, the nutritional benefits are limited because it is obtained from refined wheat flour [2,3]. Moreover, due to the large quality variation of wheat flour used in bread-making, bakery producers usually add different types of additives to wheat flour, especially chemicals, to improve product quality from a technological point of view [4]. Nowadays, one of the trends in bread making is to improve the nutritional quality of bread by substituting wheat flour with other flour types without affecting the quality of the final products. Specialist attempts are made to balance the content of vitamins, minerals, and fibers lost through wheat refining without adding any chemical compound to the bakery products [5]. Moreover, an attempt is made to improve the quality of proteins in baked goods by adding various ingredients which contain essential amino acids that are deficient in wheat flour [6,7]. To correct these nutritional deficiencies, different grain flours can be used in bread making, such as legumes, oilseeds, pseudocereals, etc. [5,7,8].

The substitution of wheat flour with different legumes types improves the nutritional value of bread by decreasing the glycemic index and by increasing the fiber, mineral content, and protein quality [9]. However, the addition of legume flours to wheat flour has the disadvantage that they contain some antinutritive factors, such as tannins, phytates, or trypsin inhibitors which may affect the nutritional value of bread [8]. Furthermore, the addition level in wheat flour is limited due to their gluten dilution effect which affects dough viscoelastic structure and its ability to retain gases during fermentation which can lead to bakery products of poor quality [7]. Thus, it is more convenient to use legume flours in a germinated form, a process that has a positive effect on the nutritional profile of legumes, but also their sensory profile. During legume germination, many positive changes occur, as follows: a decrease in antinutritive factors contained in legumes, an increase in protein content, an improvement in the availability of sodium, magnesium, iron, zinc, a decrease in lipids and carbohydrates content [10–12]. Additionally, the amount of volatile organic compounds, such as 2-methylbutanal and dimethyltrisulfite increases which leads to the intensification of legume grains' flavor. At the same time, their sweetness intensifies [13], improving the sensory characteristics. In addition, during germination the enzymatic activity of legume grains increases which may improve wheat flour quality if it is enzymatically deficient without adding other chemical additives during bread making [14].

Two legumes with a superior nutritional profile that may be used in bread making are lentil and soybean. Lentils boast a significant content of vitamins, minerals, carbohydrates, dietary fiber, and a low glycemic index [15]. Lentils are also rich in proteins (21–31%) which contain all essential amino acids (39.3 g of essential amino acids per 100 g of protein); they are a source of glutamic acid, lysine, arginine, leucine, acid aspartic [16]. Lentils also contain many micronutrients, of which vitamin B9, zinc, and iron have the highest weight. Lentils contain the highest amount of polyphenols, compared to all other vegetables [17]. Regarding health aspects, medical studies have shown that the consumption of lentils has benefits on cardiovascular diseases and cancer prevention [18], but also has implications in promoting slow and moderate postprandial blood glucose increase [19,20]. Soybean is boasted as having a significant amount of high-quality protein (38–55%), essential amino acids, lipids (20%), and carbohydrates (27%) [21–23]. The phytochemicals present in soybean are of great interest for health because studies have shown that they lower the amount of cholesterol, have an anticarcinogenic capacity, and contribute to bone health [24].

Soybean and lentil flour often appear in specialized studies due to the possibility of being used as an ingredient in various bakery products [1,7]. The use of lentil and soybean flours as partial substitutes in wheat flour is justified by both nutritional and sensory aspects. Their use, particularly in the germinated form, may improve the quality of bakery products even from a technological point of view, especially if the wheat flour used has enzymatic deficiencies. According to the results obtained by Zhang et al. [25], native red lentil flour incorporation in wheat dough led to higher water absorption and mechanical weakening, while dough development time, stability and minimum torque, and cooking stability were lower compared to the control and increased with the addition level, due to the influence of the chemical compounds of the ingredient added. In the study of Marchini et al. [26] on the effects of lentil flour in the wheat dough, it is stated that the addition level increase caused water absorption rises, dough stability reduction, delayed protein weakening, and worsening of dough pasting consistency which could be related to the lower swelling power of pulses compared to wheat. The incorporation of germinated lentil flour in Sangak bread determined water absorption, dough development time, and softness degree increases compared to the control, while the stability of dough did not differ significantly. Wheat bread fortification with defatted soybean led to higher water absorption and dough extensibility and lower dough stability compared to the control according to results presented by Mashayekh et al. [27]. The addition of native and germinated soybean flours in wheat dough increased water absorption, maximum consistency time, and dough stability, while dough maximum resistance to extension and extensibility was not significantly affected [28]. The rise of the 7S protein fraction extracted from native and germinated soybean in wheat dough was related to the increment of water absorption and extensographic maximum resistance to extension, especially in the case of protein extracted from native soybeans [29].

This study aimed to optimize the formulation of germinated soybean and lentil flours that can be added to refined wheat flour of low alpha-amylase activity to improve dough rheological properties. For the optimal combination between the soybean and lentil germinated flour, the dough microstructure was analyzed by using epifluorescence light microscopy (EFLM). To our knowledge, no other studies have examined soybean and lentil germinated flour addition to wheat flour in a combined form. The importance of their use in bread making derives from the valuable nutritional composition of these legumes, but also the technological advantages of their use in a germinated form.

2. Materials and Methods

2.1. Materials

Refined wheat flour of the 650 type (harvest 2020) provided by the S.C. Dizing S.R.L. company (Brusturi, Neamt, Romania) was used. Germinated legume flours were obtained from lentil (*Lens culinaris Merr*) and soybean (*Glycine max* L.) which were germinated for 4 days, lyophilized and milled before they were used in the wheat flour according to the method reported in our previous studies [12,14].

The flours were analyzed according to the international ICC standard methods: ash content (ICC 104/1), moisture content (ICC 110/1), fat content (ICC 136), protein content (ICC 105/2) [30]. To be certain that the germinated soybean (SGF) and lentil (LGF) flour may be used in bread making, they were analyzed also from a microbiological point of view according to the following methods: molds and yeast according to SR ISO 7954:2001 [31], mycotoxins by using an ELISA kit (Prognosis Biotech, Larissa, Greece) and *Bacillus cereus* according to the SR EN ISO 7932:2005 [32]. The wheat flour has been also analyzed for its wet gluten content and gluten deformation index according to SR 90:2007 method [33].

Soybean germinated flour (SGF) at ratios of 5, 10, 15, and 20% and lentil germinated flour (LGF) at ratios of 2.5, 5, 7.5, and 15% were mixed with wheat flour and coded as SGF 5, SGF 10, SGF 15, SGF 20 and LGF 2.5, LGF 5, LGF 7.5 and LGF 10, respectively. Wheat flour without SGF or LGF was used as a control (C).

2.2. Dough Rheological Properties

2.2.1. Empirical Dough Rheological Properties during Mixing and Extension

Dough rheological properties during mixing and extension were analyzed by using an Alveo Consistograph (Chopin Technologies, Cedex, France) according to ICC 171 and ICC 121 standards [30] respectively. The Consistograph test was made to determine dough rheological properties during mixing: water absorption capacity (WA), tolerance to kneading (Tol), consistency of the dough after 250 s (D250), and 450 s (D450). The Alveograph test was performed to determine dough rheological properties during extension: maximum pressure (P), dough extensibility (L), baking strength (W), and configuration ratio of the Alveograph curve (P/L).

2.2.2. Empirical Dough Rheological Properties during Fermentation and Falling Number

Empirical dough rheological properties during fermentation were analyzed using the Rheofermentometer device (Chopin Rheo, type F3, Villeneuve-La-GarenneCedex, France) according to the standard method AACC89–01.01 [34]. The Rheofermentometer parameters analyzed for the dough samples obtained by kneading of 250 g mixed flours, 7 g compressed yeast of the *Saccharomyces cerevisiae* type, and 5 g salt according to the Consistograph water absorption value were: the total CO_2 volume production (VT, mL), the maximum height of gaseous production (H'm, mm), volume of the gas retained in the dough at the end of the test (VR, mL) and retention coefficient (CR, %). The falling number values expressed in s were determined by using a Falling number device (FN 1305, Perten Instruments AB, Stockholm, Sweden) according to ICC 107/1 method [30].

2.2.3. Dynamic Dough Rheological Properties

The dynamic dough rheological properties were obtained with a HAAKE MARS 40 device (Termo-HAAKE, Karlsruhe, Germany) with a 2 mm gap and parallel plate geometry of 40 mm diameter, according to previous works [14,35,36]. The dough samples were placed between rheometer plates and analyzed for the storage modulus (G′), loss modulus (G″), and loss tangent (tan δ) at a frequency of 1 Hz. Additionally, the maximum gelatinization temperatures were analyzed for the dough samples during heating from 25 to 100 °C at a rate of 4 °C per min at a fixed strain of 0.001 and a frequency of 1 Hz.

2.3. Dough Microstructure

The epifluorescence light microscopy (EFLM) images of dough with and without the best combination between the soybean and lentil germinated flour addition in wheat flour were analyzed with a Motic AE 31 (Motic, Optic Industrial Group, Xiamen, China) equipped with catadioptric objectives LWD PH 203 (N.A. 0.4). The images and dough samples preparation were obtained according to methods reported in our previous studies [14,37,38]. The dough sample was immersed in a fixing solution made of 1% rhodamine B for protein coloring and 0.5% fluorescein for starch coloring for at least 1 h.

2.4. Statistical Analysis

All the measurements were done in duplicate. Analysis of variance (ANOVA) was applied to compare mean values of the samples with SGF and LGF respectively, at different addition levels. Statistically significant differences were considered at $p < 0.05$ by the Tukey test. For this purpose, XLSTAT for Excel 2021 version (Addinsoft, New York, NY, USA) software was used.

Then, an experimental design was performed to identify single and combined effects of factors on the responses. The study of SGF and LGF addition levels effects on wheat dough characteristics and the optimization were performed on a trial version of Design Expert software (Stat-Ease, Inc., Minneapolis, MN, USA). A full factorial design with two factors varied at five levels, SGF addition at 0, 5, 10, 15, and 20% and LGF addition at 0, 2.5, 5, 7.5, and 10%, and Response Surface Methodology (RSM) with a two-factor interaction (2FI) model were used. The responses considered were the following: FN—falling number, WA—water absorption, Tol—tolerance to kneading, D250—dough consistency after 250 s, D450—dough consistency after 450 s, P—dough tenacity, L—dough extensibility, W—baking strength, P/L—curve configuration ratio, H′m—maximum height of gaseous production, VT—total CO_2 volume production, VR—the volume of the gas retained in the dough at the end of the test, CR—retention coefficient, G′—elastic modulus, G″—viscous modulus, tan δ—loss tangent, T_i—initial gelatinization temperature, T_{max}—maximum gelatinization temperature.

The effects of SGF and LGF addition levels on dough properties were evaluated through mathematical modeling. The most suitable model to predict data variation for each response was selected according to F-test results, coefficient of determination (R^2), and adjusted coefficients of determination (*Adj.-R^2*). The effects of factors and their interactions were underlined using Analysis of Variance (ANOVA), considering a significance level of 95%.

SGF and LGF addition levels optimization was done by applying the desirability function. The coded and real values of factors are listed in Table 1.

Table 1. Coded vs. real values of factors.

Run	Coded Values		Real Values	
	A	B	SGF (%)	LGF (%)
1	−1.00	−1.00	0.00	0.00
2	−1.00	−0.50	0.00	2.50
3	−1.00	0.00	0.00	5.00
4	−1.00	0.50	0.00	7.50
5	−1.00	1.00	0.00	10.00
6	−0.50	−1.00	5.00	0.00
7	−0.50	−0.50	5.00	2.50
8	−0.50	0.00	5.00	5.00
9	−0.50	0.50	5.00	7.50
10	−0.50	1.00	5.00	10.00
11	0.00	−1.00	10.00	0.00
12	0.00	−0.50	10.00	2.50
13	0.00	0.00	10.00	5.00
14	0.00	0.50	10.00	7.50
15	0.00	1.00	10.00	10.00
16	0.50	−1.00	15.00	0.00
17	0.50	−0.50	15.00	2.50
18	0.50	0.00	15.00	5.00
19	0.50	0.50	15.00	7.50
20	0.50	1.00	15.00	10.00
21	1.00	−1.00	20.00	0.00
22	1.00	−0.50	20.00	2.50
23	1.00	0.00	20.00	5.00
24	1.00	0.50	20.00	7.50
25	1.00	1.00	20.00	10.00

A: SGF—soybean germinated flour (%), B: LGF—lentil germinated flour (%).

The goals established for the factors and responses considered, along with their lower and upper limits are presented in Table 2. The differences among the optimal and control sample were tested using the Student-*t*-test, at a significance level of 95%, by using XLSTAT for Excel 2021 version (Addinsoft, New York, NY, USA) software.

Table 2. Factors and responses goals established for optimization.

Variable	Goal	Lower Limit	Upper Limit	Importance
A: SGF (%)	is in range	0.00	20.00	3
B: LGF (%)	is in range	0.00	10.00	3
WA (%)	is in range	50.70	54.30	3
Tol (s)	maximize	128.00	232.00	3
D250 (mb)	minimize	270.00	644.00	3
D450 (mb)	minimize	819.00	1117.00	3
P (mm)	is in range	88.00	132.00	3
L (mm)	is in range	25.00	75.00	3
W (10^{-4} J)	is in range	141.00	301.00	3
P/L (adim.)	is in range	1.38	5.04	3
G′ (Pa)	is in range	29,290.00	72,310.00	3
G″ (Pa)	is in range	10,780.00	31,460.00	3
tan δ (adim.)	minimize	0.34	0.50	3
T_i (°C)	is in range	47.60	53.50	3
T_{max} (°C)	is in range	73.40	77.40	3
FN (s)	minimize	185.00	350.00	3

Table 2. *Cont.*

Variable	Goal	Lower Limit	Upper Limit	Importance
H′m (mL)	maximize	62.60	77.00	3
VT (mL)	maximize	1268.00	1886.00	3
VR (mL)	maximize	1070.00	1369.00	3
CR (%)	maximize	64.30	86.90	3

A: SGF—soybean germinated flour (%), B: LGF—lentil germinated flour (%), FN—falling number, WA—water absorption, Tol—tolerance to kneading, D250—dough consistency after 250 s, D450—dough consistency after 450 s, P—dough tenacity, L—dough extensibility, W—baking strength, P/L—curve configuration ratio, H′m—maximum height of gaseous production, VT—total CO_2 volume production, VR—the volume of the gas retained in the dough at the end of the test, CR—retention coefficient, G′—elastic modulus, G″—viscous modulus, tan δ—loss tangent, T_i—initial gelatinization temperature, T_{max}—maximum gelatinization temperature.

3. Results

3.1. Flour Characteristics

The wheat flour used in this study presented the following characteristics: moisture content of 14.6%, an ash content of 0.66%, the protein content of 12.3%, the fat content of 1.12%, wet gluten content of 30.4%, and gluten deformation index of 3 mm. The falling number value of the wheat flour was 356 s which indicates that it has a low α amylase activity [39]. The germinated lentil (LGF) presented 19.5% protein, 1.0% fat, 3.1% ash, 8.8% moisture, whereas the germinated soybean (SGF) presented 40.2% protein, 17.9% fat, 5.1% ash and 10.5% moisture. From a microbiological point of view, the germinated and lyophilized legumes samples were free of Bacillus cereus and presented 1 UFC/g yeast and molds. Mycotoxins values for SGF and LGF were the following: for zearalenone of 28.18 and 63.02 ppb respectively, for ochratoxin of 24.64 ppb and 18.53 ppb, and for aflatoxin less than 1.4 ppb. The microbiological data obtained recommend the use of germinated legume flours as ingredients in food products [40,41].

3.2. Effects of SGF and LGF Levels on Falling Number and Dough Rheology

SGF addition to wheat flour resulted in a significant decrease ($p < 0.05$) of Falling Number values as the level was higher and compared to the control, a similar trend was observed for LGF incorporation (Table 3). Dough mixing behavior in terms of water absorption, and tolerance to kneading showed significant reduction as the amount of SGF raised, while dough consistency parameters varied irregularly. Similar reduction trends of water absorption and tolerance to kneading were observed for LGF samples, while dough consistency parameters increased proportionally. Significant decreases ($p < 0.05$) of dough extensibility and baking strength were obtained as the levels of SGF or LGF were higher compared to the control (Table 3). Dough tenacity increased as the amount of SGF was raised, while LGF determined an opposite change, except for LGF 5. The curve configuration ratio values also increased in proportion with the SGF or LGF addition level, except for LGF 7.5. All the parameters listed above were influenced significantly ($p < 0.05$) by SGF or LGF incorporation.

Dough rheological parameters during fermentation, viscoelastic moduli and gelatinization temperatures were affected significantly ($p < 0.05$) by SGF or LGF level (Table 4).

The maximum height of gaseous production, total CO_2 volume production, and volume of the gas retained in the dough at the end of the test was reduced as the amount of SGF raised, the retention coefficient being changed irregularly. On the other hand, LGF caused an increase of dough maximum height of gaseous production and total CO_2 volume production, except for LGF 10 and a decrease in retention coefficient values, while the volume of the gas retained in the dough at the end of the test parameter exhibited an irregular trend. The elastic and viscous moduli increased significantly ($p < 0.05$) as the addition levels of SGF or LGF increased, while the loss tangent changes were irregular (Table 4). The maximum gelatinization temperature registered an increasing trend proportional to the SGF or LGF amount, while the initial gelatinization temperature raised only with LGF level, the opposite trend was observed for SGF.

Table 3. Falling number and rheological properties during mixing and extension of wheat dough with different addition levels of soybean germinated flour (SGF) and lentil germinated flour (LGF).

Sample	FN (s)	WA (%)	Tol (s)	D250 (mb)	D450 (mb)	P (mm)	L (mm)	W ($10-4$ J)	P/L (adim.)
C	350 ± 2.83 [aA]	54.3 ± 0.14 [aA]	214 ± 2.83 [bA]	394 ± 1.41 [aD]	943 ± 4.24 [aC]	104 ± 1.41 [dB]	72 ± 2.83 [aA]	301 ± 4.24 [aA]	1.44 ± 0.04 [dB]
SGF 5	323 ± 2.83 [b]	54.0 ± 0.14 [ab]	223 ± 5.66 [ab]	293 ± 2.83 [c]	881 ± 1.41 [b]	115 ± 1.41 [c]	53 ± 2.83 [b]	241 ± 4.24 [b]	2.17 ± 0.08 [c]
SGF 10	305 ± 2.83 [c]	53.7 ± 0.14 [ab]	232 ± 2.83 [a]	270 ± 5.66 [d]	819 ± 4.24 [d]	119.5 ± 0.71 [bc]	46 ± 1.41 [b]	219 ± 4.24 [c]	2.59 ± 0.09 [c]
SGF 15	275 ± 4.24 [c]	53.4 ± 0.28 [bc]	217 ± 4.24 [ab]	272 ± 2.83 [d]	858 ± 4.24 [c]	124 ± 2.83 [ab]	35 ± 2.83 [c]	186 ± 2.83 [d]	3.54 ± 0.21 [b]
SGF 20	243 ± 1.41 [e]	52.8 ± 0.14 [c]	191 ± 5.66 [c]	319 ± 2.83 [b]	878 ± 1.41 [b]	128 ± 2.83 [a]	31 ± 1.41 [c]	170 ± 4.24 [d]	4.15 ± 0.07 [a]
One-way ANOVA *p* values	$p < 0.0001$	$p < 0.003$	$p < 0.002$	$p < 0.0001$	$p < 0.0001$	$p < 0.0001$	$p < 0.0001$	$p < 0.0001$	$p < 0.0001$
LGF 2.5	295 ± 4.24 [B]	53.7 ± 0.14 [B]	191 ± 5.66 [B]	406 ± 7.07 [CD]	1000 ± 7.07 [B]	109 ± 0.00 [B]	75 ± 1.41 [A]	285 ± 2.83 [B]	1.45 ± 0.03 [B]
LGF 5	274 ± 4.24 [C]	53.1 ± 0.14 [C]	177 ± 4.24 [BC]	418 ± 5.66 [BC]	1015 ± 7.07 [AB]	115 ± 1.41 [A]	68 ± 1.41 [AB]	269 ± 4.24 [C]	1.69 ± 0.01 [A]
LGF 7.5	252 ± 2.83 [D]	52.6 ± 0.14 [CD]	166 ± 5.66 [C]	435 ± 4.24 [B]	1020 ± 4.24 [AB]	91 ± 1.41 [C]	63 ± 1.41 [B]	183 ± 2.83 [D]	1.44 ± 0.06 [B]
LGF 10	229 ± 4.24 [E]	52.2 ± 0.00 [D]	161 ± 2.83 [C]	571 ± 4.24 [A]	1029 ± 5.66 [A]	88 ± 1.41 [C]	50 ± 1.41 [C]	173 ± 2.83 [D]	1.76 ± 0.08 [A]
One-way ANOVA *p* values	$p < 0.0001$	$p < 0.0001$	$p < 0.0001$	$p < 0.0001$	$p < 0.0001$	$p < 0.0001$	$p < 0.0001$	$p < 0.0001$	$p < 0.002$

FN—falling number, WA—water absorption, Tol—tolerance to kneading, D250—dough consistency after 250 s, D450—dough consistency after 450 s, P—dough tenacity, L—dough extensibility, W—baking strength, P/L—curve configuration ratio. Soybean germinated flour (SGF) containing samples: a–e, mean values in the same column followed by different letters are significantly different ($p < 0.05$); Lentil germinated flour (LGF) containing samples: A–E, mean values in the same column followed by different letters are significantly different ($p < 0.05$).

Table 4. Empirical and dynamic rheological properties of wheat dough with different addition levels of soybean germinated flour (SGF) and lentil germinated flour (LGF).

Sample	H'm (mL)	VT (mL)	VR (mL)	CR (%)	G' (Pa)	G" (Pa)	tan δ (adim.)	T_i (°C)	T_{max} (°C)
C	65.9 ± 0.14 [cE]	1532 ± 4.24 [cE]	1228 ± 2.83 [bC]	80.15 ± 0.07 [bA]	29,290 ± 5.66 [eE]	10,780 ± 4.24 [eE]	0.368 ± 0.00 [bD]	51.9 ± 0.14 [aB]	73.4 ±0.28 [dD]
SGF 5	68.7 ± 0.14 [a]	1665 ± 5.66 [a]	1335 ± 1.41 [a]	80.15 ± 0.21 [b]	39,190 ± 4.24 [d]	13,440 ± 2.83 [d]	0.343 ± 0.00 [e]	51.4 ± 0.14 [a]	74.1 ± 0.28 [cd]
SGF 10	67.3 ± 0.14 [b]	1567 ± 2.83 [b]	1200 ± 2.83 [c]	76.55 ± 0.07 [c]	44,120 ± 5.66 [c]	16,670 ± 2.83 [c]	0.378 ± 0.00 [a]	49.7 ± 0.14 [b]	74.8 ± 0.14 [bc]
SGF 15	65.9 ± 0.14 [c]	1534 ± 4.24 [c]	1235 ± 2.83 [b]	80.45 ± 0.07 [b]	55,060 ± 2.83 [b]	19,750 ± 2.83 [b]	0.359 ± 0.00 [c]	48.9 ± 0.28 [c]	75.5 ± 0.14 [ab]
SGF 20	62.6 ± 0.14 [d]	1360 ± 7.07 [d]	1176 ± 2.83 [d]	86.45 ± 0.21 [a]	64,920 ± 2.83 [a]	23,050 ± 4.24 [a]	0.355 ± 0.00 [d]	47.6 ± 0.14 [d]	76.3 ± 0.14 [a]
One-way ANOVA *p* values	*p* < 0.0001	*p* < 0.0001	*p* < 0.0001	*p* < 0.0001	*p* < 0.0001	*p* < 0.0001	*p* < 0.0001	*p* < 0.0001	*p* < 0.0001
LGF 2.5	69.1 ± 0.14 [D]	1631 ± 2.83 [D]	1281 ± 4.24 [B]	78.55 ± 0.35 [B]	31,480 ± 2.83 [D]	11,810 ± 4.24 [D]	0.375 ± 0.00 [B]	50.3 ± 0.00 [C]	74.6 ± 0.14 [C]
LGF 5	73.3 ± 0.14 [B]	1836 ± 5.66 [B]	1369 ± 5.66 [A]	74.50 ± 0.57 [C]	32,160 ± 5.66 [C]	12,500 ± 2.83 [C]	0.389 ± 0.00 [A]	52.3 ± 0.14 [B]	75.3 ± 0.14 [B]
LGF 7.5	77.0 ± 0.28 [A]	1886 ± 4.24 [A]	1282 ± 4.24 [B]	67.95 ± 0.35 [D]	40,600 ± 2.83 [A]	14,700 ± 5.66 [A]	0.362 ± 0.00 [E]	52.7 ±0.28 [AB]	75.8 ± 0.00 [B]
LGF 10	70.7 ± 0.14 [C]	1799 ± 4.24 [C]	1172 ± 5.66 [D]	65.10 ± 0.14 [E]	38,100 ± 2.83 [B]	14,130 ± 4.24 [B]	0.371 ± 0.00 [C]	53.5 ± 0.28 [A]	76.9 ± 0.00 [A]
One-way ANOVA *p* values	*p* < 0.0001	*p* < 0.0001	*p* < 0.0001	*p* < 0.0001	*p* < 0.0001	*p* < 0.0001	*p* < 0.0001	*p* < 0.0001	*p* < 0.0001

H'm—maximum height of gaseous production, VT—total CO_2 volume production, VR—volume of the gas retained in the dough at the end of the test, CR—retention coefficient, G'—elastic modulus, G"—viscous modulus, tan δ—loss tangent, T_i—initial gelatinization temperature, T_{max}—maximum gelatinization temperature. Soybean germinated flour (SGF) containing samples: a–e, mean values in the same column followed by different letters are significantly different ($p < 0.05$); Lentil germinated flour (LGF) containing samples: A–E, mean values in the same column followed by different letters are significantly different ($p < 0.05$).

3.3. Optimization of LGF and SGF Addition Levels

3.3.1. Diagnostic Checking of the Models

The data for falling number (FN), dough tolerance to kneading (Tol), dough consistency after 250 s (D250), dough consistency after 450 s (D450), baking strength (W), and curve configuration ratio (P/L) properties were successfully fitted ($p < 0.05$) to the quadratic model which explained 96, 74, 66, 75, 86 and 74% respectively of the variations, as the ANOVA results showed (Table 5). The 2FI mathematical model chosen for water absorption (WA) and dough extensibility (L) data fitting explained 74 and 87% respectively of the variation and it was significant at $p < 0.05$ in both cases. Dough tenacity (P) alveographic results were fitted to the cubic model which was significant at $p < 0.05$ and explained 77% of data variation.

Table 5. ANOVA results of the models fitted for FN and dough rheological properties during mixing and extension data.

Factors	Parameters								
	FN (s)	WA (%)	Tol (s)	D250 (mb)	D450 (mb)	P (mm)	L (mm)	W (10^{-4} J)	P/L (adim.)
Constant	254.50	52.72	191.11	372.23	933.41	114.85	46.72	196.66	2.49
A	−27.20 ***	−0.64 ***	−7.48	−18.12	−19.36	−10.66	−17.84 ***	−46.28 ***	1.09 ***
B	−47.40 ***	−0.92 ***	−34.60 ***	116.88 ***	77.04 ***	−7.93	−4.96 **	−30.56 ***	0.48 *
A × B	11.48 **	−0.12	−6.44	42.60	37.64 *	3.00	2.56	26.08 **	0.65 *
A^2	−0.6857		−13.31	4.97	37.03	−3.37		1.20	0.08
B^2	−2.80		−5.31	75.77 *	13.60	2.23		−0.9143	0.19
A^2B						−5.54			
AB^2						12.06 *			
A^3						114.85 *			
B^3						−10.66			
Model evaluation									
R^2	0.96	0.74	0.74	0.66	0.75	0.77	0.87	0.86	0.74
Adj.-R^2	0.95	0.70	0.67	0.57	0.69	0.64	0.85	0.82	0.68
p-value	0.0001	0.0001	0. 0001	0.0005	0.0001	0.0017	0.0001	0.0001	0.0001

*** $p < 0.001$, ** $p < 0.01$, * $p < 0.05$, A—soybean germinated flour (%), B—lentil germinated flour (%), R^2, *Adj.-R^2*—measures of model fit, FN—falling number, WA—water absorption, Tol—tolerance to kneading, D250—dough consistency after 250 s, D450—dough consistency after 450 s, P—dough tenacity, L—dough extensibility, W—baking strength, P/L—curve configuration ratio.

The quadratic model successfully fitted ($p < 0.05$) the data for the maximum height of gaseous production (H'm), the volume of gas retained in the dough at the end of the test (VR), elastic modulus (G′), loss tangent (tan δ) and initial gelatinization temperature (T_i), the variations were explained in proportions of 62 to 98% (Table 6). For total CO_2 volume production (VT) and maximum gelatinization temperature (T_{max}) data prediction, the cubic model was found to be adequate ($p < 0.05$) with an explained variation of 78 and 96% respectively, while the retention coefficient (CR) values were fitted to the modified cubic model which was significant at $p < 0.05$ and explained 61% of data variation (Table 6).

Table 6. ANOVA results of the mathematic models fitted for dough empirical and dynamic rheological properties data.

Factors	Parameters								
	H'm (mL)	VT (mL)	VR (mL)	CR (%)	G′ (Pa)	G″ (Pa)	tan δ (adim.)	T_i (°C)	T_{max} (°C)
Constant	69.57	1609.22	1283.41	79.77	49460.86	20268.40	0.4143	50.77	75.57
A	−3.22 ***	−151.59	−68.28 ***	−0.3634	16590.80 ***	7117.60 ***	0.0104	−1.65 ***	1.94 ***
B	0.50	−104.84	−19.32	3.84	4064.80 ***	3569.60 ***	0.0391 ***	0.92 ***	0.51 *
A × B	−1.06	−66.40	−3.52	2.55	−688.80	1040.00 **	0.0115	0.15	−0.44 ***
A^2	0.21	41.77	−50.57 *	79.77 *	1932.00 *		−0.0288 *	−0.33	0.45 **
B^2	−3.29 *	−101.66 *	−43.60	−0.3634	194.29		−0.0008	0.12	−0.09
A^2B		185.14 **		3.84 ***					0.39 *
AB^2		−39.43		2.55					75.57
A^3		36.40							1.94 ***
B^3		11.47							0.51
Model evaluation									

Table 6. *Cont.*

Factors	Parameters								
	H′m (mL)	**VT (mL)**	**VR (mL)**	**CR (%)**	**G′ (Pa)**	**G″ (Pa)**	**tan δ (adim.)**	**T$_i$ (°C)**	**T$_{max}$ (°C)**
R^2	0.62	0.78	0.62	0.61	0.98	0.98	0.71	0.83	0.96
Adj.-R^2	0.53	0.66	0.53	0.45	0.98	0.98	0.64	0.79	0.93
p-value	0.0013	0.0010	0.0012	0.0113	<0.0001	<0.0001	0.0001	<0.0001	<0.0001

*** $p < 0.001$, ** $p < 0.01$, * $p < 0.05$, A—soybean germinated flour (%), B—lentil germinated flour (%), R^2, *Adj.*-R^2—measures of model fit, H′m—maximum height of gaseous production, VT—total CO_2 volume production, VR—volume of the gas retained in the dough at the end of the test, CR—retention coefficient, G′—elastic modulus, G″—viscous modulus, tan δ—loss tangent, T$_i$—initial gelatinization temperature, T$_{max}$—maximum gelatinization temperature.

3.3.2. Effects of SGF and LGF on Falling Number and Dough Rheological Properties during Mixing and Extension

Flour properties and dough behavior during processing stages are influenced by the ingredients added, depending on their proportions and chemical composition. Falling number (FN) values decreased significantly ($p < 0.05$) when the SGF addition level increased (Figure 1), a similar trend was observed for LGF (Table 5). The interaction between factors significantly affected flour FN variation in a positive way.

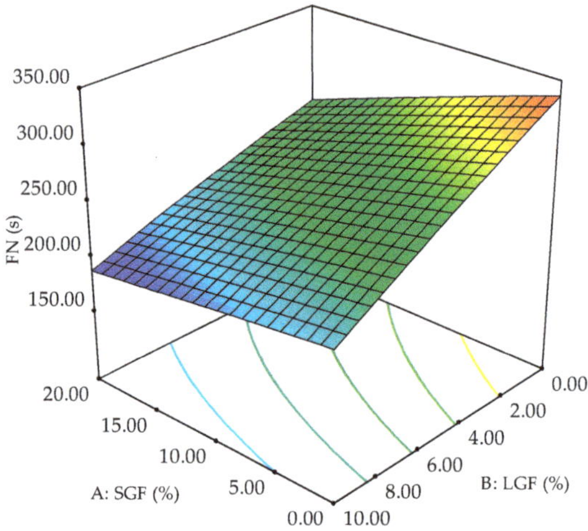

Figure 1. Three-dimensional response surface graphic presenting the interaction between SGF and LGF addition levels on flour falling number (FN).

Dough behavior during mixing was influenced by SGF and LGF addition in wheat flour. Water absorption registered a significant ($p < 0.05$) decrease (Figure 2a) as SGF and LGF addition levels raised, the interaction between factors had a non-significant ($p > 0.05$) effect (Table 5). Dough kneading tolerance showed significant ($p < 0.05$) decreases (Figure 2b) with LGF addition level increase, with SGF showing a non-significant effect ($p > 0.05$).

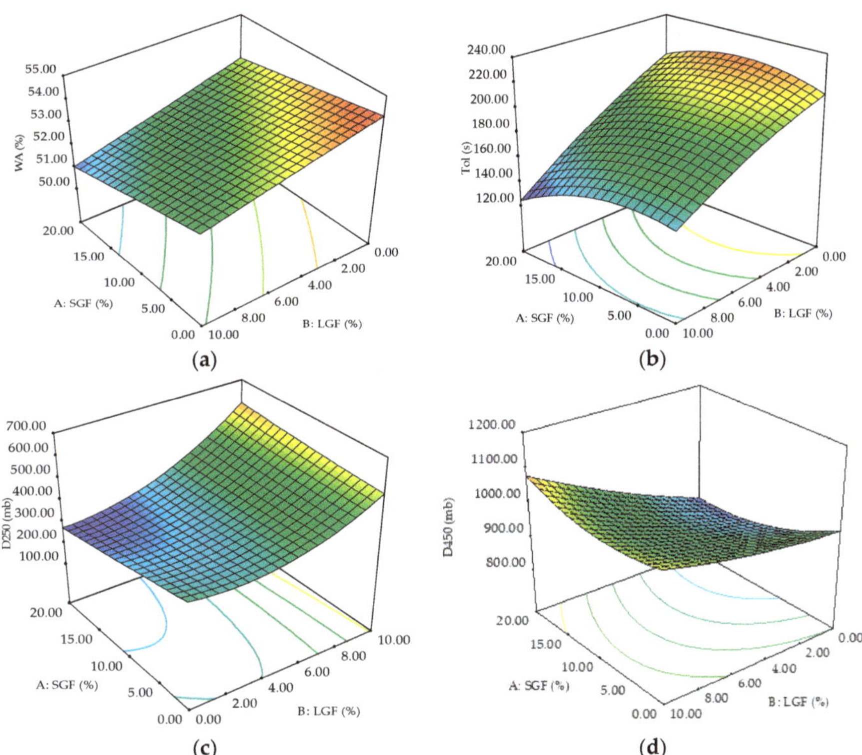

Figure 2. Three-dimensional response surface graphic presenting the interaction between SGF and LGF addition levels on dough properties during mixing: (**a**) water absorption (WA), (**b**) tolerance to kneading (Tol), (**c**) dough consistency after 250 s (D250), and (**d**) dough consistency after 450 s (D450).

Dough consistencies after 250 and 450 s respectively were significantly ($p < 0.05$) affected by the LGF factor (Table 5), while SGF and the interaction between factors had significant influence only on the D450 parameter. The rise in LGF amounts led to proportionally higher dough consistency parameters (Figure 2c,d).

The effects of factors on dough extension properties are presented in Figure 3.

Dough tenacity showed an irregular trend, rising with SGF levels and decreasing with LGF at levels higher than 2% (Figure 3), the effects being significant for the interaction between SGF and the quadratic term of LGF and for the cubic term on SGF. On the other hand, SGF increases in wheat flour caused a strong decrease ($p < 0.001$) of dough extensibility, LGF and SGF presented a significant effect (Figure 3b). Dough baking strength was significantly ($p < 0.05$) influenced by both factors and their interaction (Table 5), with decreases of its values being observed with the addition levels of germinated legumes (Figure 3c). The curve configuration ratio showed increases with raised SGF and LGF amounts (Figure 3d), with both factors and their interaction being significant ($p < 0.05$).

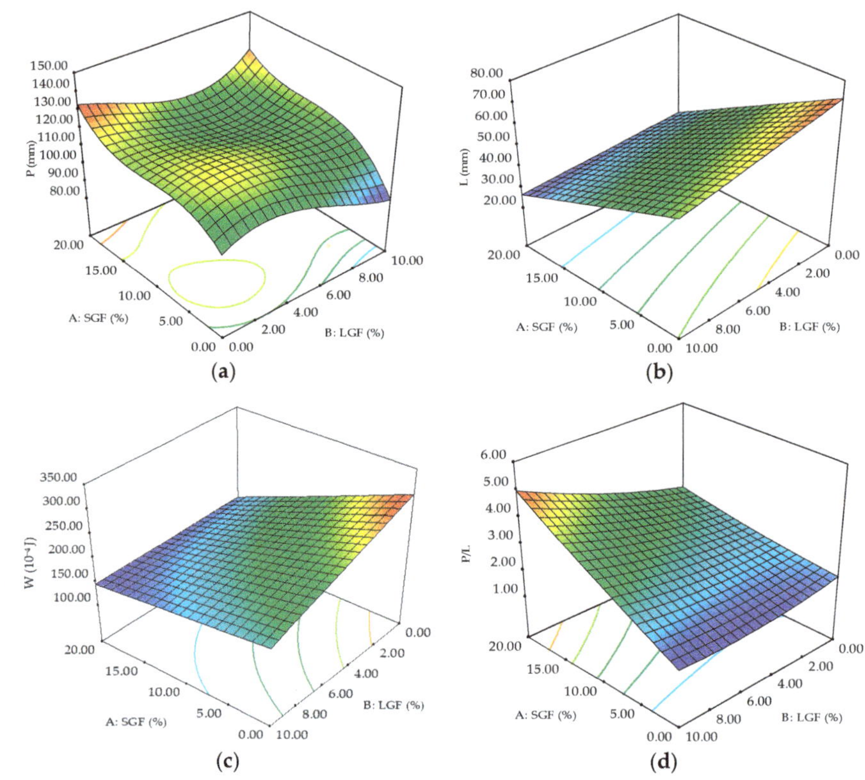

Figure 3. Three-dimensional response surface graphic presenting the interaction between SGF and LGF addition levels on dough properties during extension: (**a**) dough tenacity (P), (**b**) dough extensibility (L), (**c**) baking strength (W) and (**d**) curve configuration ratio (P/L).

3.3.3. Effects of SGF and LGF on Dough Fermentation and Dynamic Rheological Properties

Dough rheological properties are important for baked product processing optimization since they could predict dough behavior during mixing, fermentation, and handling. The maximum height of gaseous production showed a significant ($p < 0.05$) decrease as the addition level of SGF was higher, while LGF and the interaction between factors did not exert significant effects (Figure 4a). Only the interactions between the quadratic term of SGF with LGF factor and the quadratic term of LGF presented significant ($p < 0.05$) influence on the total CO_2 volume production (Table 6), while the volume of gas retained in the dough at the end of the test significantly decreased with increased SGF addition (Figure 4c). SGF quadratic term and the interaction between SGF quadratic term and LGF factor had significant ($p < 0.05$) effects (Table 6) on the dough retention coefficient, with a slightly decreasing trend being observed with LGF amount raise (Figure 4d), while in the case of SGF a reduction of up to 10% was observed, then the values increased.

The dynamic rheological properties in terms of elastic modulus, viscous modulus, and loss tangent were influenced by SGF and LGF addition in the wheat dough as follows: G' increased significantly ($p < 0.05$) as the amounts of SGF and LGF were higher (Figure 5a), while the interactions between them presented a non-significant ($p > 0.05$) effect (Table 6); G'' was affected by both factors and their interaction ($p < 0.05$), an increasing trend was observed with the increasing addition levels (Figure 5b); the loss tangent rose with SGF levels, increased up to 10% and increased as the amount of LGF was higher (Figure 5c), but only the quadratic term of SGF exerted a significant effect.

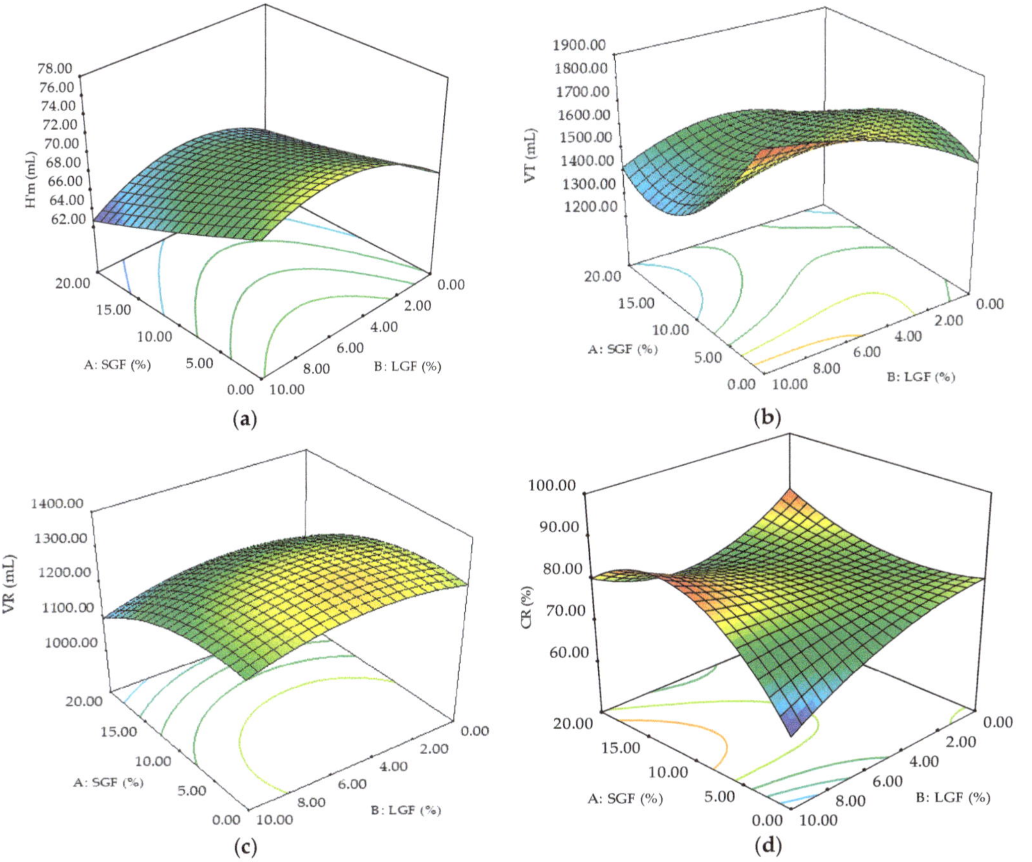

Figure 4. Three-dimensional response surface graphic presenting the interaction between SGF and LGF addition levels on dough rheological properties during fermentation: (**a**) maximum height of gaseous production (H′m), (**b**) total CO_2 volume production (VT), (**c**) volume of the gas retained in the dough at the end of the test (VR) and (**d**) retention coefficient (CR).

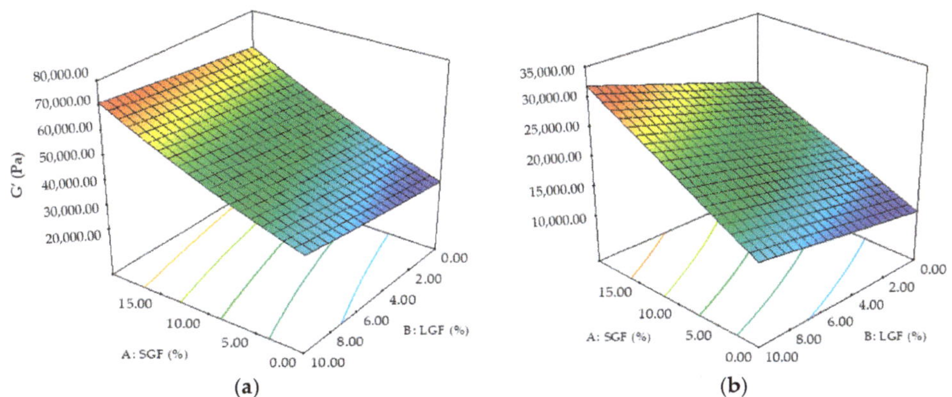

Figure 5. *Cont.*

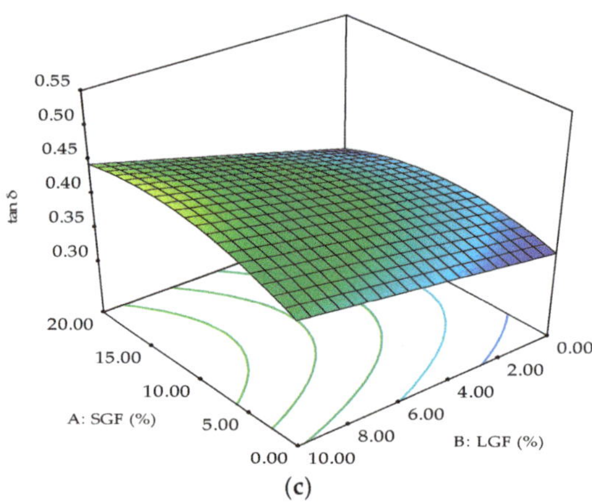

Figure 5. Three-dimensional response surface graphic presenting the interaction between SGF and LGF addition levels on dough dynamic rheological properties: (**a**) elastic modulus (G′), (**b**) viscous modulus (G″) and (**c**) loss tangent (tan δ).

Composite flour dough elastic and viscous moduli variations during heating provide valuable information on starch gelatinization which could be helpful in the prediction of dough behavior in the baking stage. The initial gelatinization temperature (T_i) is determined at the minimum value of G′, while the maximum gelatinization temperature (T_{max}) is considered at the maximum value of G″ [42]. The initial gelatinization temperature was significantly ($p < 0.05$) affected by SGF and LGF addition (Table 6), a decreasing trend being observed for SGF and the opposite trend for LGF as the addition level was higher (Figure 6a), while LGF and the interaction between factors showed a non-significant influence ($p > 0.05$). SGF and LGF terms exerted significant effects on the maximum gelatinization temperature, an increasing tendency being obtained as the germinated legume flour amounts increased (Figure 6b). A reverse trend on the maximum gelatinization temperature was given by the interaction of SGF with LGF.

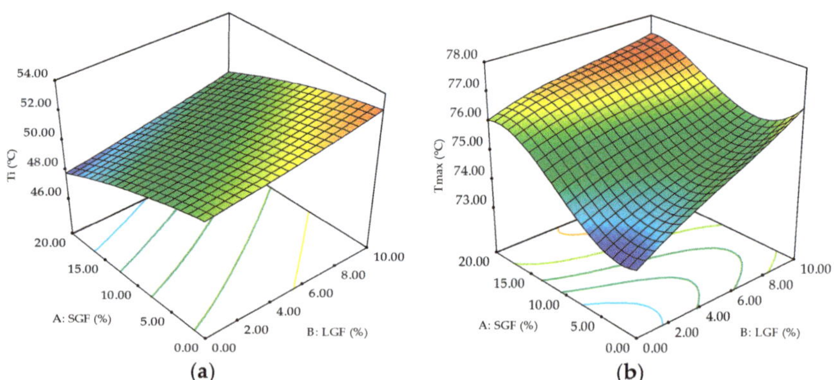

Figure 6. Three-dimensional response surface graphic presenting the interaction between SGF and LGF addition levels on dough rheological properties during heating: (**a**) initial gelatinization temperature (T_i), (**b**) maximum gelatinization temperature (T_{max}).

3.3.4. Optimal and Control Samples Properties

The optimal addition levels of SGF and LGF in wheat flour and the predicted values of the responses are presented in Table 7. The results of the optimization of the considered response revealed that the optimal formulation contains 5.60% SGF, 3.62% LGF, and 90.76%

wheat flour. The falling number and the rheological properties of the optimal sample showed significantly different ($p < 0.05$) values compared to the control, except for the loss tangent (Table 7). The FN, WA, Tol, D250, D450, L, W, CR, and T_i values of the optimal sample were lower compared to the control, while in the case of P, P/L, H'm, VT, VR, G', G'' and T_{max} higher values were obtained.

Table 7. Optimal vs. control sample properties.

Variable	Optimal Sample	Control
A: SGF (%)	5.60	0.00
B: LGF (%)	3.62	0.00
FN (s)	280.51 [b]	350.00 [a]
WA (%)	53.25 [b]	54.30 [a]
Tol (s)	200.19 [b]	214.00 [a]
D250 (mb)	359.80 [b]	394.00 [a]
D450 (mb)	933.32 [b]	943.00 [a]
P (mm)	119.95 [a]	104.00 [b]
L (mm)	56.20 [b]	72.00 [a]
W (10^{-4} J)	228.64 [b]	301.00 [a]
P/L (adim.)	1.99 [a]	1.43 [b]
H'm (mL)	70.50 [a]	65.90 [b]
VT (mL)	1684.98 [a]	1532.00 [b]
VR (mL)	1305.19 [a]	1228.00 [b]
CR (%)	78.78 [b]	80.10 [a]
G' (Pa)	41,384.57 [a]	29,290.00 [b]
G'' (Pa)	16,296.27 [a]	10,780.00 [b]
tan δ (adim.)	0.39 [a]	0.37 [a]
T_i (°C)	51.20 [b]	51.90 [a]
T_{max} (°C)	74.67 [a]	73.40 [b]

A: SGF—soybean germinated flour (%), B: LGF—lentil germinated flour (%), R^2, FN—falling number, WA—water absorption, Tol—tolerance to kneading, D250—dough consistency after 250 s, D450—dough consistency after 450 s, P—dough tenacity, L—dough extensibility, W—baking strength, P/L—curve configuration ratio, H'm—maximum height of gaseous production, VT—total CO_2 volume production, VR—the volume of the gas retained in the dough at the end of the test, CR—retention coefficient, G'—elastic modulus, G''—viscous modulus, tan δ—loss tangent, T_i—initial gelatinization temperature, T_{max}—maximum gelatinization temperature, [a,b] values followed by distinct letters in the same row are significantly different ($p < 0.05$).

3.4. Optimal and Control Dough Microstructure

Dough microstructures obtained for dough samples with and without germinated soybean and lentil addition are shown in Figure 7.

The images obtained show a dough structure with red areas interconnected with green areas in a homogeneous and continuous matrix. The red-colored areas indicate the presence of protein, whereas the green-colored areas depict the presence of starch. These different colors for dough compounds were determined by the two fluorochromes used in the EFLM technique namely rhodamine B and fluorescein. Rhodamin B is labeling in red the protein present in the dough system, whereas the fluorescein is labeling the starch granules in green [43]. From both images obtained it may be seen that starch granules are surrounded by a continuous protein network forming a fine dough matrix structure. For the optimal dough sample, it may be seen a slightly higher red area compared to the control due to the high protein content of this dough sample.

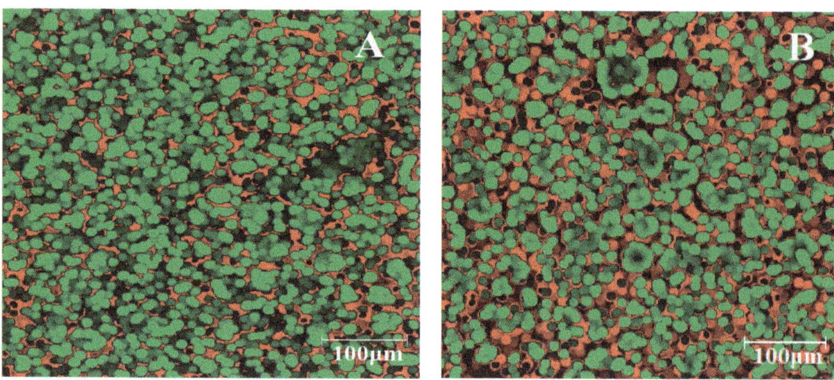

Figure 7. Microstructure taken by EFLM of wheat dough: (**A**) control sample and (**B**) optimal sample. Red is protein and green are starch granules.

4. Discussion

4.1. Effects of SGF and LGF on Falling Number and Dough Rheology

The rheological properties of dough could provide information about its behavior during mixing, extension, and fermentation, underlying also the influence of the ingredients added to the chemical composition that can inhibit or promote molecular interactions in the dough matrix [44]. The falling number is a measure of the α-amylase activity of flour and could be defined as the time necessary to stir and to allow the viscometer stirrer to fall an established distance through the aqueous flour gel undergoing liquefaction [43]. The decrease of falling number values with increased LGF and SGF addition indicated an increase of the α-amylase activity which could be related to the intake of calcium that stabilizes α-amylase [45]. Furthermore, these changes could be due to the enhanced activities of endogenous amylases found in the germinated legume flours which can promptly denature starch grains during heating along with wheat flour amylases, explaining the decrease in falling number values.

Water absorption was expressed as the quantity of water necessary to center the highest part of the mixing curve on the arbitrary 500 BU (Brabender units) [46]. The addition of SGF and LGF caused a reduction of water absorption probably due to the germ enzyme activities on starch grains, causing their hydrolyzation to dextrins which presents low water binding capacity and/or to the proteins de-polymerization as a result of the intense protease activity in germinated flours, similar findings were reported by Hejri-Zarifi et al. [46] and by Marti et al. [47]. Water absorption decrease could be related to the lower falling number values since it is known that high α-amylase activity could give lower water absorption [48]. The decrease in water absorption could be due to protein de-polymerization as a consequence of the intense protease activity in germinated wheat [48]. Dough kneading tolerance showed significant decreases ($p < 0.05$) as the addition level of LGF was higher, similar results being reported by Eissa et al. [49] for Egyptian Balady Bread and biscuits supplemented with germinated legume seeds flours. Dough kneading tolerance increased with increases of SGF up to 10%. Shorter stability of the wheat flour supplemented with germinated soy flour was found by Rosales-Juarez [28], while Sadowska et al. [50] reported dough stability prolongation for the wheat flour with different doses of germinated pea flour added. The decrease of kneading tolerance can be caused by peptidase formed during germination which determines the advanced disruption of the protein network. Kneading tolerance reduction showed a weakening of the gluten matrix structure that could be attributed to a noticeable incompatibility between the protein spectrum of legume flours and wheat gluten protein [51]. It is supposed that with the increase of germinated legumes quantity in the composite flour, the energy required for the optimal development of dough consistency raised, which was related to an increased mechanical agitation need, caused by the non-gluten proteins from the dough system [52]. Dough consistency increases proportionally to the LGF amounts and could be explained by the chemical composition of LGF [14], these results underlying their positive effects on dough rheological properties and confirming the possibility to improve low-protein wheat flours for breadmaking. Similar results were reported by Mohammed et al. [53] for wheat flour enriched with chickpea. The presence of fiber and proteins of germinated legume flours could lead to more intense interactions with water which will contribute to the formation of a more consistent gluten network [54]. Dough consistency changes could be related to the hydrolysis products resulting from germination [55].

The addition of SGF in wheat flour caused the increase of dough tenacity, while LGF determined the decrease of this parameter at levels higher than 10%. The increase of dough tenacity with SGF addition can be related to the ascorbic acid content which increases during germination [56]. Dough extensibility was reduced as SGF and LGF addition levels rose, while the P/L ratio registered an opposite trend, similar to the results reported by Hernandez-Aguilar et al. [57] for wheat dough supplemented with germinated lentil flour. These results supported the data obtained for the viscoelastic moduli which increased as the amounts of SGF and LGF were higher, indicating a dough with greater rigidity which is not easy to handle due to its low extensibility. SGF and LGF factors and their interaction determined the decrease of dough baking strength, suggesting a weakening of the gluten matrix. The decrease of the alveographic parameters could be due to the dietary fiber compositions of SGF and LGF which led to dough strength and stability changes, probably as a result of the small numbers of hydroxyl groups of fiber that can interact with water through hydrogen bonding, which will impact gluten network compactness [58]. The decrease in dough compactness could be related to the disruption of the well-defined protein–starch complex in wheat flour dough by the exogenous proteins, as previously stated [58].

Dough behavior during fermentation can be evaluated using rheofermentometer parameters. Gluten networks developed through mixing properties are essential for gas retention and the final structure of bread [1]. The maximum height of gaseous production and volume of gas retained in the dough at the end of the test decreased with the increase of SGF added in wheat flour, except for the sample with 5% SGF. This behavior could suggest the collapse of dough structures due to the reduction in the ability of the gluten network to withstand the physical stresses as a result of proteolytic activity [47]. When LGF was incorporated into wheat flour, the maximum height of gaseous production increased significantly, up to 7.5% and then decreased, but the value remained higher compared to the control. Total CO_2 volume production was higher in the dough with LGF compared to dough with SGF. On the other hand, the retention coefficient registered higher values in samples with SGF and the best final bread volume could be expected. The addition of SGF led to the decrease of the maximum height of gaseous production and volume of gas retained in the dough at the end of the test as the amount was higher, probably as a result of the gluten matrix dilution effect [59]. The gas retention coefficient expressed as the ratio between the volume of gas retained by the dough and the total volume of gas produced during the test decreased with SGF up to 10% and LGF addition levels increased. Gas retention reduction led to higher dough permeability due to the gluten matrix weakening caused by amylose and amylopectin hydrolysis and could be affected by the enzyme's activities during germination [60]. Furthermore, protease enzymes could hydrolyze peptide bonds, which could promote the partial denaturation of the protein network and thus reduce the dough's ability to enclose air [47]. Our results were in agreement with those reported by Suarez-Estrella et al. [61] for wheat dough enriched with germinated quinoa flour. Probably, the increasing availability of mono- and disaccharides as substrates for yeast due to legume germination enhanced the carbon dioxide produced during fermentation [47].

All the samples included in this study showed a solid-like behavior since $G' > G''$, the visco-elastic moduli increased with frequency. Dough elastic and viscous moduli presented higher values as the SGF and LGF amounts raised. The loss tangent increased at SGF levels up to 10%, then it was reduced and raised as the amount of LGF was higher, confirming the positive effects of germinated legume flours for wheat flour since dough from stronger flour has G' values higher compared to weaker ones [62]. The proteins found in legumes can influence water distribution within the dough matrix with significant implications in components interactions [63]. The incorporation of higher amounts of germinated flours could alter the starch-gluten matrix, influencing the viscoelastic behavior of dough and cumbering its handling, similar observations being made by Hernandez-Aguilar et al. [57] for wheat dough enriched with germinated lentil. Germination causes the decrease of wheat starch crystallinity, the enzymes activated during germination preferentially hydrolyze the amorphous starch areas which led to the raised double-helical ordered structure, contributing to the increase in the formation of the gel structure [64]. The loss tangent could be a measure of the structural order (molecular interactions) of dough, with low tan δ values suggesting a rigid and stiff mass, while higher values led to a moist and slack dough [65,66]. Loss tangent increase indicated the depletion of the elastic character of dough, probably as a result of the incorporation of non-gluten flours, such as SGF and LGF. These changes could be possibly due to the presence of low molecular mass molecules caused by de-polymerization during germination of soybean and lentil which will contribute to the increase of the viscous character of dough samples [63]. Legume flours led to the increase of dough fiber proportion, the effect on the rheological behavior of dough being possibly also attributed to interactions between the fiber structure and wheat proteins [1].

The decrease of the initial gelatinization temperature T_i with SGF addition level increase and the rise of T_i in the case of LGF and T_{max} in the case of SGF could be related to the starch structure which was proven to influence dough behavior during heating [67]. Furthermore, the amylose and amylopectin ratio, the degree of heterogeneity, and the amounts of amylase-lipid complexes could have been impacted the gelatinization temperatures [68]. Probably, these results could be also explained by the activation of enzymes during germination, increasing the α-amylase, proteolytic and lipolytic activities [55]. The decrease of the maximum gelatinization temperature could suggest that germination altered soybean and lentil starch granule surface, determining higher resistance to temperature changes, complying with findings reported by Frias et al. [69].

4.2. Optimal Addition Levels of SGF and LGF

The optimization of SGF and LGF addition in wheat flour resulted in an optimal combination of 5.60% SGF, 3.62% LGF, and 90.76% wheat flour. The differences regarding the falling number and rheological properties between the optimal and control sample could be related to the intake of proteins, lipids, carbohydrates, and minerals of the germinated flours and their interactions within the dough matrix [8]. According to the obtained results, both optimal and control flours are considered strong ones, the enriched sample presenting moderate extension properties (P/L > 0.5) and higher α-amylase activity (lower FN) [48]. Higher dough total CO_2 volume production and volume of the gas retained in the dough at the end of the test of the optimal sample compared to the control could be due to the increased amount of fermentable sugars along with the activation of amylase during germination of legumes, leading to enhanced CO_2 production [14].

The images obtained of the dough microstructure showed slight differences among the structures of dough samples. The addition of germinated soybean and lentil led to a lower green area and a higher red one compared to the control due to the higher level of protein from the enriched dough. This ratio color change from the dough structure is due to the high level of protein from the LGF and SGF compared with wheat flour which is partially replaced by them. Both dough samples' structures appeared compact, homogenous, in which starch granules were enveloped by proteins, being glued together. No black regions are present in these images, meaning that LGF and SGF at these addition levels did not affect in a negative way dough structure which presents good rheological characteristics, such as elasticity, gas holding capacity, and extensibility.

5. Conclusions

Legume flour potential can be increased by applying a germination process, allowing the attainment of high nutritive bakery products, with minimum impairment of quality attributes. The results obtained in this study revealed that SGF and LGF influenced dough behavior during mixing, extension, and fermentation. The mathematical modeling of data allowed the interpretation of the effects of SGF and LGF factors, along with their interactions, on flour and dough properties, the explanation rate of the models proposed varied between 61 and 96%.

LGF led to the decrease of falling number, water absorption, kneading tolerance, dough extensibility, and baking strength, while dough consistency, configuration ratio of the Alveograph curve, visco-elastic moduli, loss tangent, initial and maximum gelatinization temperature increased proportionally with the amount used. SGF increases induced lower values of falling number, water absorption, dough extensibility, baking strength, the maximum height of gaseous production, the volume of gas retained in the dough at the end of the test, and initial gelatinization temperature, while the configuration ratio of the Alveograph curve, elastic and viscous moduli and maximum gelatinization temperature was raised with increased SGF. The interactions between SGF and LGF exerted significant ($p < 0.05$) influences on the falling number, dough consistency after 450 s, dough baking strength, viscous modulus, and maximum gelatinization temperature.

The optimal combination of SGF and LGF in wheat flour was found to be 5.60 and 3.62% respectively. Compared to the control, the optimal sample showed lower falling number, water absorption, tolerance to kneading, consistency of dough, extensibility, and initial gelatinization temperature, while for dough tenacity, the maximum height of gaseous production, total CO_2 volume production, the volume of the gas retained in the dough at the end of the test, visco-elastic moduli and maximum gelatinization temperatures higher values were obtained. Dough rheological property variations when germinated legume flours are added to wheat flour knowledge could help producers to optimize the production and recipes of improved bakery products, according to the consumers and technologies requirements. There is a scarcity of papers underlying the effects of germinated legumes, such as soybean and lentil on wheat dough rheological behavior and a lack of information regarding their combined effects. Thus, the results presented in this work bring useful information about the simultaneous effects of two germinated legume flours in the wheat dough, fulfilling the state of art regarding germination application and incorporation of legume flours in wheat bread production. Further research regarding bread quality parameters as influenced by SGF and LGF should be performed.

Author Contributions: The authors contributed equally to this research. All authors have read and agreed to the published version of the manuscript.

Funding: This work was funded for its research by Romanian Ministry of Education and Research, CNCS–UEFISCDI, project number PN-III-P1-1.1-TE-2019-0892.

Institutional Review Board Statement: Not applicable.

Informed Consent Statement: Not applicable.

Data Availability Statement: Not applicable.

Acknowledgments: This work was supported by a grant from the Romanian Ministry of Education and Research, CNCS–UEFISCDI, project number PN-III-P1-1.1-TE-2019-0892, within PNCDI III.

Conflicts of Interest: The authors declare no conflict of interest.

References

1. Bojňanská, T.; Musilová, J.; Vollmannová, A. Effects of adding legume flours on the rheological and breadmaking properties of dough. *Foods* **2021**, *10*, 1087. [CrossRef] [PubMed]
2. Ciudad-Mulero, M.; Barros, L.; Fernandes, Â.; Ferreira, I.C.F.R.; Callejo, M.J.; Matallana-González, M.C.; Fernández-Ruiz, V.; Morales, P.; Carrillo, J.M. Potential health claims of durum and bread wheat flours as functional ingredients. *Nutrients* **2020**, *12*, 504. [CrossRef] [PubMed]
3. Mironeasa, S. Mironeasa Costel Dough bread from refined wheat flour partially replaced by grape peels: Optimizing the rheological properties. *J. Food Process Eng.* **2019**, *42*, e13207. [CrossRef]
4. Yu, L.; Nanguet, A.L.; Beta, T. Comparison of antioxidant properties of refined and whole wheat flour and bread. *Antioxidants* **2013**, *2*, 370–383. [CrossRef] [PubMed]
5. Brennstuhl, M.J.; Martignon, S.; Tarquinio, C. Diet and mental health: Food as a path to happiness? *Nutr. Clin. Metab.* **2021**, *35*, 168–183. [CrossRef]
6. Bresciani, A.; Marti, A. Using Pulses in Baked Products: Lights, Shadows, and Potential Solutions. *Foods* **2019**, *8*, 451. [CrossRef] [PubMed]
7. Belc, N.; Duta, D.E.; Culetu, A.; Stamatie, G.D. Type and amount of legume protein concentrate influencing the technological, nutritional, and sensorial properties of wheat bread. *Appl. Sci.* **2021**, *11*, 436. [CrossRef]
8. Atudorei, D.; Codina, G.G. Perspectives on the use of germinated legumes in the bread making process, a review. *Appl. Sci.* **2020**, *10*, 6244. [CrossRef]
9. Erbersdobler, H.; Barth, C.; Jahreis, G. Grain Legumes in the Human Nutrition Nutrient Content and Protein Quality of Pulses. *Ernährungs-Umsch. Forsch. Prax.* **2017**, *64*, 550–554. [CrossRef]
10. Ma, Z.; Boye, J.I.; Hu, X. Nutritional quality and techno-functional changes in raw, germinated and fermented yellow field pea (*Pisum sativum* L.) upon pasteurization. *LWT Food Sci. Technol.* **2018**, *92*, 147–154. [CrossRef]
11. Ohanenye, I.C.; Tsopmo, A.; Ejike, C.E.C.C.; Udenigwe, C.C. Germination as a bioprocess for enhancing the quality and nutritional prospects of legume proteins. *Trends Food Sci. Technol.* **2020**, *101*, 213–222. [CrossRef]
12. Atudorei, D.; Stroe, S.G.; Codină, G.G. Impact of germination on the microstructural and physicochemical properties of different legume types. *Plants* **2021**, *10*, 592. [CrossRef] [PubMed]
13. Kaczmarska, K.T.; Chandra-Hioe, M.V.; Frank, D.; Arcot, J. Enhancing wheat muffin aroma through addition of germinated and fermented Australian sweet lupin (*Lupinus angustifolius* L.) and soybean (*Glycine max* L.) flour. *LWT* **2018**, *96*, 205–214. [CrossRef]
14. Atudorei, D.; Atudorei, O.; Codină, G.G. Dough Rheological Properties, Microstructure and Bread Quality of Wheat-Germinated Bean Composite Flour. *Foods* **2021**, *10*, 1542. [CrossRef] [PubMed]
15. Wang, N.; Hatcher, D.W.; Toews, R.; Gawalko, E.J. Influence of cooking and dehulling on nutritional composition of several varieties of lentils (*Lens culinaris*). *LWT Food Sci. Technol.* **2009**, *42*, 842–848. [CrossRef]
16. Monnet, A.F.; Laleg, K.; Michon, C.; Micard, V. Legume enriched cereal products: A generic approach derived from material science to predict their structuring by the process and their final properties. *Trends Food Sci. Technol.* **2019**, *86*, 131–143. [CrossRef]
17. Romano, A.; Gallo, V.; Ferranti, P.; Masi, P. Lentil flour: Nutritional and technological properties, in vitro digestibility and perspectives for use in the food industry. *Curr. Opin. Food Sci.* **2021**, *40*, 157–167. [CrossRef]
18. Asif, M.; Rooney, L.W.; Ali, R.; Riaz, M.N. Application and Opportunities of Pulses in Food System: A Review. *Crit. Rev. Food Sci. Nutr.* **2013**, *53*, 1168–1179. [CrossRef]
19. García-Alonso, A.; Goñi, I.; Saura-Calixto, F. Resistant starch and potential glycaemic index of raw and cooked legumes (lentils, chickpeas and beans). *Eur. Food Res. Technol.* **1998**, *206*, 284–287. [CrossRef]
20. Perri, G.; Coda, R.; Rizzello, C.G.; Celano, G.; Ampollini, M.; Gobbetti, M.; De Angelis, M.; Calasso, M. Sourdough fermentation of whole and sprouted lentil flours: In situ formation of dextran and effects on the nutritional, texture and sensory characteristics of white bread. *Food Chem.* **2021**, *355*, 129638. [CrossRef]
21. Dhingra, S.; Jood, S. Organoleptic and nutritional evaluation of wheat breads supplemented with soybean and barley flour. *Food Chem.* **2002**, *77*, 479–488. [CrossRef]
22. Osundahunsi, O.F.; Amosu, D.; Ifesan, B.O.T. Quality evaluation and acceptability of soy-yoghurt with different colours and fruit flavours. *Am. J. Food Technol.* **2007**, *2*, 273–280. [CrossRef]
23. Bolarinwa, I.F.; Oyesiji, O.O. Gluten free rice-soy pasta: Proximate composition, textural properties and sensory attributes. *Heliyon* **2021**, *7*, e06052. [CrossRef]
24. Hammond, E.G.; Murphy, P.A.; Johnson, L.A. SOY (SOYA) BEANS | Properties and Analysis. *Encycl. Food Sci. Nutr.* **2003**, 5389–5392. [CrossRef]
25. Zhang, Y.; Hu, R.; Tilley, M.; Siliveru, K.; Li, Y. Effect of pulse type and substitution level on dough rheology and bread quality of whole wheat–based composite flours. *Processes* **2021**, *9*, 1687. [CrossRef]
26. Marchini, M.; Carini, E.; Cataldi, N.; Boukid, F.; Blandino, M.; Ganino, T.; Vittadini, E.; Pellegrini, N. The use of red lentil flour in bakery products: How do particle size and substitution level affect rheological properties of wheat bread dough? *LWT* **2021**, *136*, 110299. [CrossRef]
27. Mashayekh, M.; Mahmoodi, M.R.; Entezari, M.H. Effect of fortification of defatted soy flour on sensory and rheological properties of wheat bread. *Int. J. Food Sci. Technol.* **2008**, *43*, 1693–1698. [CrossRef]

28. Rosales-Juárez, M.; González-Mendoza, B.; López-Guel, E.C.; Lozano-Bautista, F.; Chanona-Pérez, J.; Gutiérrez-López, G.; Farrera-Rebollo, R.; Calderón-Domínguez, G. Changes on dough rheological characteristics and bread quality as a result of the addition of germinatedand non-germinated soybean flour. *Food Bioprocess Technol.* **2008**, *1*, 152–160. [CrossRef]

29. López-Guel, E.C.; Lozano-Bautista, F.; Mora-Escobedo, R.; Farrera-Rebollo, R.R.; Chanona-Pérez, J.; Gutiérrez-López, G.F.; Calderón-Domínguez, G. Effect of Soybean 7S Protein Fractions, Obtained from Germinated and Nongerminated Seeds, on Dough Rheological Properties and Bread Quality. *Food Bioprocess Technol.* **2012**, *5*, 226–234. [CrossRef]

30. ICC. *Standard Methods of the International Association for Cereal Chemistry. Methods 104/1, 110/1, 136, 105/2, 171, 121, 107/1*; International Association for Cereal Chemistry: Vienna, Austria, 2010.

31. SR ISO 7954:2001. *Microbiology—General Guidance for Enumeration of Yeasts and Moulds—Colony Count Technique at 25 Degrees C*; ANSI: New York, NY, USA, 2001.

32. SR EN ISO 7932:2005. *Microbiology of Food and Feed. Horizontal Method for Counting Presumptive Bacillus Cereus. Colony Counting Technique at 30 Degrees C*; National Standards Authority of Ireland: Dublin, Ireland, 2005.

33. SR 90:2007. *Wheat Flour. Analysis Method*; Romanian Standards Association: Bucharest, Romania, 2007.

34. AACC. *Approved Methods of Analysis, 11th Edition—AACC Method 89-01.01. Cereals & Grains Association. Yeast Activity, Gas Production*; Cereals & Grains Association: St. Paul, MN, USA, 2000.

35. Codină, G.G.; Zaharia, D.; Stroe, S. Influence of calcium ions addition from gluconate and lactate salts on refined wheat flour dough rheological properties. *CyTA J. Food* **2018**, *16*, 884–891. [CrossRef]

36. Iuga, M.; Mironeasa, S. Grape seeds effect on refined wheat flour dough rheology: Optimal amount and particle size. *Ukr. Food J.* **2019**, *8*, 799–814. [CrossRef]

37. Mironeasa, S.; Codină, G.G. Dough Rheological Behavior and Microstructure Characterization of Composite Dough with Wheat and Tomato Seed Flours. *Foods* **2019**, *8*, 626. [CrossRef] [PubMed]

38. Mironeasa, S.; Codină, G. Influence of Mixing Speed on Dough Microstructure and Rheology. *Food Technol. Biotechnol.* **2013**, *51*, 509–519.

39. Codină, G.G.; Dabija, A.; Oroian, M. Prediction of pasting properties of dough from mixolab measurements using artificial neuronal networks. *Foods* **2019**, *8*, 447. [CrossRef] [PubMed]

40. European Comission Recommendation (EC) 2006/576. *Off. J. Eur. Union* **2006**, *L285*, 30–50.

41. European Comission Regulation (EC) No. 1881/2006. *Off. J. Eur. Union* **2006**, *364*, 5.

42. Pereira, P.M.; Oliveira, J.C. Measurement of glass transition in native wheat flour by dynamic mechanical thermal analysis (DMTA). *Int. J. Food Sci. Technol.* **2000**, *35*, 183–192. [CrossRef]

43. Mudawi, H.A.; Mohamed, S.; Abdelrahim, K.; Musa, R.; Elawad, O.; Yang, T.A.; Halim, A.; Ahmed, R.; Elnur, K.; Ishag, A. Chemical composition and functional properties of wheat bread containing wheat and legumes bran. *Int. J. Food Sci. Nutr.* **2016**, *1*, 2455–4898.

44. Chinachoti, P.; Vodovotz, Y. *Bread Staling*; CRC Press: Boca Raton, FL, USA, 2018; ISBN 1351078798.

45. Mironeasa, S.; Codină, G.G.; Mironeasa, C. The effects of wheat flour substitution with grape seed flour on the rheological parameters of the dough assessed by mixolab. *J. Texture Stud.* **2012**, *43*, 40–48. [CrossRef]

46. Hejri-Zarifi, S.; Ahmadian-Kouchaksaraei, Z.; Pourfarzad, A.; Khodaparast, M.H.H. Dough performance, quality and shelf life of flat bread supplemented with fractions of germinated date seed. *J. Food Sci. Technol.* **2014**, *51*, 3776–3784. [CrossRef]

47. Marti, A.; Cardone, G.; Pagani, M.A.; Casiraghi, M.C. Flour from sprouted wheat as a new ingredient in bread-making. *LWT Food Sci. Technol.* **2018**, *89*, 237–243. [CrossRef]

48. Dapcevic, T.; Pojic, M.; Hadnaev, M.; Torbic, A. The Role of Empirical Rheology in Flour Quality Control. *Wide Spectra Qual. Control* **2011**, 335–360. [CrossRef]

49. Eissa, H.; Hussein, A.; Mostafa, B. Rheological Properties and Quality Evaluation of Egyptian Balady Bread and Biscuits Supplemented With Flours of Ungerminated and Germinated Legume Seeds or Mushroom. *Pol. J. Food Nutr. Sci.* **2007**, *57*, 487–496.

50. Sadowska, J.; Błaszczak, W.; Fornal, J.; Vidai-Valverde, C.; Frias, J. Changes of wheat dough and bread quality and structure as a result of germinated pea flour addition. *Eur. Food Res. Technol.* **2003**, *216*, 46–50. [CrossRef]

51. Kohajdová, Z.; Karovičová, J.; Magala, M. Effect of lentil and bean flours on rheological and baking properties of wheat dough. *Chem. Pap.* **2013**, *67*, 398–407. [CrossRef]

52. Mohammed, I.; Ahmed, A.R.; Senge, B. Effects of chickpea flour on wheat pasting properties and bread making quality. *J. Food Sci. Technol.* **2014**, *51*, 1902–1910. [CrossRef]

53. Mohammed, I.; Ahmed, A.R.; Senge, B. Dough rheology and bread quality of wheat-chickpea flour blends. *Ind. Crops Prod.* **2012**, *36*, 196–202. [CrossRef]

54. Guardianelli, L.; Puppo, M.C.; Salinas, M.V. Influence of pistachio by-product from edible oil industry on rheological, hydration, and thermal properties of wheat dough. *LWT* **2021**, *150*, 111917. [CrossRef]

55. Ouazib, M.; Garzon, R.; Zaidi, F.; Rosell, C.M. Germinated, toasted and cooked chickpea as ingredients for breadmaking. *J. Food Sci. Technol.* **2016**, *53*, 2664–2672. [CrossRef]

56. Bau, H.M.; Villaume, C.; Nicolas, J.P.; Méjean, L. Effect of germination on chemical composition, biochemical constituents and antinutritional factors of soya bean (*Glycine max*) seeds. *J. Sci. Food Agric.* **1997**, *73*, 1–9. [CrossRef]

57. Hernandez-Aguilar, C.; Dominguez-Pacheco, A.; Palma Tenango, M.; Valderrama-Bravo, C.; Soto Hernández, M.; Cruz-Orea, A.; Ordonez-Miranda, J. Lentil sprouts: A nutraceutical alternative for the elaboration of bread. *J. Food Sci. Technol.* **2020**, *57*, 1817–1829. [CrossRef] [PubMed]

58. Ficco, D.B.M.; Muccilli, S.; Padalino, L.; Giannone, V.; Lecce, L.; Giovanniello, V.; Del Nobile, M.A.; De Vita, P.; Spina, A. Durum wheat breads 'high in fibre' and with reduced in vitro glycaemic response obtained by partial semolina replacement with minor cereals and pulses. *J. Food Sci. Technol.* **2018**, *55*, 4458–4467. [CrossRef]

59. Kumar, D.; Mu, T.; Ma, M. Effects of potato flour on dough properties and quality of potato-wheat-yogurt pie bread. *Nutr. Food Sci.* **2020**, *50*, 885–901. [CrossRef]

60. Sanz Penella, J.M.; Collar, C.; Haros, M. Effect of wheat bran and enzyme addition on dough functional performance and phytic acid levels in bread. *J. Cereal Sci.* **2008**, *48*, 715–721. [CrossRef]

61. Suárez-Estrella, D.; Cardone, G.; Buratti, S.; Pagani, M.A.; Marti, A. Sprouting as a pre-processing for producing quinoa-enriched bread. *J. Cereal Sci.* **2020**, *96*, 103111. [CrossRef]

62. Singh, H.; Singh, N.; Kaur, L.; Saxena, S.K. Effect of sprouting conditions on functional and dynamic rheological properties of wheat. *J. Food Eng.* **2001**, *47*, 23–29. [CrossRef]

63. Patrascu, L.; Vasilean, I.; Turtoi, M.; Garnai, M.; Aprodu, I. Pulse germination as tool for modulating their functionality in wheat flour sourdoughs. *Qual. Assur. Saf. Crop. Foods* **2019**, *11*, 269–282. [CrossRef]

64. Kun, G.; Yanxiang, L.; Bin, T.; Xiaohong, T.; Duqin, Z.; Liping, W. An insight into the rheology and texture assessment: The influence of sprouting treatment on the whole wheat flour. *Food Hydrocoll.* **2021**, 107248. [CrossRef]

65. Iuga, M.; Boestean, O.; Ghendov-Mosanu, A.; Mironeasa, S. Impact of Dairy Ingredients on Wheat Flour Dough Rheology and Bread Properties. *Foods* **2020**, *9*, 828. [CrossRef]

66. Weipert, D. 13 Fundamentals of Rheology and Spectrometry. *Fundam. Rheol. Spectrom.* **1980**, 117–146. Available online: https://www.scribd.com/document/312893026/Bra%C5%A1no-Rheology (accessed on 29 November 2021).

67. He, H.; Hoseney, R.C. Differences in gas retention, protein solubility, and rheological properties between flours of different baking quality. *Cereal Chem.* **1991**, *68*, 526–530.

68. Jan, R.; Saxena, D.C.; Singh, S. Comparative study of raw and germinated Chenopodium (*Chenopodium album*) flour on the basis of thermal, rheological, minerals, fatty acid profile and phytocomponents. *Food Chem.* **2018**, *269*, 173–180. [CrossRef] [PubMed]

69. Frias, J.; Fornal, J.; Ring, S.G.; Vidal-Valverde, C. Effect of germination on physico-chemical properties of lentil starch and its components. *LWT Food Sci. Technol.* **1998**, *31*, 228–236. [CrossRef]

applied sciences

Article

Synergistic Effects of Heat-Moisture Treatment Regime and Grape Peels Addition on Wheat Dough and Pasta Features

Mădălina Iuga *, Ana Batariuc and Silvia Mironeasa *

Faculty of Food Engineering, Stefan Cel Mare University of Suceava, 13 Universitatii Street, 720229 Suceava, Romania; ana.batariuc@usm.ro
* Correspondence: madalina.iuga@usm.ro (M.I.); silviam@fia.usv.ro (S.M.)

Abstract: Heat moisture treatment (HMT) can be a useful method of wheat flour functionality modification, improving the nutritional value of pasta along with grape peels (GPF) addition. The aim of this study was to investigate the combined effects of HMT temperature, time, and moisture and GPF level on dough and pasta properties. Dough rheology and texture, pasta color, texture, total polyphenols (TPC), dietary fiber (DF), and resistant starch (RS) contents were evaluated. Furthermore, an optimization was performed based on Response Surface Methodology (RSM) and desirability function. The results showed that HMT regime and GPF determined proportional dough viscoelastic moduli and firmness increase. On the other hand, cooked pasta firmness and gumminess decreased with HMT conditions and GPF level rise. Higher pasta RS and DF content was promoted by HMT and GPF components. The reduction effect of HMT on TPC was countered by the incorporation of GPF, a rich source of polyphenols. The optimization revealed that the recommended wheat flour treatment regime would be 87.56 °C, 3 h, and 26.01% moisture, while the quantity of GPF that could be added was 4.81%. For these values, the maximum functional and nutritional values would be achieved with minimum negative impact on pasta quality.

Keywords: wheat flour; grape peels; heat-moisture treatment; pasta; functional ingredients

Citation: Iuga, M.; Batariuc, A.; Mironeasa, S. Synergistic Effects of Heat-Moisture Treatment Regime and Grape Peels Addition on Wheat Dough and Pasta Features. *Appl. Sci.* **2021**, *11*, 5403. https://doi.org/10.3390/app11125403

Academic Editor: Francisco Artés-Hernández

Received: 10 May 2021
Accepted: 7 June 2021
Published: 10 June 2021

Publisher's Note: MDPI stays neutral with regard to jurisdictional claims in published maps and institutional affiliations.

1. Introduction

Wheat flour is the basic ingredient for many staple foods, such as bread, pasta, muffins, biscuits, etc., consumed worldwide. Food industry evolution implies the development of novel products with high nutritional value that could capture consumers' attention. In countries such as Romania, where the cultivated area of durum wheat (*Triticum durum*) is small, only 1% of the total wheat cultivated area [1], pasta producers usually use common (*Triticum aestivum*) wheat flour for short pasta manufacturing, trying to achieve the quality of durum wheat pasta by employing special processing technologies such as vacuum dough mixing and extrusion.

The nutritional and technological behavior of wheat flour can be changed by applying various physical treatments, one of them being heat-moisture treatment (HMT) which implies flour heating at higher level than the gelatinization temperature and under water restriction (<35%) [2]. HMT is an easy to control process and it is eco-friendly compared to chemical modifications of starch-based matrices. The intensities of component change in wheat flour during treatment depends on the regime applied (temperature, time, and moisture level), but also on the interactions between molecules, especially starch-protein and starch-lipids [3]. The main effects of HMT consist of starch crystalline area disruption, double helices destruction, and a reorganization of the crystallites and interactions between wheat polymer chains [4]. These molecular changes lead to nutritional value enhancement, but also to different technological properties of the final product. Some studies have revealed positive effects of HMT, including improving starch digestibility, increasing resistant and slowly digestible starch, and lowering rapid digestible starch values [4,5]. Dough rheological properties and final product texture are affected by flour treatment, depending on the amylose content, starch shape deformation that may occur, starch swelling and interactions between continuous and dispersed phases [3,6]. Liao et al. [7] reported higher elastic and viscous moduli of vermicelli pasta dough from HMT potato starch as the moisture content applied was higher, with the textural parameters in terms of hardness, springiness and chewiness also increasing with treatment moisture increase.

The wine industry generates high amounts of valuable by-products that can be successfully incorporated in food formulations as functional ingredients due to their raised quantity of dietary fibers and bioactive compounds with antioxidant properties [8,9]. Grape peels are rich in dietary fibers, 98.5% being represented by insoluble fractions, the chemical composition depending on the variety and vinification process [10]. The potential of grape peels to inhibit oxidative processes is given by their high contents of polyphenols such as anthocyanins, hydroxycinnamic acids, catechins, and flavonols [10,11]. Therefore, grape peels can be used to increase the nutritional and functional value of foods. Gaita et al. [12] reported higher polyphenolics and antioxidant activities of pasta enriched with grape peels. According to the data presented by Savla and Yardi [13], the addition of 5% grape pomace to gluten free pasta caused the increase of fiber content 13.50 times compared to the control. Tolve et al. [14] obtained increases of total polyphenolics content of pasta with grape pomace of 200–500% compared to the control. Gaceu et al. [15] stated that wheat flour supplemented by 15% grape peels presented 90% more fiber than flour without addition, while the calcium and potassium contents were improved by more than 150%. Sant'Anna et al. [16] reported higher cooking loss (5.38–6.35%) of fettuccini pasta with 25–75 g/kg grape marc incorporated, compared to the control (5.45%).

The rheological tests can give valuable information about the interactions between grape peels and wheat flour components and gluten matrices, and could differentiate the rheological behavior of supplemented flour dough during mixing, modelling and drying [17]. Frequency sweep tests can provide information about the differences of dough structures between samples with different ingredients added [18]. For this purpose, the deformation frequency is increased progressively at a constant amplitude of the strain. The results obtained at low frequencies led to the behavior of dough at slow changes of stress, while at high frequencies information of dough response to fast load is given [18]. Dough rheology and final product texture could be negatively affected by the addition of fiber-rich ingredients, depending on the amount and particle size. In order to minimize the negative effects of grape peels addition on dough rheological and pasta texture properties due to the gluten dilution, a small particle size could be used since a smaller impact was observed in previous studies [19]. According to the literature, wheat dough with grape peels added presented higher hardness compared to the control [20], with the elastic and viscous moduli being higher compared to the control [17]. The data presented by Tolve et al. [14] showed an increase in durum wheat pasta firmness of 30% and an increase in adhesiveness caused by the addition of grape pomace, while uncooked pasta luminosity decreased from 65.95 to 43.55 [14].

In Europe, pasta is usually made from durum semolina, while for Asian noodle manufacturing soft wheat flour is used [21]. Soft wheat pasta is usually processed by sheeting and cutting, while pasta from durum wheat is cold-extruded. Thus, compared to durum wheat pasta, soft wheat pasta has a softer and more elastic structure, with a color that ranges from "white to creamy white to moderately yellow". Durum wheat products are characterized as having a harder structure, intense yellow nuance, nutty flavor, stability to overcooking, and a particular eating quality [22]. The intrinsic properties of wheat flour and the processing conditions are the most important factors that influence pasta cooking quality [23].

There are a few researches presenting the impact of HMT applied to wheat. Furthermore, even if there are some papers that showed the influence of grape peels on dough and final product properties, to our knowledge, no studies have been published regarding the combined effect of HMT and grape peels addition on common wheat dough and pasta quality. Thus, the aim of this investigation was to evaluate the synergistic effects of HMT regime in terms of temperature, time, and moisture content and grape peels addition level on dough rheology and texture and on pasta's functional, physical, and textural properties in order to optimize the production process.

2. Materials and Methods

2.1. Materials and Treatment Regime

The wheat flour used for investigation belonged to the *Triticum aestivum* species and was produced in 2019 by Dizing S.R.L. (Brusturi, Neamt, Romania). Grape peels flour (GPF) of the Feteasca Regala variety coming from Iasi Research and Development Center for Viticulture and Vinification (Iasi, Romania) was obtained from grinding in a Kitchen Aid mill (Whirlpool Corporation, Benton Harbor, MI, USA), after manual separation from dried pomace. The particle size of <180 μm was achieved by sieving the resultant flour on a Retsch Vibratory Sieve Shaker AS 200 basic (Retsch GmbH, Haan, Germany). The magnitude of changes caused by heat moisture treatment (HMT) depend on wheat botanical origin, amylose and amylopectin contents, and treatment conditions such as temperature, moisture, and time [6]. The conditions of HMT should be selected in function of cereal type, since native starches present different properties, especially different amylose-amylopectin ratio, and have different processing requirements [24].

HMT of wheat flour was done according to a process outlined in a previous study [25]. The desired moisture was achieved by calculating the appropriate amount of water, according to the native flour moisture previously determined (Figure 1). Water was incorporated in small portions into wheat flour samples by continuous mixing in a sealed system of a Kitchen Aid mixer (Whirlpool Corporation, Benton Harbor, MI, USA), and the mix was placed in hermetically sealed glass containers in 2 cm layers (about 200 g). After 24 h of resting at 20 °C for moisture equilibration, the samples were placed in a convection oven for the given time in agreement with the experimental matrix, which was calculated after 30 min of sample thermalization. After 30 min of cooling, the treated flour was dried at 40 °C for 12 h, ground, and sieved to obtain a particle size of <300 μm.

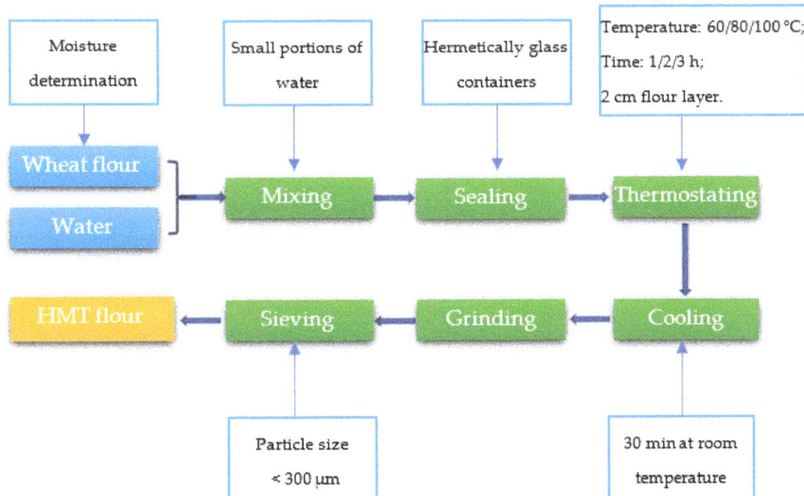

Figure 1. Heat moisture treatment (HMT) graphical representation.

2.2. Dough and Pasta Manufacturing

Composite flours from treated wheat flour and grape peels were obtained by 15 min mixing the appropriate amounts of ingredients in a Yucebas Y21 machine (Izmir, Turkey).

Pasta dough with a moisture of 40% was produced by adding the corresponding water to the flour (according to the moisture previously established) and mixing for 5 min in a

Kitchen Aid mixer (Whirlpool Corporation, Benton Harbor, MI, USA). Moisture contents were checked with a Kern DBS thermobalance with infrared emission quartz heaters (Kern, Balingen, Germany). The dough was put for at least 15 min in sealed containers for moisture equilibration, then the pasta was modeled using a rigatoni mold accessory of the Kitchen Aid mixer. Pieces of dough were subjected to rheological and texture analysis before pasta modelling. Samples were dried first for 30 min at 20 °C, then 60 min at 40 °C, 120 min at 80 °C, and 120 min at 40 °C [26]. After cooling, pasta was packed in polyethylene bags and kept in a dry place until analysis.

2.3. Synergistic Effects of HMT and GPF on Dough Properties

2.3.1. Dough Rheological Behavior

Dynamic rheological testing was carried out by means of frequency sweep analysis on a Thermo-HAAKE, MARS 40 (Karlsruhe, Germany) device. Laminated dough samples rested for at least 15 min [27] were placed between the parallel plates at a gap of 3 mm, and the edges were sealed with Vaseline to avoid water evaporation. The elastic (G′) and viscous (G″) moduli variation with frequency were registered at 20 °C in triplicate by increasing the range from 0.1 to 20 Hz, at a strain of 15 Pa, in the linear viscoelastic region previously established. For further optimization, the values of G′ and G″ at 1 Hz frequency were considered.

2.3.2. Dough Texture

Texture profile analysis (TPA) was used to evaluate dough firmness in triplicate by double cycle compression on a Perten TVT-6700 texturometer (Perten Instruments, Hägersten, Sweden). For this purpose, a spheric piece of dough of 50 g was subjected to analysis at a testing speed of 5.0 mm/s and a trigger force of 20 g [28], with a 35 mm cylindrical probe. The measurements were performed the same day as pasta manufacturing, at 20 °C.

2.4. Synergistic Effects of HMT and GPF on Dry Pasta Properties

2.4.1. Dry Pasta Color

CIE Lab system was used for pasta color property evaluation, the measurements being done on a Konica Minolta CR-400 (Konica Minolta, Tokyo, Japan) colorimeter. The calibration of the device was performed with a standard white [28]. Sample luminosity (L^*) was reported as mean of three determinations. Durum wheat pasta of good quality has a bright yellow color given by the presence of natural carotenoid pigments [29], while soft wheat pasta has a "white to creamy white to moderately yellow" appearance [21].

2.4.2. Total Polyphenolics Content (TPC)

The method described by Melilli et al. [30] was carried out to prepare pasta extracts for TPC determination. Mixes of 1:10 of ground sample with methanol (80% v/v) were vortexed and put on a sonication water bath at 37 °C and 45 Hz for 40 min. After filtration on a Whatman paper with 125 μm pores dimensions, portions of 200 μL were mixed with 800 μL distilled water, 500 μL of Folin–Ciocalteu reagent (1N) and 2500 μL of sodium carbonate (20% v/v). Triplicate samples were prepared and were left to rest in a dark place for 40 min, then the absorbance at 725 nm was measured on a UV–VIS–NIR Shimadzu 3600 (Tokyo, Japan) device. The calibration curve was made with gallic acid and the coefficient of determination was $R^2 = 0.99$. The results were expressed as mg GAE/100 g of pasta as it is.

2.4.3. Dietary Fiber Content (DF)

An Infrared FOSS 6500 NIR (Foss, Silver Springs, FL, USA) device was employed to estimate the DF content. Ground pasta was subjected to analysis at room temperature and the spectrometer was calibrated by using INGOT calibrations (AUNIR, Towcester, UK).

Three measurements were performed for each experiment and the results were reported on dry pasta as it is.

2.5. Synergistic Effects of HMT and GPF on Cooked Pasta Properties

Pasta was boiled according to the optimum cooking time considered when the white core of uncooked starch disappeared.

2.5.1. Cooked Pasta Texture

Pasta texture was evaluated by TPA, firmness and gumminess parameters being registered in triplicate. For this purpose, one piece of pasta was subjected to double compression with a cylindrical probe of 35 mm, at a height of 50%, a speed of 5.0 mm/s, and a trigger force of 20 g [28]. Measurements were performed on a Perten TVT-6700 device (Perten Instruments, Hägersten, Sweden).

2.5.2. Resistant Starch (RS) Determination

Fresh boiled pasta resistant starch content was determined by using a Megazyme kit (Cat. No. K-DSTRS; Megazyme, Bray, Ireland), according to the standard method AOAC 2017.16. The principle of the method was to digest the sample with α-amylase and a-myloglucosidase for 4 h and to quantify spectrophotometrically (at 510 nm) the resultant glucose by using GOPOD reagent.

2.6. Optimization of HMT Regime and GPF Addition

The Design Expert software (Stat-Ease, Inc., Minneapolis, MN, USA) trial version was employed to create the experimental matrix in order to evaluate the effects of factors (temperature, time, moisture, GPF level) on the considered responses (dough G′, G″, firmness, pasta color, TPC, DF, firmness, gumminess and RS content). The study was carried out on 30 resulting experiments (Table 1). Response surface methodology (RSM) with a central composite design (CCD) was used for experiment planning and data processing. The option of face centered at $\alpha = 1$ with six points at the center and three repetitions for each experiment was selected. Mathematical model fittings were evaluated trough F sequential test, coefficient of determination (R^2), and adjusted coefficients of determination ($Adj.$-R^2). Analysis of Variance (ANOVA) performed by Design Expert software was used to establish the significant effects at $p < 0.05$ of the factors and their interactions on the responses. The optimization of HMT regime and GPF level was done by means of desirability function. The constraints applied consisted of maximization of G′, G″, TPC, RS, DF, and pasta firmness, while L^*, dough hardness, and pasta gumminess were kept within the range.

RSM is a well-known tool successfully employed in different industries, with various applications in food formulation, quality, and processing technologies optimization. RSM is based on simple algebraic equations although they present an advantage regarding ease of use [31].

Table 1. Codded vs. real values of the factors and values of the responses.

Exp. No	Coded Values of Factors				Real Values of Factors				Values of the Responses								
	A	B	C	D	Temperature (°C)	Time (h)	Moisture (%)	GPF Level (%)	G' (Pa)	G'' (Pa)	Dough Firmness (g)	L^*	RS (%)	TPC (mg GAE/100 g)	Pasta Firmness (g)	Pasta Gumminess (g)	DF (%)
1	−1.00	−1.00	−1.00	−1.00	60.00	1.00	14.00	1.00	65,243.33 ± 4125.49	21,620.00 ± 1230.49	2394.67 ± 80.64	72.50 ± 0.20	1.35 ± 0.03	10.40 ± 0.14	5582.33 ± 139.52	4582.05 ± 64,67	0.04 ± 0.02
2	−1.00	−1.00	−1.00	1.00	60.00	1.00	14.00	6.00	115,366.67 ± 6107.65	36,690.00 ± 1839.59	2851.67 ± 98.29	65.25 ± 0.14	1.75 ± 0.02	12.85 ± 0.08	5077.00 ± 30.05	3979.43 ± 103.13	1.10 ± 0.00
3	−1.00	−1.00	1.00	−1.00	60.00	1.00	30.00	1.00	92,510.00 ± 4133.63	28,150.00 ± 1192.64	2317.33 ± 39.80	73.37 ± 0.19	2.21 ± 0.03	10.42 ± 0.36	4769.00 ± 99.83	3702.50 ± 88.17	0.85 ± 0.05
4	−1.00	−1.00	1.00	1.00	60.00	1.00	30.00	6.00	96,053.33 ± 7110.29	29,543.33 ± 2363.90	2529.00 ± 87.54	65.09 ± 0.86	2.39 ± 0.04	14.38 ± 0.41	5532.67 ± 38.73	3475.66 ± 133.51	1.95 ± 0.05
5	−1.00	0.00	0.00	0.00	60.00	2.00	22.00	3.50	96,845.00 ± 985.00	30,820.00 ± 750.00	2382.33 ± 85.65	68.44 ± 0.40	2.36 ± 0.06	11.70 ± 0.23	5646.67 ± 161.38	3571.40 ± 32.50	1.10 ± 0.00
6	−1.00	1.00	−1.00	−1.00	60.00	3.00	14.00	1.00	62,435.00 ± 2745.00	20,715.00 ± 1025.00	2087.67 ± 43.82	71.44 ± 0.29	1.69 ± 0.02	8.85 ± 0.08	5257.67 ± 149.31	3917.56 ± 188.13	0.10 ± 0.00
7	−1.00	1.00	−1.00	1.00	60.00	3.00	14.00	6.00	110,300.00 ± 4500.00	36,245.00 ± 1305.00	2973.67 ± 179.62	63.42 ± 0.33	2.09 ± 0.03	14.09 ± 0.35	4749.00 ± 209.16	3640.87 ± 114.96	1.35 ± 0.05
8	−1.00	1.00	1.00	−1.00	60.00	3.00	30.00	1.00	111,300.00 ± 9139.47	31,683.33 ± 2632.38	2593.67 ± 81.86	74.33 ± 74.33	2.14 ± 0.03	10.22 ± 0.18	5171.19 ± 41.56	3319.90 ± 78.63	1.10 ± 0.00
9	−1.00	1.00	1.00	1.00	60.00	3.00	30.00	6.00	123,733.33 ± 7958.85	36,056.67 ± 2643.56	2830.00 ± 28.35	68.60 ± 0.13	2.32 ± 0.02	14.20 ± 0.24	5509.38 ± 256.63	3703.64 ± 230.56	2.20 ± 0.10
10	0.00	−1.00	0.00	0.00	80.00	1.00	22.00	3.50	179,900.00 ± 12,019.98	48,253.33 ± 2805.54	2607.00 ± 181.35	72.51 ± 0.07	1.41 ± 0.04	11.83 ± 0.33	4690.00 ± 115.57	3439.52 ± 45.24	2.10 ± 0.00
11	0.00	0.00	−1.00	0.00	80.00	2.00	14.00	3.50	90,680.00 ± 6131.86	29,120.00 ± 1993.66	2592.33 ± 125.01	69.31 ± 0.21	1.30 ± 0.01	12.21 ± 0.08	4569.00 ± 51.51	3826.95 ± 117.56	1.30 ± 0.10
12	0.00	0.00	0.00	−1.00	80.00	2.00	22.00	1.00	192,200.00 ± 6773.48	46,030.00 ± 1203.37	2623.67 ± 70.74	77.52 ± 0.15	1.35 ± 0.03	9.64 ± 0.05	4782.67 ± 192.73	3758.35 ± 194.28	1.65 ± 0.05
13	0.00	0.00	0.00	0.00	80.00	2.00	22.00	3.50	171,123.33 ± 85,406.97	49,411.17 ± 4023.58	3036.04 ± 148.87	69.57 ± 1.58	1.40 ± 0.02	12.60 ± 0.26	4471.00 ± 160.08	3478.30 ± 78.55	2.25 ± 0.04
14	0.00	0.00	0.00	1.00	80.00	2.00	22.00	6.00	243,933.33 ± 7211.33	59,853.33 ± 2311.54	3365.33 ± 116.98	66.32 ± 0.04	1.46 ± 0.01	12.70 ± 0.70	4676.67 ± 188.60	3260.48 ± 14.93	2.95 ± 0.15
15	0.00	0.00	1.00	0.00	80.00	2.00	30.00	3.50	311,533.33 ± 18,779.33	65,976.67 ± 4810.64	4694.33 ± 444.08	73.20 ± 0.65	1.62 ± 0.05	11.06 ± 0.22	4589.67 ± 257.41	3151.64 ± 173.89	2.15 ± 0.05
16	0.00	1.00	0.00	0.00	80.00	3.00	22.00	3.50	259,600.00 ± 15,661.42	58,203.33 ± 4250.84	3443.67 ± 110.21	72.36 ± 0.51	1.58 ± 0.01	11.00 ± 0.22	5504.00 ± 184.46	3438.89 ± 94.96	2.25 ± 0.05
17	1.00	−1.00	−1.00	−1.00	100.00	1.00	14.00	1.00	146,400.00 ± 13,596.69	38,243.33 ± 3639.13	3697.67 ± 53.61	77.52 ± 0.06	1.01 ± 0.00	9.73 ± 0.19	3750.33 ± 168.00	3913.92 ± 149.77	1.70 ± 0.00
18	1.00	−1.00	−1.00	1.00	100.00	1.00	14.00	6.00	194,733.33 ± 6165.50	50,006.67 ± 1664.52	3856.00 ± 131.64	63.91 ± 0.37	1.30 ± 0.02	13.67 ± 0.22	4415.67 ± 214.31	3599.77 ± 71.85	2.90 ± 0.10
19	1.00	−1.00	1.00	−1.00	100.00	1.00	30.00	1.00	244,766.67 ± 5784.75	41,600.00 ± 1398.71	6012.67 ± 374.68	77.72 ± 0.18	1.84 ± 0.03	8.46 ± 0.38	4232.00 ± 118.44	3328.82 ± 152.72	1.80 ± 0.00
20	1.00	−1.00	1.00	1.00	100.00	1.00	30.00	6.00	345,150.00 ± 750.00	57,360.00 ± 1120.00	6505.33 ± 104.58	71.21 ± 0.15	1.87 ± 0.02	12.24 ± 0.08	4699.33 ± 95.35	3229.02 ± 24.20	2.20 ± 0.10
21	1.00	0.00	0.00	0.00	100.00	2.00	22.00	3.5	321,033.33 ± 11,834.84	59,920.00 ± 2044.70	5869.00 ± 156.60	73.53 ± 0.37	1.72 ± 0.02	7.72 ± 0.25	4225.00 ± 189.81	3413.43 ± 63.80	2.15 ± 0.25
22	1.00	1.00	−1.00	−1.00	100.00	3.00	14.00	1.00	195,550.00 ± 8150.00	38,915.00 ± 1865.00	5956.33 ± 50.82	77.34 ± 0.13	1.39 ± 0.03	6.94 ± 0.03	4445.67 ± 105.70	3718.83 ± 166.23	2.20 ± 0.10
23	1.00	1.00	−1.00	1.00	100.00	3.00	14.00	6.00	343,300.00 ± 11,200.00	63,195.00 ± 525.00	6502.67 ± 70.01	71.28 ± 0.15	1.88 ± 0.04	9.99 ± 0.03	3405.33 ± 335.50	3451.07 ± 151.60	3.00 ± 0.10
24	1.00	1.00	1.00	−1.00	100.00	3.00	30.00	1.00	566,866.67 ± 23,071.70	85,683.33 ± 3634.37	3522.33 ± 264.08	74.43 ± 0.20	1.81 ± 0.01	7.04 ± 0.03	3598.33 ± 56.70	3365.23 ± 47.14	1.60 ± 0.10
25	1.00	1.00	1.00	1.00	100.00	3.00	30.00	6.00	441,700.00 ± 42,785.63	67,473.33 ± 6410.43	3423.33 ± 58.71	71.53 ± 0.34	2.10 ± 0.02	10.13 ± 0.20	3500.67 ± 47.17	2645.07 ± 67.37	2.90 ± 0.00

G'—elastic modulus, G''—viscous modulus, L^*—luminosity, DF—dietary fiber, TPC—total polyphenolics content, RS—resistant starch.

3. Results

Dough elastic and viscous moduli, firmness, pasta color, polyphenolics contents, dietary fiber content, pasta firmness, gumminess, and resistant starch content responses were fitted to the quadratic polynomial regression model. The quadratic models (Equation (1)) were selected because they presented the highest *Adj.-R^2* values compared to other mathematical models proposed:

$$Y = x_0 + x_1A + x_2B + x_3C + x_4D + x_5AB + x_6AC + x_7BC + x_8AD + x_9CD + x_{10}BD + x_{11}A^2 + x_{12}B^2 + x_{13}C^2 + x_{14}D^2 \quad (1)$$

where Y is the response, x_0-x_9 are the regression coefficients, and A, B, C, and D are the factors.

3.1. Pasta Dough Properties

Dough pasta rheological properties are affected by HMT and the addition of fiber-rich ingredients such as GPF. The variation on the elastic modulus G′ was explained by the quadratic model ($R^2 = 0.93$, $p < 0.01$), temperature, time, moisture, and GPF level factors significantly influencing this parameter (Table 2).

Table 2. ANOVA results for quadratic model fitted to pasta dough parameters.

Factor	G′ (Pa)	G″ (Pa)	Dough Firmness (g)
Constant	211,420.21	51,275.41	3248.00
A	106,984.07 **	12,826.30 **	1243.63 **
B	40,814.54 **	4816.85 **	31.22
C	56,089.17 **	6043.15 **	84.19
D	18,722.13 **	4654.63 **	201.72 *
A × B	36,110.73 **	3710.00 **	−66.21
A × C	40,515.52 **	3224.58 **	−31.96
A × D	3583.44	−173.33	−43.29
B × C	17,204.90 *	3233.33 **	−607.21 **
B × D	−7468.85	−1125.83	15.63
C × D	−18,929.90 **	−3957.92 **	−75.37
A^2	−4938.95	−6526.83 **	807.01 **
B^2	5871.89	1331.50	−293.32
C^2	−12,771.45	−4348.50	324.68
D^2	4188.55	1044.84	−324.16
Model diagnostic			
p-value	<0.01	<0.01	<0.01
R^2	0.93	0.86	0.76
Adj.-R^2	0.91	0.83	0.71

A—temperature, B—time, C—moisture, D—GPF level, G′—elastic modulus, G″—viscous modulus, *—significant at $p < 0.05$, **—significant at $p < 0.01$.

The response surface plot showed (Figure 2a,b) that G′ significantly ($p < 0.01$) increased with an increase in factor levels, indicating a strengthening effect of HMT and GPF on wheat dough. The biggest significant ($p < 0.01$) positive influence was observed for temperature factor, while the interaction between moisture and GPF level presented the highest significant negative effect on G′. The variation of the viscous modulus (G″) was successfully described by the quadratic model ($R^2 = 0.86$, $p < 0.01$). Temperature, time, moisture, and GPF level rise led to higher G″ values (Figure 2c,d), with all the considered factors presenting significant ($p < 0.01$) influence (Table 2).

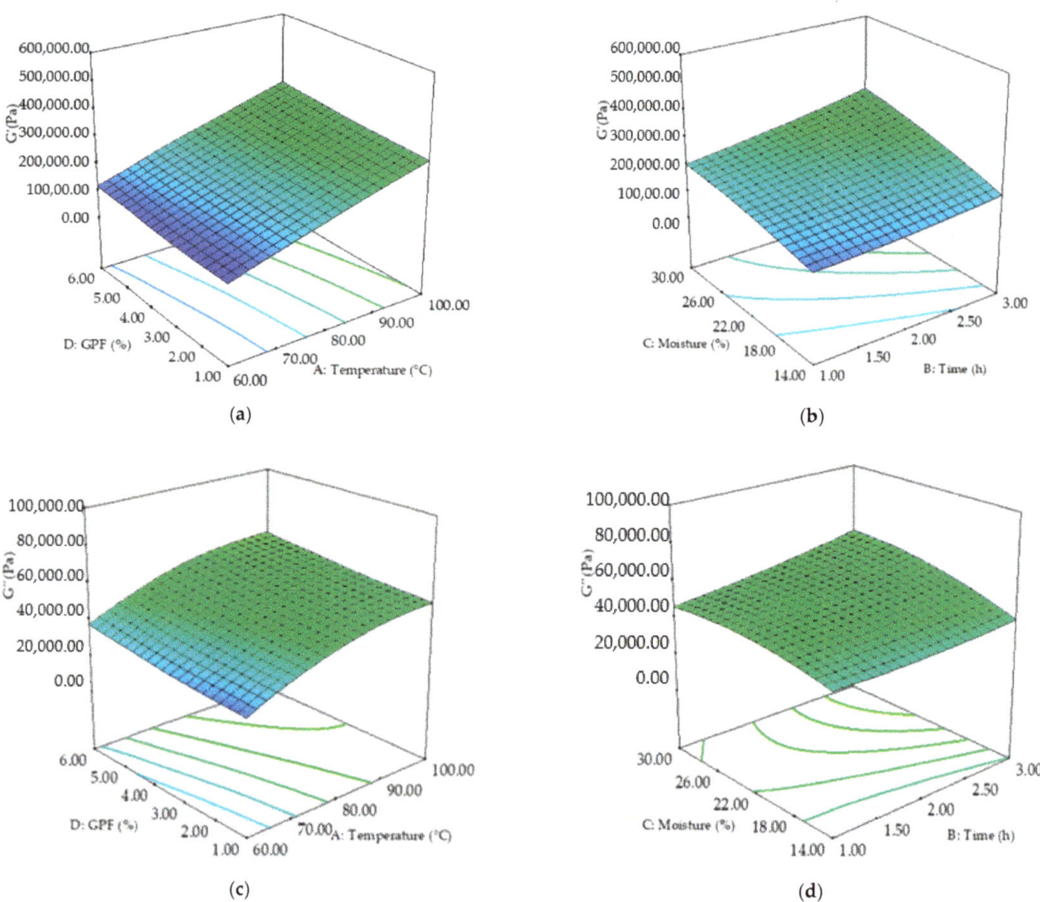

Figure 2. Three-dimensional response surface plots presenting the synergistic effects of factors on dough elastic G′ (**a,b**) and viscous G″ (**c,d**) moduli.

HMT temperature showed the biggest positive influence, while the most important negative effect on G″ was observed for the quadratic term of temperature, considering a significance level of $p < 0.01$. Higher elastic and viscous moduli are desirable for pasta shape keeping. G′ and G″ values (Figure 2) were significantly higher compared to the data obtained by Fanari et al. [32] for semolina dough, probably due to the difference in dough moisture used.

Wheat flour HMT and GPF addition affected dough firmness, the quadratic model describing 71% of data variation at $p < 0.01$. Only temperature and GPF level factors influenced significantly ($p < 0.05$) dough firmness (Table 2). The highest positive effect was observed for the linear term of temperature, while the interaction between HMT time and moisture showed a significant ($p < 0.01$) negative effect. Dough firmness increase was directly proportional with temperature level increase, while GPF determined higher dough firmness at additions up to 5% (Figure 3). Dough firmness is an important technological property that is directly related to pasta handling and modeling. Soft doughs are not desirable for short pasta due to their low capacity to keep the shape and the issues that may appear during drying. These results showed that both HMT and GPF addition had beneficial effects on dough firmness and on further pasta quality.

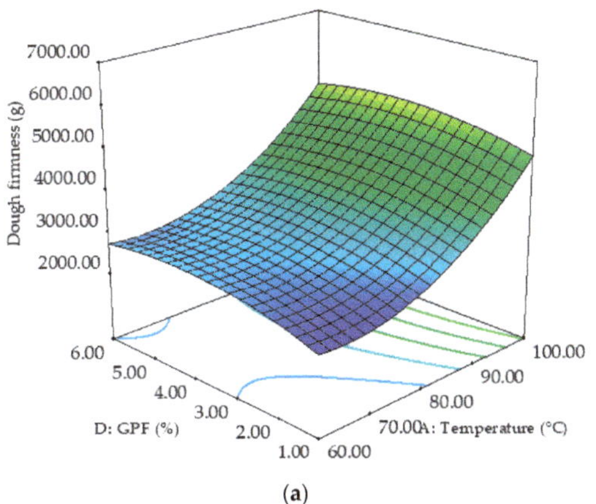
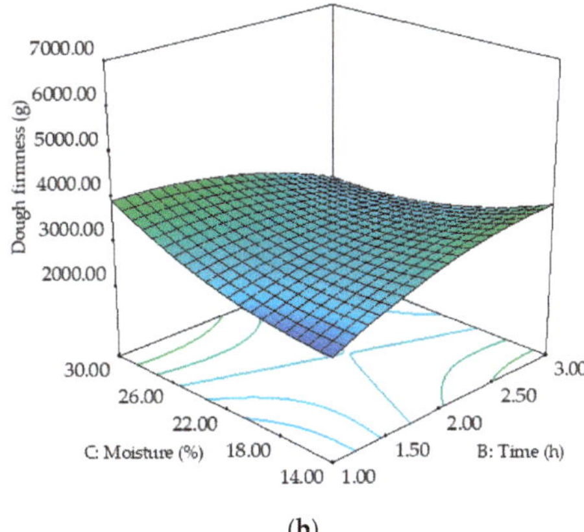

Figure 3. Three-dimensional response surface plots presenting the synergistic effects of factors: GPF level–temperature (**a**) and moisture-time (**b**) on dough firmness.

3.2. Dry Pasta Properties

Pasta color is one of the quality characteristics of pasta and impacts directly the consumer perception and purchase intention. The ANOVA results for the quadratic models fitted to the dry pasta properties are presented in Table 3. The quadratic regression model explained 87% of L^* parameter variation, being significant at $p < 0.01$ (Table 3). Temperature, moisture, and GPF level showed significant ($p < 0.01$) influence on L^*, while time factor had a non-significant influence ($p > 0.05$).

Table 3. ANOVA results for quadratic model fitted to dry pasta parameters.

Factor	L^*	DF (%)	TPC (mg GAE/100 g)
Constant	71.10	2.11	11.52
A	2.00 **	0.59 **	−1.18 **
B	0.31	0.11 **	−0.64 **
C	0.97 **	0.17 **	−0.03
D	−3.86 **	0.53 **	1.81 **
A × B	0.16	0.02	−0.58 **
A × C	−0.25	−0.30 **	−0.34 **
A × D	0.01	−0.05 *	−0.11
B × C	−0.17	0.01	0.18
B × D	0.81 **	0.04	0.08
C × D	0.72 **	−0.03	0.01
A^2	−0.62	−0.44 **	−1.45 **
B^2	0.82	0.11	0.25
C^2	−0.36	−0.34 **	0.47 *
D^2	0.30	0.24 **	0.01
Model diagnostic			
p-value	<0.01	<0.01	<0.01
R^2	0.87	0.97	0.92
Adj.-R^2	0.84	0.96	0.90

A—temperature, B—time, C—moisture, D—GPF level, L^*—luminosity, DF—dietary fiber, TPC—total polyphenolics content, *—significant at $p < 0.05$, **—significant at $p < 0.01$.

The highest positive effect was observed for the linear term of temperature, while the most important negative effect ($p < 0.01$) was that of GPF level. The decrease of L^* was proportional with GPF level increase, while the opposite trend was obtained with temperature (Figure 4).

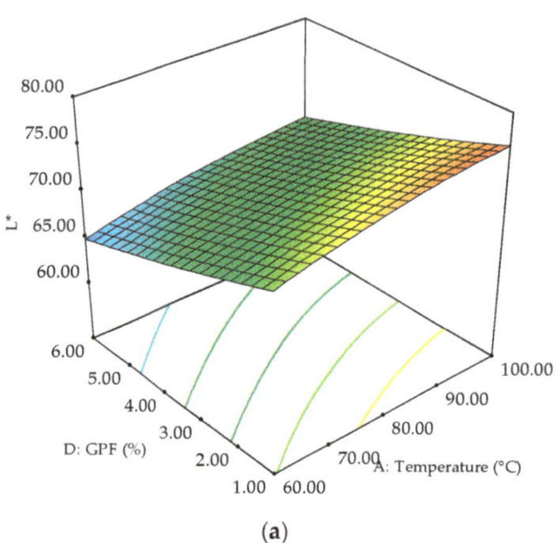 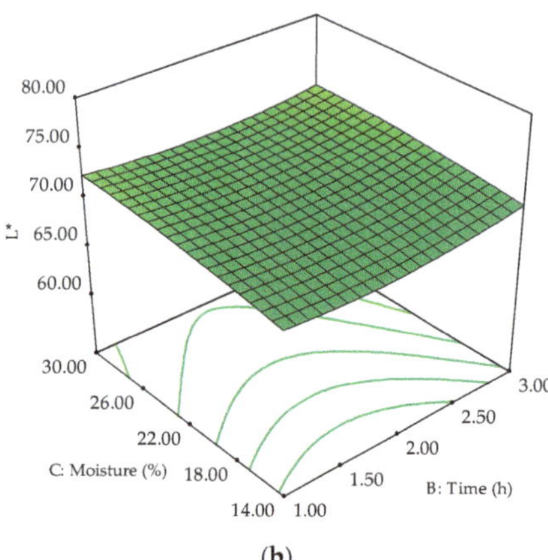

(a)   (b)

Figure 4. Three-dimensional response surface plots presenting the synergistic effects of factors: GPF level–temperature (**a**) and moisture-time (**b**) on dry pasta luminosity (L^*).

Grape peels are known to increase DF content of foods since they are an important source of insoluble and soluble fibers. The quadratic model was correctly chosen ($p < 0.01$) to describe 97% of DF content data variation. HMT conditions and GPF addition level showed significant effects ($p < 0.01$) on DF. The linear term of temperature presented the biggest positive influence, while its quadratic term had the highest negative effect on DF. The response surface plots describing the synergistic effects of factors on pasta DF are given in Figure 5. These results are confirming the enhancement of pasta nutritional value by HMT and GPF addition.

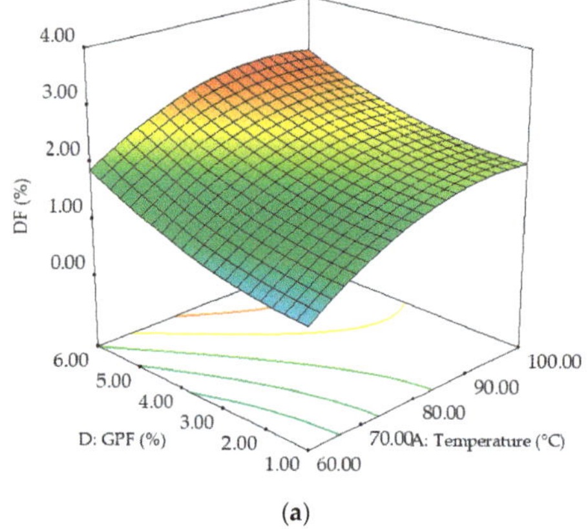 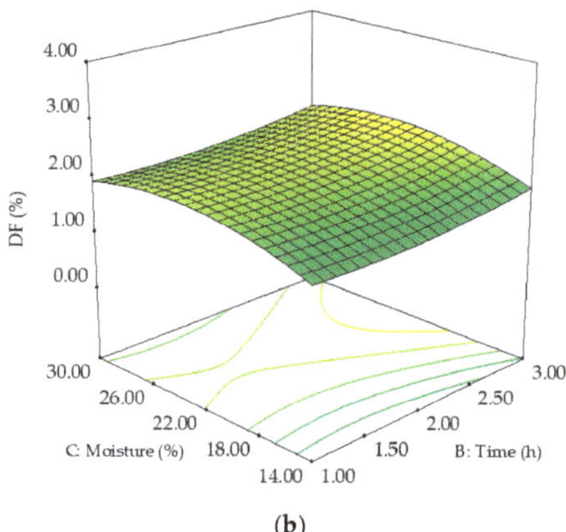

(a)   (b)

Figure 5. Three-dimensional response surface plots presenting the synergistic effects of factors: GPF level–temperature (**a**) and moisture-time (**b**) on dietary fiber content (DF).

Pasta polyphenolics content is affected by HMT regime and the addition of GPF, which is a rich source of bioactive compounds. The quadratic model was adequate ($p < 0.01$) to describe 92% of TPC data variation. HMT temperature and time and GPF addition level showed significant ($p < 0.01$) effects on TPC (Table 3). GPF level linear term had the highest positive influence, while the quadratic term of temperature showed the most important negative effect at $p < 0.01$. The negative impact of HMT given by the decrease of TPC with temperature and time increase was countered by the addition of GPF, which determined raised TPC values as the level was higher (Figure 6).

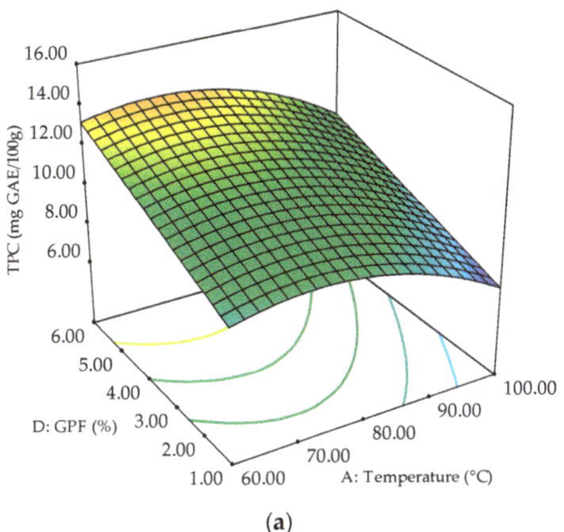

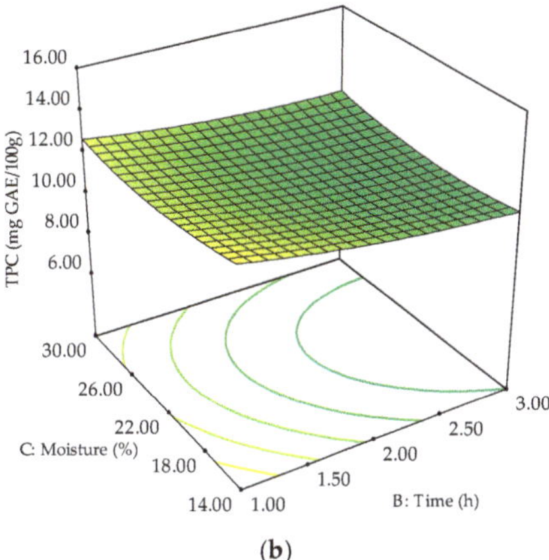

(a)   (b)

Figure 6. Three-dimensional response surface plots presenting the synergistic effects of factors: GPF level–temperature (**a**) and moisture-time (**b**) on total polyphenolics content (TPC).

3.3. Cooked Pasta Properties

Pasta texture could be a predictor for some of the sensory characteristics of the product. Cooked pasta firmness is another important quality parameter that was significantly influenced ($p < 0.05$) by wheat HMT temperature and time (Table 4).

Table 4. ANOVA results for quadratic model fitted to cooked pasta parameters.

Factor	Pasta Firmness (g)	Pasta Gumminess (g)	RS (%)
Constant	4773.37	3454.78	1.49
A	−612.37 **	−179.33 **	−0.19 **
B	−89.28 *	−113.87 **	0.10 **
C	19.46	−261.61 **	0.25 **
D	−1.30	−145.68 **	0.13 **
A × B	−117.10 *	16.65	0.04 **
A × C	−18.93	−12.33	−0.01
A × D	−5.82	−42.47	−0.01
B × C	−30.36	40.29	−0.10 **
B × D	−168.72 **	22.66	0.03 *
C × D	178.78 **	49.88	−0.06 **
A^2	61.68	45.48	0.52 **
B^2	222.84	−7.73	−0.02
C^2	−294.82 *	42.35	−0.05
D^2	−144.49	62.48	−0.11 **
Model diagnostic			
p-value	<0.01	<0.01	<0.01
R^2	0.79	0.79	0.96
Adj.-R²	0.75	0.74	0.95

A—temperature, B—time, C—moisture, D—GPF level, RS—resistant starch, *—significant at $p < 0.05$, **—significant at $p < 0.01$.

The quadratic regression model explained 79% of data variation at $p < 0.01$. The highest negative effect was that of the linear term of temperature, while the interaction between moisture and GPF level showed the larger positive effect ($p < 0.01$). As can be seen in Figure 7, pasta firmness decreased with temperature and time increase, while moisture and GPF level had a non-significant effect ($p > 0.05$).

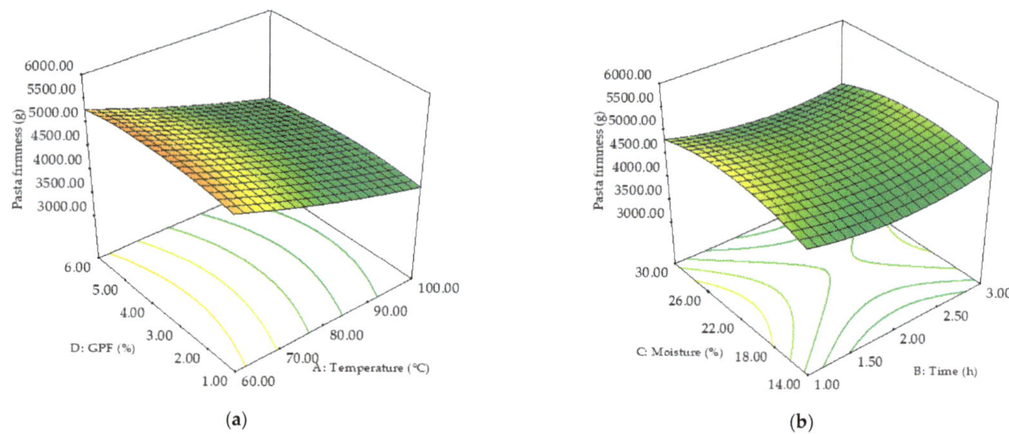

(a)

(b)

Figure 7. Three-dimensional response surface plots presenting the synergistic effects of factors: GPF level–temperature (**a**) and moisture-time (**b**) on pasta firmness.

Pasta gumminess data were fitted to the quadratic regression model, which presented a coefficient of determination of 0.79 and a significance of $p < 0.01$. All the considered factors showed significant effects ($p < 0.01$) on pasta gumminess, while their interactions and quadratic terms had a non-significant effect ($p > 0.05$). The highest negative influence was obtained for HMT moisture factor (Table 4). HMT regime in terms of temperature, time, moisture, and the GPF level determined a proportional decrease of pasta gumminess, as it is shown in Figure 8.

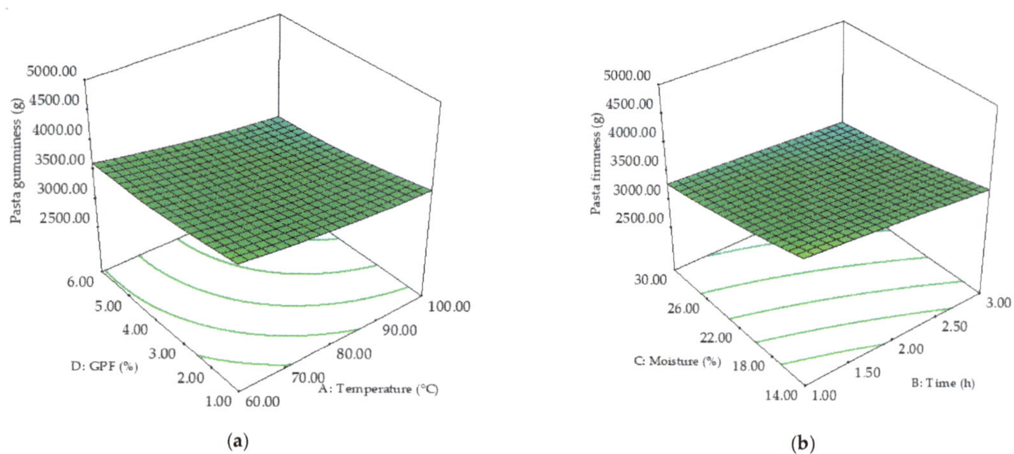

(a)

(b)

Figure 8. Three-dimensional response surface plots presenting the synergistic effects of factors: GPF level–temperature (**a**) and moisture-time (**b**) on pasta gumminess.

The formation of RS during HMT led to improved nutritional and functional value of pasta. The quadratic model chosen to describe RS content data variation was suitable ($p < 0.01$), since a high determination coefficient ($R^2 = 0.96$) was obtained. All the factors presented significant ($p < 0.01$) influence on past RS content (Table 3). The quadratic term of temperature had the biggest positive influence at $p < 0.01$, while its linear term had the highest negative effect on RS. GPF addition level, HMT time, and moisture increase determined increases of RS content, while temperature showed an opposite trend up to 80 °C (Figure 9).

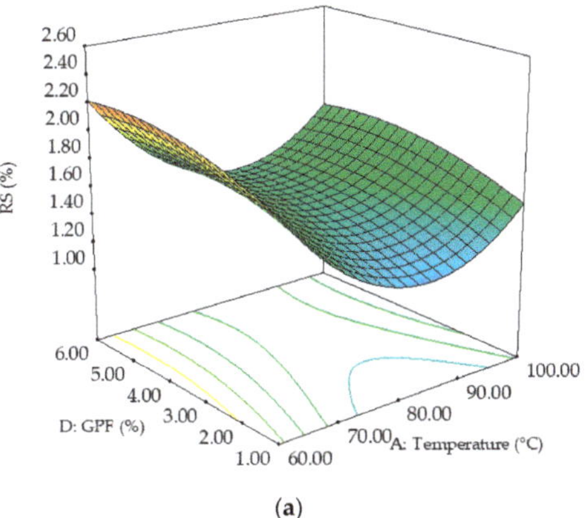

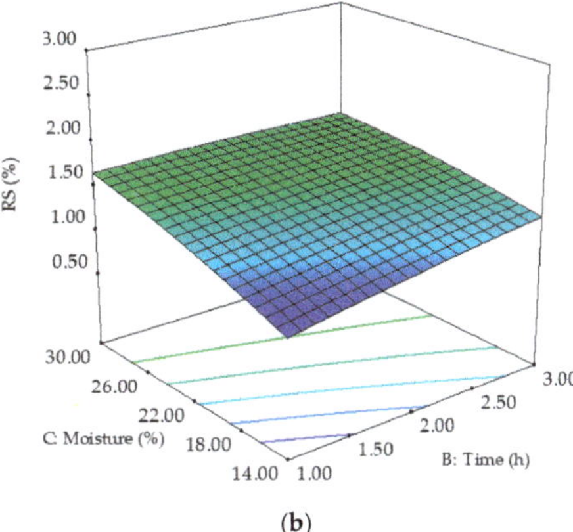

(a)

(b)

Figure 9. Three-dimensional response surface plots presenting the synergistic effects of factors: GPF level–temperature (**a**) and moisture-time (**b**) on pasta resistant starch (RS).

HMT temperature increase up to 80 °C determined RS increase. A similar trend was observed for time factor and moisture up to 22%. GPF addition level increase resulted in a proportional RS contents rise (Figure 9).

3.4. Optimization

The optimal values of the factors and the predicted values of the responses are shown in Table 5.

Table 5. Minimum, maximum, and optimal values of the variables.

Variable	Min	Max	Optimal
A—Temperature (°C)	60.00	100.00	87.56
B—Time (h)	1.00	3.00	3.00
C—Moisture (%)	14.00	30.00	26.01
D—GPF level (%)	1.00	6.00	4.81
G′ (Pa)	59,690.00	586,600.00	355,471.01
G″ (Pa)	19,690.00	89,870.00	67,972.84
Dough firmness (g)	2058.00	6622.00	3355.70
*L**	63.06	77.93	71.90
RS (%)	1.01	2.43	1.69
TPC (mg GAE/100 g)	6.90	14.85	11.36
Pasta firmness (g)	3028.00	5798.00	4473.80
Pasta gumminess (g)	2575.70	4649.91	3132.90
DF (%)	0.02	3.10	2.80

G′—elastic modulus, G″—viscous modulus, *L**—luminosity, DF—dietary fiber, TPC—total polyphenolics content, RS—resistant starch.

The results of the optimization revealed a recommended HMT regime for wheat flour of 87.56 °C temperature applied for 3 h at a moisture of 26.01%. GPF can be incorporated in wheat pasta at a level of 4.81% without major negative impact on quality parameters and with considerable nutritional and functional benefits.

3.5. Correlations between Dough and Pasta Properties

Correlations between dough and pasta properties were obtained and are presented in Table 6. Pasta luminosity L^* was significantly ($p < 0.05$) negatively correlated with TPC and RS contents. Dough firmness was positively correlated ($p < 0.01$) with pasta firmness, G′, G″, TPC, and DF contents. Pasta gumminess showed a negative correlation at $p < 0.01$ with G′, G″, and DF, while with RS the correlation was significant at $p < 0.05$. The elastic (G′) and viscous (G″) moduli were significantly ($p < 0.01$) correlated ($p < 0.01$) with TPC and DF contents.

Table 6. Correlations between dough and pasta properties.

Variables	L^*	Dough Firmness (g)	Pasta Firmness (g)	Pasta Gumminess (g)	G′ (Pa)	G″ (Pa)	TPC (mg GAE/100 g)	DF (%)	RS (%)
L^*	1.00	0.32 **	−0.31 *	−0.02	0.26 *	0.13	−0.81 **	−0.12	−0.27 *
Dough firmness (g)		1.00	−0.51 **	−0.33 **	0.54 **	0.48 **	−0.39 **	0.51 **	−0.09
Pasta firmness (g)			1.00	0.43 **	−0.68 **	−0.63 **	0.42 **	−0.49 **	0.27 *
Pasta Gumminess (g)				1.00	−0.64 **	−0.64 **	0.03	−0.65 **	−0.31 *
G′ (Pa)					1.00	0.95 **	−0.41 **	0.57 **	−0.01
G″ (Pa)						1.00	−0.23 *	0.68 **	−0.11
TPC (mg GAE/100 g)							1.00	0.17	0.19
DF (%)								1.00	−0.10
RS (%)									1.00

L^*—luminosity, G′—elastic modulus, G″—viscous modulus, TPC—total polyphenolics content, DF—dietary fiber, RS—resistant starch, *—significant at $p < 0.05$, **—significant at $p < 0.01$.

4. Discussion

The rheological testing of pasta dough can give information about the interactions of wheat and GPF components and could predict dough behavior during handling and modelling. One of the advantages of using a small amplitude oscillatory shear test is that no damages on the structure occurs and that it provides information about the structural changes to the system. All the tested samples presented G′ > G″, suggesting the solid-like nature of the material, which was expected. The increase of HMT temperature, time, and moisture determined the increase of G′ and G″ dynamic moduli probably as a result of the starch amylose and amylopectin chains reorganizations and/or interactions between phases during treatment [4]. Higher dough viscosity could be explained by the disulfide bonds formed by amino acids and starch structure reorganization [25]. During HMT, starch aggregates are formed which lead to changes of starch and gluten interactions [33]. Lazaridou et al. [34] also reported higher G′ and G″ values for wheat-barley treated flour, probably due to the plasticization process that HMT promoted causing starch amorphous-glassy cell wall stiffening. Dynamic moduli of pasta dough increase revealed a structure reinforcement, similar findings being reported for vermicelli dough from sweet potato starch treated at moisture levels up to 30% [7]. On the other hand, G′ and G″ moduli rise could be attributed also to the addition of GPF, a fiber-rich ingredient. This trend may be caused by the competition for water between gluten and GPF and/or to the high fiber content that can behave as filler in dough networks, a fact supported also by the significant ($p < 0.01$) correlations of DF content. Similar results were reported by Mironeasa et al. [19]. The interactions between flour mix components may be improved or inhibited by GPF presence, compounds such as organic acids causing dough strengthening and consequently increases of G′ and G″ values [17]. Fanari et al. [35] revealed the dependence of dough rheological properties with the water content added in the system. Lindahl et al. [36] reported no significant differences of G′ values (>10,000 Pa at 1 Hz) between durum and common wheat dough, underlying at the same time the importance of water. Peressini et al. [37] reported G′ values between 43,000 and 145,700 Pa for durum wheat pasta dough enriched with different fibers.

Dough firmness increased as HMT temperature and GPF addition level increased. Lazaridou et al. [34] also reported higher firmness of dough made of treated wheat-barley flour, which indicated a stiffer dough related to the macromolecular rearrangements of flour molecules during HMT. The chemical composition of GPF could also be responsible for dough firmness increase, with the strong correlation with DF content ($r = 0.51$, $p < 0.01$) supporting this hypothesis. Polyphenols presence may possibly contribute to the viscoelastic characteristics of pasta dough. Mironeasa et al. [17] also reported higher firmness values with GPF addition level increase at particle size <180 µm.

The synergistic effect of HMT and GPF addition was observed on dry pasta luminosity. HMT caused an increase of L^* with temperature and moisture level increase, which may be related to the modification of starch crystalline architecture which led to physicochemical property changes [38]. On the other hand, GPF decreased L^* values as the addition level was higher due to its pigments and to the promotion of nonenzymatic browning reactions that would determine darker products. Similar observations were made by Aksoylu et al. [39] for grape seeds enriched biscuits. TPC from GPF may also have contributed to L^* decrease, with a very strong correlation ($r = -0.81$, $p < 0.01$) being observed.

Pasta DF content was significantly increased ($p < 0.01$) as the HMT moisture and time and GPF level increased. There results may be related to the aggregate formation and amylose/amylopectin chain structure changes under heat energy and water molecule migrations, with similar results being reported by Zheng et al. [40]. Furthermore, starch may form complexes with other composite flour components such as lipids, proteins and polyphenols, intensifying the effects of aggregations, static, and hydrogen linkages [41]. In addition, GPF is a rich source of soluble and insoluble DF, which contributed significantly to the pasta DF content increase.

The polyphenolics content was negatively affected by HMT, but this decrease was countered by the addition of GPF, which resulted in a TPC increase. HMT may possibly cause bioactive compound damage at high temperatures. GPF is known to be a source of polyphenols such as malvidin-3-O-glucoside, peonidin-3-O-glucoside, cyanidin-3-O-glucoside, and catechin [42–44]. Our results are in agreement with those of Sant'Anna et al. [16], who reported higher TPC for fettuccini pasta enriched with grape pomace, and of Gaita et al. [12], who studied wheat pasta supplemented with grape peels.

Pasta texture is a good predictor of the sensory profile and represents a key factor in consumer acceptance. The increase of HMT temperature and duration determined a decrease in pasta firmness, which may be related to the denaturation of gluten proteins. Liu et al. [45] reported lower tapioca starch gel hardness, probably due to the interactions between phases that occur during HMT. Galvez and Resurreccion [46] stated that pasta should be neither too hard nor too soft. The interaction between moisture and GPF level factor significantly ($p < 0.01$) positively influenced pasta firmness. Tolve et al. [14] reported an increase in pasta firmness when grape pomace was incorporated. On the other hand, pasta firmness decrease may be related to GPF ability to bind water, similar to gluten proteins. Polyphenols may also contribute to pasta texture behavior as a result of interactions with other components, a fact supported by the positive correlation ($r = 0.42$, $p < 0.01$) between firmness and TPC. Our data were in agreement with those reported by Li et al. [47] for noodles with treated wheat flour, which presented a firmness of 3397.76 g. The gumminess parameter was significantly decreased ($p < 0.01$) by HMT conditions and GPF level increase. Chandla et al. [48] also obtained lower gumminess values for pasta made of HMT amaranth starch noodles. Using a small particle size of GPF (<180 µm) might have favorable effects on pasta texture by lowering gumminess values, similar to findings being reported by Chen et al. [49] for Chinese noodles with wheat bran addition.

RS is a fraction resistant to hydrolysis in the digestive system under the action of enzymes, having a physiological behavior similar to DF [50]. HMT and GPF addition significantly ($p < 0.01$) increased RS content of wheat pasta. Starch resistance to enzyme action could have been determined by the crystalline areas' perfection and more dense amorphous areas caused by HMT [5]. Similar to our results, Wang et al. [5] observed that RS formation is directly proportional to the moisture content of the sample. The interactions between flour components, especially starch, lipids, and proteins, could possibly be responsible for the higher RS content, since proteins films could have been formed on starch surface and lead to difficult amylase attack [4]. In addition, the presence of polyphenols from GPF could have a decisive role in starch digestion. It was stated that polyphenols may interact with starch to form complexes through non-covalent linkages, which are not accessible for hydrolysis [51]. Furthermore, polyphenols may inhibit the digestive enzymes by means of proteins-polyphenols interactions [52]. Thus, the synergistic effects of HMT and GPF led to an improved functional value of pasta by increasing RS content.

The optimization of HMT regime and GPF level allowed the establishment of a formulation with maximum nutritional and functional benefits and with minimum quality impairment. Thus, at 87.56 °C with a moisture of 26.01% for 3 h and an addition level of 4.81% GPF, pasta presented high TPC content (11.36 mg/100 g) and RS content (2.80%). Chen et al. [4] reported an RS content of HMT wheat starch at 25% moisture of 2.64%. Nakov et al. [53] obtained a TPC value of 9.41 mg/100 g of cakes with 4% grape pomace powder.

5. Conclusions

HMT is a useful tool for physical modification of wheat flour that enhanced its nutritional and functional value. The synergistic effect of HMT and GPF addition resulted in high DF, TPC, and RS contents of pasta due to the intake of GPF and the starch and proteins modifications during treatment. Doug firmness and viscoelastic moduli increased with HMT temperature, time, moisture, and GPF level increases. Pasta firmness and gumminess lowering was proportional to HMT regime and GPF addition increase.

The optimal wheat flour HMT regime was found to be a temperature of 87.56 °C, a moisture content of 26.01%, and a time of 3 h. GPF can be incorporated at a quantity of 4.81% in order to achieve the best pasta quality with the most health benefits. These results could be helpful for the development of novel functional pasta by applying an easy, controllable process and by using an inexpensive fiber-rich ingredient like GPF. The data presented in this study could be helpful for further pasta processing automatization.

Author Contributions: Conceptualization, M.I. and S.M.; methodology, M.I., S.M. and A.B.; software, M.I.; validation, M.I., S.M. and A.B.; formal analysis, M.I.; investigation, M.I. and A.B.; resources, M.I., S.M. and A.B.; data curation, M.I.; writing—original draft preparation, M.I.; writing—review and editing, M.I. and S.M.; visualization, M.I. and S.M.; supervision, S.M. All authors have read and agreed to the published version of the manuscript.

Funding: This work was supported by the Romania National Council for Higher Education Funding, CNFIS, project number CNFIS-FDI-2021-0357.

Institutional Review Board Statement: Not applicable.

Informed Consent Statement: Not applicable.

Data Availability Statement: Not applicable.

Conflicts of Interest: The authors declare no conflict of interest.

References

1. Borcean, I. Durum Wheat: Domestic Production, not Import. Available online: https://www.revista-ferma.ro/articole/agronomie/graul-durum-productie-autohtona-nu-import (accessed on 23 February 2021).
2. Collar, C.; Armero, E. Functional and Thermal Behaviours of Heat-Moisture-Treated Starch-Rich Wheat-Based Blended Matrices: Impact of Treatment of Non-wheat Flours. *Food Bioprocess Technol.* **2019**, *12*, 599–612. [CrossRef]
3. Iuga, M.; Mironeasa, S. A review of the hydrothermal treatments impact on starch based systems properties. *Crit. Rev. Food Sci. Nutr.* **2020**, *60*, 3890–3915. [CrossRef] [PubMed]
4. Chen, X.; He, X.; Fu, X.; Huang, Q. In vitro digestion and physicochemical properties of wheat starch/flour modified by heat-moisture treatment. *J. Cereal Sci.* **2015**, *63*, 109–115. [CrossRef]
5. Wang, H.; Zhang, B.; Chen, L.; Li, X. Understanding the structure and digestibility of heat-moisture treated starch. *Int. J. Biol. Macromol.* **2016**, *88*, 1–8. [CrossRef] [PubMed]
6. Zavareze, E.D.R.; Dias, A.R.G. Impact of heat-moisture treatment and annealing in starches: A review. *Carbohydr. Polym.* **2011**, *83*, 317–328. [CrossRef]
7. Liao, L.; Liu, H.; Gan, Z.; Wu, W. Structural properties of sweet potato starch and its vermicelli quality as affected by heat-moisture treatment. *Int. J. Food Prop.* **2019**, *22*, 1122–1133. [CrossRef]

8. Iuga, M.; Mironeasa, C.; Mironeasa, S. Oscillatory Rheology and Creep-Recovery Behaviour of Grape Seed-Wheat Flour Dough: Effect of Grape Seed Particle Size, Variety and Addition Level. *Bull. Univ. Agric. Sci. Veter. Med. Cluj Napoca. Food Sci. Technol.* **2019**, *76*, 40–51. [CrossRef]

9. Yu, J.; Ahmedna, M. Functional components of grape pomace: Their composition, biological properties and potential applications. *Int. J. Food Sci. Technol.* **2013**, *48*, 221–237. [CrossRef]

10. Deng, Q.; Penner, M.H.; Zhao, Y. Chemical composition of dietary fiber and polyphenols of five different varieties of wine grape pomace skins. *Food Res. Int.* **2011**, *44*, 2712–2720. [CrossRef]

11. Iuga, M.; Ropciuc, S.; Mironeasa, S. Antioxidant Activity and Total Phenolic Content of Grape Seeds and Peels from Romanian Varieties. *Food Environ. Saf. J.* **2017**, *16*, 276–282.

12. Gaita, C.; Alexa, E.; Moigradean, D.; Filomena, C.; Poiana, M.-A. Designing of high value-added pasta formulas by incorporation of grape pomace skins. *Rom. Biotechnol. Lett.* **2020**, *25*, 1607–1614. [CrossRef]

13. Savla, H.; Yardi, V. Development of Gluten-Free Pasta (Sevaiya) Using Grape Pomace and Assessing its Quality And Acceptability. *Int. Proc. Chem. Biol. Environ. Eng.* **2016**, *95*, 44–49. [CrossRef]

14. Tolve, R.; Pasini, G.; Vignale, F.; Favati, F.; Simonato, B. Effect of Grape Pomace Addition on the Technological, Sensory, and Nutritional Properties of Durum Wheat Pasta. *Foods* **2020**, *9*, 354. [CrossRef] [PubMed]

15. Gaceu, L.; Gruia, R.; Oprea, O.B. Researches Regarding the Superior Valorisation of By-Products From The Winery Industry, and the Obtaining of Bakery Products. *Ann. Acad. Rom. Sci. Ser. Agric. Silvic. Vet. Med. Sci.* **2020**, *9*, 25–36.

16. Sant'Anna, V.; Christiano, F.D.P.; Marczak, L.D.F.; Tessaro, I.C.; Thys, R.C.S. The effect of the incorporation of grape marc powder in fettuccini pasta properties. *LWT Food Sci. Technol.* **2014**, *58*, 497–501. [CrossRef]

17. Mironeasa, S.; Iuga, M.; Zaharia, D.; Mironeasa, C. Rheological Analysis of Wheat Flour Dough as Influenced by Grape Peels of Different Particle Sizes and Addition Levels. *Food Bioprocess Technol.* **2019**, *12*, 228–245. [CrossRef]

18. Fanari, F.; Desogus, F.; Scano, E.A.; Carboni, G.; Grosso, M. The effect of the relative amount of ingredients on the rheological properties of semolina doughs. *Sustainability* **2020**, *12*, 2705. [CrossRef]

19. Mironeasa, S.; Iuga, M.; Zaharia, D.; Mironeasa, C. Optimization of grape peels particle size and flour substitution in white wheat flour dough. *Sci. Study Res. Chem. Chem. Eng. Biotechnol. Food Ind.* **2019**, *20*, 29–42.

20. Aghamirzaei, M.; Peighambardoust, S.H.; Azadmard-Damirchi, S.; Majzoobi, M. Effects of grape seed powder as a functional ingredient on flour physicochemical characteristics and dough rheological properties. *J. Agric. Sci. Technol.* **2015**, *17*, 365–373.

21. Serna-Saldívar, S.O. *Cereal Grains Laboratory Reference and Procedures Manual*; CRC Press: Boca Raton, FL, USA, 2012; ISBN 9781466555631.

22. Sicignano, A.; Di Monaco, R.; Masi, P.; Cavella, S. From raw material to dish: Pasta quality step by step. *J. Sci. Food Agric.* **2015**, *95*, 2579–2587. [CrossRef]

23. Ciccoritti, R.; Scalfati, G.; Cammerata, A.; Sgrulletta, D. Variations in content and extractability of durum wheat (Triticum turgidum L. var durum) arabinoxylans associated with genetic and environmental factors. *Int. J. Mol. Sci.* **2011**, *12*, 4536–4549. [CrossRef] [PubMed]

24. Hoover, R. The impact of heat-moisture treatment on molecular structures and properties of starches isolated from different botanical sources. *Crit. Rev. Food Sci. Nutr.* **2010**, *50*, 835–847. [CrossRef] [PubMed]

25. Collar, C.; Armero, E. Value-added of heat moisture treated mixed flours in wheat-based matrices: A functional and nutritional approach. *Food Bioprocess Technol.* **2018**, *11*, 1536–1551. [CrossRef]

26. Bergman, C.; Gualberto, D.; Weber, C. Development of a high-temperature-dried soft wheat pasta supplemented with cowpea (Vigna unguiculata (L) Walp)—Cooking quality, color, and sensory evaluation. *Cereal Chem.* **1994**, *71*, 523–527.

27. Marchetti, L.; Cardós, M.; Campaña, L.; Ferrero, C. Effect of glutens of different quality on dough characteristics and breadmaking performance. *LWT Food Sci. Technol.* **2012**, *46*, 224–231. [CrossRef]

28. Ungureanu-Iuga, M.; Dimian, M.; Mironeasa, S. Development and quality evaluation of gluten-free pasta with grape peels and whey powders. *LWT Food Sci. Technol.* **2020**, *130*, 1–9. [CrossRef]

29. Troccoli, A.; Borrelli, G.M.; De Vita, P.; Fares, C.; Di Fonzo, N. Durum wheat quality: A multidisciplinary concept. *J. Cereal Sci.* **2000**, *32*, 99–113. [CrossRef]

30. Melilli, M.G.; Pagliaro, A.; Scandurra, S.; Gentile, C.; Di Stefano, V. Omega-3 rich foods: Durum wheat spaghetti fortified with Portulaca oleracea. *Food Biosci.* **2020**, *37*, 100730. [CrossRef]

31. Banga, J.R.; Balsa-Canto, E.; Moles, C.G.; Alonso, A.A. Improving food processing using modern optimization methods. *Trends Food Sci. Technol.* **2003**, *14*, 131–144. [CrossRef]

32. Fanari, F.; Desogus, F.; Scano, E.A.; Carboni, G.; Grosso, M. The rheological properties of semolina doughs: Influence of the relative amount of ingredients. *Chem. Eng. Trans.* **2019**, *76*, 703–708. [CrossRef]

33. Mann, J.; Schiedt, B.; Baumann, A.; Conde-Petit, B.; Vilgis, T.A. Effect of heat treatment on wheat dough rheology and wheat protein solubility. *Food Sci. Technol. Int.* **2014**, *20*, 341–351. [CrossRef]

34. Lazaridou, A.; Marinopoulou, A.; Biliaderis, C.G. Impact of flour particle size and hydrothermal treatment on dough rheology and quality of barley rusks. *Food Hydrocoll.* **2019**, *87*, 561–569. [CrossRef]

35. Fanari, F.; Carboni, G.; Grosso, M.; Desogus, F. Thermal properties of semolina doughs with different relative amount of ingredients. *Sustainability* **2020**, *12*, 2235. [CrossRef]

36. Lindahl, L.; Eliasson, A.C. A comparison of some rheological properties of durum and wheat flour doughs. *Cereal Chem.* **1992**, *69*, 30–34.
37. Peressini, D.; Cavarape, A.; Brennan, M.A.; Gao, J.; Brennan, C.S. Viscoelastic properties of durum wheat doughs enriched with soluble dietary fibres in relation to pasta-making performance and glycaemic response of spaghetti. *Food Hydrocoll.* **2020**, *102*, 105613. [CrossRef]
38. Chavez-Murillo, C.E.; Orona-Padilla, J.L.; de la Rosa Millan, J. Physicochemical, functional properties and ATR-FTIR digestion analysis of thermally treated starches isolated from black and bayo beans. *Starch/Staerke* **2019**, *71*, 1–25. [CrossRef]
39. Aksoylu, Z.; Çağindi, Ö.; Köse, E. Effects of blueberry, grape seed powder and poppy seed incorporation on physicochemical and sensory properties of biscuit. *J. Food Qual.* **2015**, *38*, 164–174. [CrossRef]
40. Zheng, B.; Zhong, S.; Tang, Y.; Chen, L. Understanding the nutritional functions of thermally-processed whole grain highland barley in vitro and in vivo. *Food Chem.* **2020**, *310*, 125979. [CrossRef] [PubMed]
41. Wang, H.; Liu, Y.; Chen, L.; Li, X.; Wang, J.; Xie, F. Insights into the multi-scale structure and digestibility of heat-moisture treated rice starch. *Food Chem.* **2018**, *242*, 323–329. [CrossRef]
42. Biniari, K.; Gerogiannis, O.; Daskalakis, I.; Bouza, D.; Stavrakaki, M. Study of some qualitative and quantitative characters of the grapes of indigenous Greek grapevine varieties (Vitis vinifera L.) using HPLC and spectrophotometric analyses. *Not. Bot. Horti Agrobot. Cluj-Napoca* **2018**, *46*, 97–106. [CrossRef]
43. Iuga, M.; Mironeasa, S. Potential of grape byproducts as functional ingredients in baked goods and pasta. *Compr. Rev. Food Sci. Food Saf.* **2020**, *19*, 2473–2505. [CrossRef]
44. Mironeasa, S. *Valorisation of Secondary Products from Wine Making*; Publishing House Performantica: Iasi, Romania, 2017; p. 56.
45. Liu, Y.F.; Laohasongkram, K.; Chaiwanichsiri, S. Effects of Heat-Moisture Treatment on Molecular Interactions and Physicochemical Properties of Tapioca Starch. *MOJ Food Process. Technol.* **2016**, *3*, 304–311. [CrossRef]
46. Galvez, F.C.F.; Resurreccion, A.V.A. Reliability of the Focus Group Technique in Determining the Quality Characteristics of Mungbean [Vigna Radiata (L.) Wilczec] Noodles. *J. Sens. Stud.* **1992**, *7*, 315–326. [CrossRef]
47. Li, M.; Liu, C.; Zheng, X.; Li, L.; Bian, K. Physicochemical properties of wheat flour modified by heat-moisture treatment and their effects on noodles making quality. *J. Food Process. Preserv.* **2020**, *44*, 1–9. [CrossRef]
48. Chandla, N.K.; Saxena, D.C.; Singh, S. Processing and evaluation of heat moisture treated (HMT) amaranth starch noodles; An inclusive comparison with corn starch noodles. *J. Cereal Sci.* **2017**, *75*, 306–313. [CrossRef]
49. Chen, J.S.; Fei, M.J.; Shi, C.L.; Tian, J.C.; Sun, C.L.; Zhang, H.; Ma, Z.; Dong, H.X. Effect of particle size and addition level of wheat bran on quality of dry white Chinese noodles. *J. Cereal Sci.* **2011**, *53*, 217–224. [CrossRef]
50. Barros, J.H.T.; Telis, V.R.N.; Taboga, S.; Franco, C.M.L. Resistant starch: Effect on rheology, quality, and staling rate of white wheat bread. *J. Food Sci. Technol.* **2018**, *55*, 4578–4588. [CrossRef] [PubMed]
51. Rocchetti, G.; Giuberti, G.; Busconi, M.; Marocco, A.; Trevisan, M.; Lucini, L. Pigmented sorghum polyphenols as potential inhibitors of starch digestibility: An in vitro study combining starch digestion and untargeted metabolomics. *Food Chem.* **2020**, *312*, 126077. [CrossRef]
52. McDougall, G.J.; Shpiro, F.; Dobson, P.; Smith, P.; Blake, A.; Stewart, D. Different Polyphenolic Components of Soft Fruits Inhibit α-Amylase and α-Glucosidase. *J. Agric. Food Chem.* **2005**, *53*, 2760–2766. [CrossRef]
53. Nakov, G.; Brandolini, A.; Hidalgo, A.; Ivanova, N.; Stamatovska, V.; Dimov, I. Effect of grape pomace powder addition on chemical, nutritional and technological properties of cakes. *LWT* **2020**, *134*, 109950. [CrossRef]

*applied
sciences*

MDPI

Article

Preliminary Investigations on the Use of a New Milling Technology for Obtaining Wholemeal Flours

Pavel Skřivan, Marcela Sluková *, Lucie Jurkaninová and Ivan Švec

Department of Carbohydrates and Cereals, University of Chemistry and Technology Prague, 166 28 Prague, Czech Republic; pavel.skrivan@vscht.cz (P.S.); lucie.jurkaninova@vscht.cz (L.J.); ivan.svec@vscht.cz (I.Š.)
* Correspondence: marcela.slukova@vscht.cz

Abstract: Wholemeal flours from various cereals and pseudocereals are a valuable source of nutritionally important fiber components, especially beta-glucans and arabinoxylans, as well as bioactive substances accompanying dietary fiber. Most types of whole-wheat flours have unfavorable baking and sensory properties. The finest granulation of bran particles in the flour has a significant effect on reducing or eliminating these deficiencies. Special disintegration equipment is required to achieve fine granulation of the bran particles. In this study, we have tested a special type of impact mill (originally intended for grinding of plastics) to produce special finely ground wholemeal flours with lower starch damage and higher farinographic absorption. Moisture content in the studied flours was significantly lower (7.4–9.8%) than is common in standard flour (13–14%). According to the results of flour analyses obtained from several cereal sources, it seems that especially in rye and wheat, this technology is suitable for both achieving fine granulation of bran particles and in terms of not very substantial damage of starch granules.

Keywords: wholegrain flour; wholemeal flour; milling technology; granulation; starch damage

Citation: Skřivan, P.; Sluková, M.; Jurkaninová, L.; Švec, I. Preliminary Investigations on the Use of a New Milling Technology for Obtaining Wholemeal Flours. *Appl. Sci.* **2021**, *11*, 6138. https://doi.org/10.3390/app11136138

Academic Editors: Silvia Mironeasa and Suyong Lee

Received: 6 May 2021
Accepted: 29 June 2021
Published: 1 July 2021

1. Introduction

The production of wholemeal flours is the oldest method of mill processing of cereals. It has been replaced by the production of flours containing mostly the endosperm over the centuries. The other anatomical parts of grain (hulls and germ) separate in a different way and these by-products are often traditionally used as feed ingredients for farm animals. The most efficient procedures of endosperm separation from the coating layers of the grain and the production of white flours are achieved in the case of wheat. White wheat and bread wheat flours consisting mostly of endosperm particles are the most widespread type of flours in Europe, USA, and currently also in the majority parts of the world [1,2].

The endosperm separation reduces the spectrum of flour components to starch and storage proteins. In the case of wheat, starch (a digestible polysaccharide) accounts for about 80% of the dry matter of bakery flours and the remaining components are the proteins of gliadin and glutenin fraction. Virtually all dietary fiber and fiber accompanying substances with proven health benefits then become a component of the by-product (bran) and only a very low proportion, in the order of units of percent, remains in the standard bakery flours. The higher content of fiber, minerals, vitamins, phenolic compounds, and other bioactive compounds can be found in darker (bread) flours. This also applies to rye, another bread cereal used mainly in Central, Eastern, and Northern Europe, albeit in this case, the endosperm separation is less effective than in the case of wheat and thus rye bread flour contains higher proportions of fiber (up to 10%) [3–6].

The most important components of fiber contained in the hulls of wheat, rye, and other cereals and pseudocereals in terms of nutritional benefits are cereal beta-glucans and arabinoxylans, and the most significant accompanying substances are phenolic compounds, showing for the most part antioxidant properties [7] and also some B-group vitamins and minerals [8–10].

For that reason, wholemeal flours are considerably more beneficial from the nutritional point of view than standard bakery flours. Their major problem is that for the production of basic types of bread and pastries of the Euro-American type, wholemeal flours are technologically less suitable and less sensorially acceptable [11,12]. Not to be overlooked is the fact that wholemeal flours are potentially more risky in terms of the content of contaminants, in particular mycotoxins and pesticide residues [13,14].

Standard wheat and rye wholemeal flours are produced in principle in two ways. In the first, most common case, the basis of the disintegration process is a simple impact mill that can be followed by a rolling mill for granulation treatment. The second method of production is the reconstitution of wholemeal flour from common mill products, which is used if the main production program of the mill is the standard composition of products (bakery flours) and wholemeal flour is a minor product. An example of such a product is Graham wheat flour. Even in the case of reconstituted flour, granulation treatment is usually carried out after the fractions have been assembled [1,2].

In both cases, despite the granulation treatment, insufficiently disintegrated particles of the tough outer layers remain in the flour, which are responsible for the specific sensory properties of the flour and bakery products. These particles absorb water and swell in the dough differently than the endosperm particles, making the dough denser and less homogeneous. The partially swollen bran particles then prevent the formation of a light, finely fibrous, porous dough structure during maturation and leavening, which is typical of wheat bread and pastries and is required by consumers. Even in the case of rye breads, where the structure of the dough is generally much more compact than in the case of wheat doughs, a higher proportion of common rye wholemeal flour has a negative effect. Typical Central European rye-wheat bread is as difficult to make from common wholemeal flour as wheat bread [15,16].

It turns out that granulation of particles of outer layers (bran particles) has a fundamental influence on technological behavior and sensory properties. If a sufficiently fine granulation is achieved, which is close to the average granulation of common very fine flours (below 150–200 µm), the sorption capacity (water binding) of such dough is significantly increased, but its mechanical and sensory properties change minimally. Such finely ground wholemeal flours can then replace conventional flours much better, and at the same time, thanks to the higher water binding, a significantly higher yield of dough and products is achieved.

A major problem to keep in mind in the case of production of such flours is the risk of too high a degree of damage to the starch granules. In other words, it is necessary to use a technique that reliably disintegrates the outer layers into very fine fractions, but at the same time does not cause extensive mechanical and especially thermal damage to the starch granules [15].

Damage to starch during disintegration is of two types: (a) mechanical, due to pressure, shear forces, and crash; and (b) thermal, especially due to friction. The deformation to which the grain and its internal structure are subjected during disintegration differs fundamentally depending on the type of mill and grinding parameters used. Grinding in roller mills has a different impact, depending on the parameters (advance, down pressure, specific load) and surface treatment of the rollers (grooving) and their mutual position. In the case of impact mills, other mechanical forces act, and grinding can be both very gentle and very destructive to starch granules, depending on the design of the mill and the process parameters [1,2].

When the grain is disintegrated in any way, partial deformation and damage to the starch granules always occurs. For bakery purposes, where fermentation is one of the key processes, a certain degree of starch damage is desirable because it becomes more amenable to the amylases present and a sufficient amount of substrate is created for yeast or lactic acid bacteria. In the case of flour for the production of wafers and biscuits, confectionery or pasta, damage to starch is not beneficial. However, with extensive damage, complications also arise in ordinary bakery production (stickiness of the dough, poor structure of the bread crumbs, etc.).

Starches of cereals and pseudocereals are easily broken down in the upper part of the human digestive tract. Most starches are completely resorbed in the small intestine, and only some starches may be partially resistant and are classified as indigestible polysaccharides [17,18]. Starch is gradually cleaved by amylases present in saliva and especially in pancreatic juice to oligosaccharides, which are further hydrolyzed to glucose monosaccharide. In the small intestine, glucose is actively absorbed into body fluids.

Dietary starch intake results in a significant increase in blood glucose. Glycaemia is the concentration of glucose in the blood, the value of which in fasting should not exceed 5.5 mol/L (in venous blood) according to the current approach [19]. Starch resorption is rapid, and its rate increases with the degree of damage to the native starch structure. Starch damage can occur biochemically (by enzymatic hydrolysis), chemically (by acid hydrolysis), mechanically, and thermally. The higher the starch resorption rate, the higher the glycemic index (GI) of the food in question [20].

The wholemeal flours tested in this study were produced in a special mill. Physical and chemical analyses of the flours were conducted and the impact of grinding technology on microstructure and properties of the flours was evaluated. What is essential for this work is the assessment of the extent to which the grain disintegration technique used leads to intensive comminution of the grain outer layers (bran particles) into fine granulation without significant damage to the starch granules.

2. Materials and Methods

Cereal grains (*Triticum aestivum* L., *Triticum spelta* L., *Secale cereale* L.) and a pseudocereal (*Fagopyrum esculentum* Moench) were decontaminated (cleaned) and then disintegrated in a special impact mill. Finely granulated wholemeal flours as final products were subsequently analyzed. The flours used were as follows: finely granulated wholemeal flour of wheat (wheat WM FG), rye (rye WM FG), spelt (spelt WM FG), and buckwheat (buckwheat WM FG).

2.1. Flour Production

The flours tested in this study were produced on a special mill from Mahltechnik Görgens GmbH (Dormagen, Germany). It is an impact mill with a vertical axis of rotation, which was not originally intended for grain processing, but which was included in a special production line in the company Mlýn Perner Svijany (Svijany, Czech Republic) and from which the grinding of cereals and pseudocereals into wholemeal flours achieves remarkable results.

The principle of the grinding process is the disintegration of the grist between specially shaped grinding segments rotating in several levels above each other and a specially modified inner shell of the wall of the grinding device (Figure 1). The material is kept suspended in the air stream throughout the disintegration, while it is possible to regulate the residence time of the grist in the grinding chamber and thus also its granulation. Thanks to this arrangement, the flour is not exposed to such thermal stress as when grinding on conventional grinding and roller mills.

The temperature in the grinding chamber ranges from 50 to 90 °C, while the temperature of the grist is in the range of 30–80 °C depending on the setting of the parameters of the special mill. The residence time of the grist in the grinding chamber is regulated by setting the parameters of the screen of the control plansifter located behind the mill. With sensitive control of the mill parameters, it is possible to achieve, on the one hand, very fine granulation of the bran parts, but on the other hand, there may not be more extensive damage to the starch granules either mechanically or thermally. The aim of our work was to verify these assumptions on selected samples of flour.

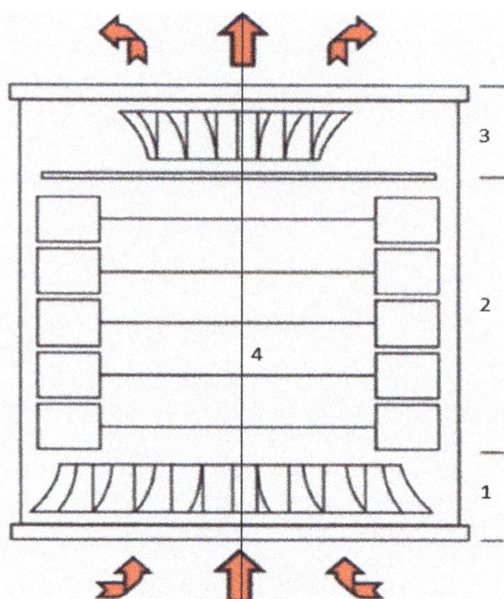

Figure 1. Scheme of grinding equipment. Description of the equipment parts: 1—grain input; 2—grinding zone; 3—sieving zone, regulation of the granulation; 4—rotation axis.

2.2. Determination of Moisture, Ash, Protein, and Fiber

The moisture content was determined by standard ICC (International Association for Cereal Science and Technology) No. 110/1, content of ash was specified by ICC No. 104/1, and protein content was measured using the Kjeldahl method (factor 5.7) (ICC No. 105/2). Soluble (SDF), insoluble (IDF), and total (TDF) fiber content was determined using a commercial Total Dietary Fiber enzyme kit from Megazyme (Ireland) according to the approved AOAC (Association of Official Agricultural Chemists) 991.43 method (determination of fiber by enzymatic-gravimetric method) on the Fibertec system (Tecator Foss, Höganäs, Sweden). The results of duplicates of moisture, ash, protein, and fiber determination were repeated three times.

2.3. Flour Granulation

The sieve analysis was performed according to previously valid Czechoslovak state standards. The sample weight amounted to 100 g, and it was sieved using a KS 1000 rotary sieve (Retsch, Haan, Germany) at a frequency of 100 rpm for 5 min. The set of sieves, equipped with chains, consisted of sieves with mesh sizes of 400, 300, 250, 200, 180, 150, 125, and 90 μm. Sieve analysis was performed in triplicate for each sample and the results were averaged.

2.4. Solvent Retention Capacity Profile

The physical-and-chemical test profile of retention capacity (SRC) was determined according to the AACC (American Association of Cereal Chemists) 56-11 methodology. The solvent retention capacity (SRC) is expressed as the weight of solvent retained by the flour after centrifugation of the flour suspension with the solvent under the given conditions. It is expressed as a percentage by weight of flour. The result is based on 14% moisture of flour. Four solvents are used independently to profile the SRC values: SRC water (demineralized water), SRC aqueous sucrose solution (50% w/w), SRC aqueous sodium carbonate solution (5% w/w), and SRC aqueous lactic acid solution (5% w/w). By combining these four SRC values, the quality profile of the flour can be determined, and its baking and technological properties can be predicted. In general, lactic acid SRC is associated with glutenin characteristics, sodium carbonate SRC with damaged starch levels, and sucrose SRCs with pentosans characteristics. The SRC of water is affected by all these components of flour. The results of duplicates of SRC profile parameters determination were repeated three times.

2.5. Determination of Falling Number

Estimation of alpha-amylase activity and degree of starch damage was performed using a Falling Number instrument (type 1400, Perten Instruments, Hägersten, Sweden). The procedure corresponded to the EN ISO 3093 standard, i.e., the weight of the flour sample corresponded to the current value of the flour moisture. An appropriate laboratory shaker of Polish origin (Wytrząsarka type SZ, biogenet, Józefów, Poland) was used to create the suspension. The results of two experiments of determination of Falling Number were verified (the results may differ not more than 5% of their mean value). There were three replicates performed in the same way for the statistical evaluation.

2.6. Farinographic Water Absorption

Farinographic binding of samples of flours and their mixtures was performed according to ISO 5530-1 with the connection of the Farinograph TS device (Brabender, Duisburg, Germany), allowing the addition of distilled water with regard to the current value of flour moisture and the repetition of the test with corrected binding. The measurement was performed in one replicate; the previously determined repeatability of the farinographic water absorption determination is 0.2 percentage points.

2.7. Amylographic Measurements

An AS-type amylograph (Brabender, Duisburg, Germany) was used to evaluate the viscosity behavior of flour suspensions during heating from 25 to 95 °C as a simulation of dough processes during baking. The test methodology was based on the international standard ICC 126/1, when the standard weight of flour and water was 80 and 450 g. The evaluated features were the onset temperatures and gelation maximum (°C) as well as the amylographic maximum (viscosity) in customary Brabender units (BU).

2.8. Scanning Electron Microscopy

The scanning electron microscope works with a stream of electrons in a vacuum in the case of conductive samples or at very low pressures for non-conductive samples. The interaction of the electron beam with the mass (sample examined) creates a visible image. The generated signals carry information about the topography of the sample and its material composition. The scanning electron microscope can provide comprehensive information about the microstructure, chemical composition, and other properties of the examined sample.

Secondary electrons (SE) and/or backscattered electrons (BSE) are most often used to monitor the microstructure of a flour sample. Images were taken on a TESCAN VEGA3 LMU microscope (tungsten cathode, SE and BSE detector) at different magnifications (500× and 2500×). The measurement was performed in UNIVAC mode (pressure 10 Pa) using a BSE detector. The accelerating voltage of the electrons was 20 kV. The flour samples were applied to a double-sided carbon tape and placed on the plates of a metal holder. Prior to scanning, the samples were covered with a continuous layer of 5 nm gold in a Quorum 150 coater.

2.9. Statistical Data Processing

The results of analytical and rheological measurements were statistically analyzed with STATISTICA software version 12.0 (StatSoft, Tulsa, OK, USA). One-way analysis of variance (ANOVA) at $p \leq 0.05$ was calculated and groups were estimated according to Duncan's new multiple range test (MRT).

3. Results and Discussion

The main research was focused on wheat flour (both sown wheat and spelt), rye, and buckwheat produced on the Görgens mill.

3.1. Chemical Analyses of Flours (Moisture, Ash, Protein, and Fiber Content)

In the case of ash, protein, and fiber content, the flours show the expected values corresponding to the individual raw materials. Moisture, ash, protein, and fiber content in the analyzed wholemeal flours are shown in Tables 1 and 2.

Table 1. Content of moisture, ash, and protein in the wholemeal finely granulated flours (WM FG). Data are the means of three replicates (±standard deviation).

Flour	Moisture (g/100 g)	Ash (g/100 g d.m.)	Protein (g/100 g d.m.)
Wheat WM FG	7.4 ± 0.1 a	1.62 ± 0.02 b	11.71 ± 0.52 b
Rye WM FG	9.4 ± 0.1 b	1.61 ± 0.03 b	8.39 ± 0.61 a
Spelt WM FG	9.0 ± 0.2 b	1.66 ± 0.03 b	13.44 ± 0.22 b
Buckwheat WM FG	9.8 ± 0.3 b	2.11 ± 0.04 c	12.70 ± 0.44 b
Wheat white (T530)	13.3 ± 0.1 c	0.56 ± 0.01 a	12.59 ± 0.11 b

Data represent the means of three replicates. Small letters in the same column denote significant differences according to Duncan's new multiple range test (MRT) ($p \leq 0.05$).

Table 2. Dietary fiber content expressed as soluble (SDF), insoluble (IDF), and total fiber (TDF) in the wholemeal finely granulated flours (WM FG). Data are the means of three replicates (±standard deviation).

Flour	SDF (g/100 g d.m.)	IDF (g/100 g d.m.)	TDF (g/100 g d.m.)
Wheat WM FG	2.1 ± 0.2 b	8.8 ± 0.5 b	12.0 ± 0.8 b
Rye WM FG	5.3 ± 0.6 c	13.0 ± 0.3 d	18.2 ± 0.8 c
Spelt WM FG	2.5 ± 0.2 b	9.6 ± 0.9 c	12.4 ± 0.9 b
Buckwheat WM FG	3.2 ± 0.2 b	6.6 ± 0.2 b	9.7 ± 0.4 b
Wheat white (T530)	0.8 ± 0.1 a	2.7 ± 0.2 a	3.3 ± 0.3 b

Data represent the means of three replicates. Small letters in the same column denote significant differences according to Duncan's new multiple range test (MRT) ($p \leq 0.05$).

It should be noted that the moisture content of the flour samples examined is significantly lower than is usual for standard flours. With standard flours, the humidity values are usually in the range of 13–14%, while flours produced on a special Görgens mill most often have a humidity in the range of 7–10%. This is due firstly to the fact that the grain is not tempered before this method of grinding and at the same time it is also due to the disintegration method, especially by keeping the flour suspended in an air stream of 60–70 °C (Figure 1). At these air temperatures, with the usual residence time of the grist in the grinding chamber for several tens of seconds, the flour is heated to temperatures of 40–50 °C. The impact of this intervention on the starch structure is the subject of our investigation. The residence time of the individual particles in the grinding chamber varies according to their size. Coarse particles that do not pass through the control sieve return and their residence time is therefore longer. It does not exceed 1 min at the parameters set during the experimental production of the flours examined.

3.2. Flour Granulation

The method of grinding and the conditions in the grinding zone, which resulted in reduced moisture of the flours, are closely related to the achieved granulations of the individual flours, which represent the subsequent granulation spectra. Granulation spectra of the studied wholemeal flours are present in Figure 2.

In our case, two basic factors have a fundamental influence on the result of disintegration in terms of particle size (granulation): mechanical properties of the disintegrated material and grinding intensity, which in this case is given by the residence time of the grist in the grinding zone of the mill. Due to the fact that on the one hand there was the finest possible granulation of flours (and especially hull particles), but on the other hand also the effort to minimize interference with the structure of the endosperm, especially starch, the above conditions were chosen (residence time and related temperature environment and material). Under these conditions, the properties of the disintegrated grain are fully manifested because the disintegration conditions are the same and relatively gentle.

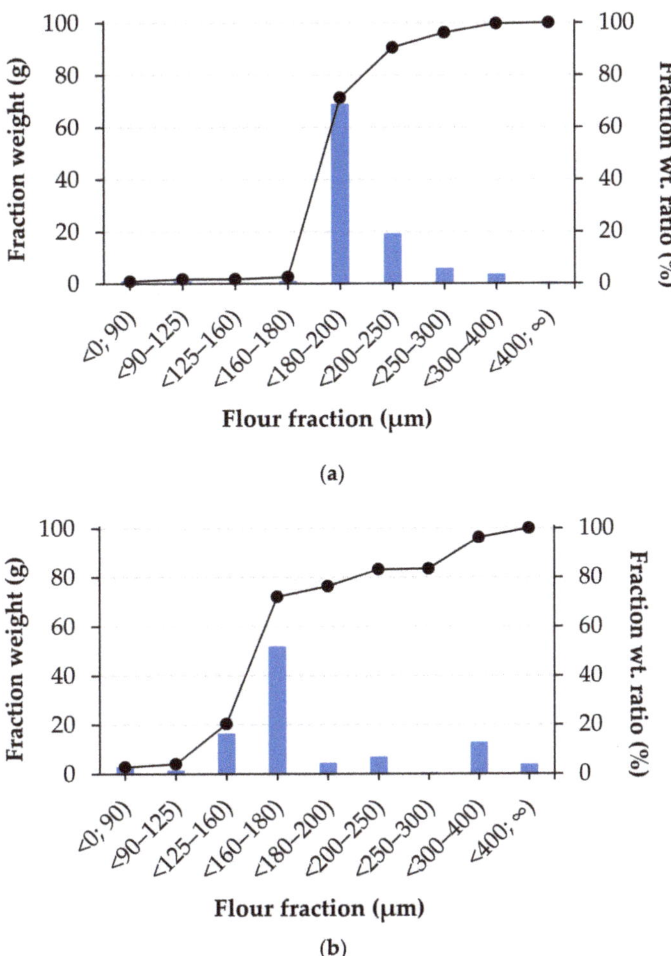

(a)

(b)

Figure 2. *Cont.*

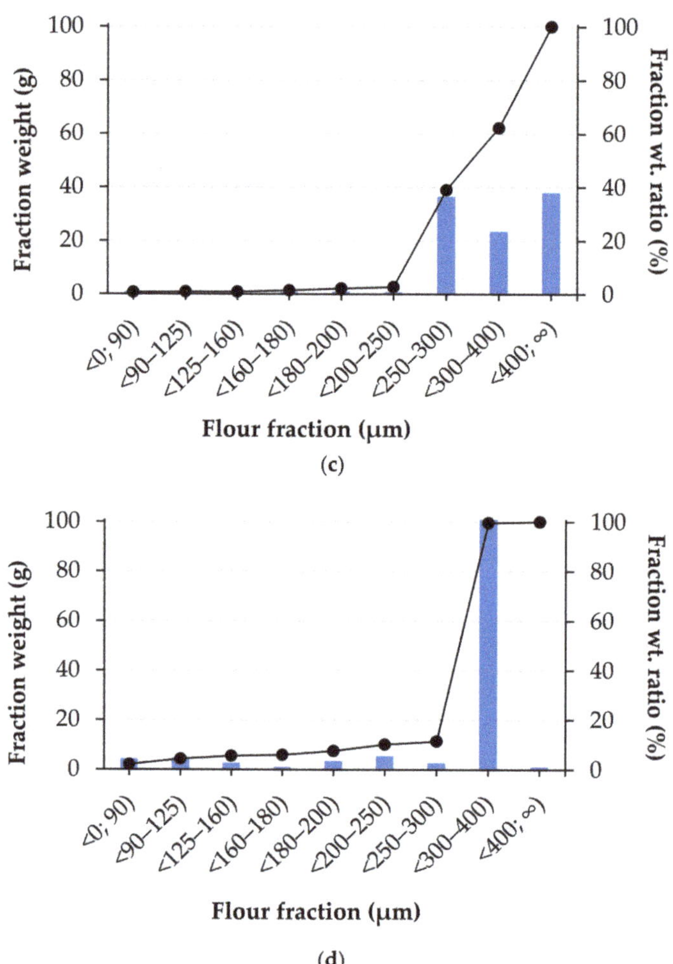

Figure 2. Granulation spectra of wheat (**a**), rye (**b**), spelt (**c**), and buckwheat (**d**) wholemeal finely granulated flours.

From this point of view, rye grain yields best to disintegration (particles 160–180 μm predominate) (Figure 2b), followed by common wheat (with a maximum frequency of particles of 200 μm without significant presence of coarser fractions) (Figure 2a). The situation is different for spelt and particularly for buckwheat (Figure 2c and 2d, respectively). A more intensive grinding process would be needed to achieve fine granulation comparable to rye or wheat flour. Especially, why buckwheat provides such a high yield of relatively coarse fraction (300–400 μm) (Figure 2d) will require further investigation.

3.3. Farinographic Water Absorption

Calculated farinographic water absorption of the analyzed wholemeal flours is shown in Table 3.

Table 3. Recalculated farinographic water absorption of the wholemeal finely granulated flours (WM FG). Data are the means of three replicates (±standard deviation).

Flour	Water Absorption (%)
Wheat WM FG	78.0 ± 0.1 b
Rye WM FG	59.2 ± 0.2 a
Spelt WM FG	70.0 ± 0.2 b
Buckwheat WM FG	63.4 ± 0.3 a
Wheat white (T530)	60.5 ± 0.1 a

Data represent the means of three replicates. Small letters in the same column denote significant differences according to Duncan's new multiple range test (MRT) ($p \leq 0.05$).

The binding of wholemeal flours was determined in a mixture with standard wheat flour of known binding (the proportion of wholemeal flours was 30% by weight) and recalculated according to the mixing tolerance index. The bindings of wholemeal flours are unsurprisingly higher, which is due to their composition (presence of a higher proportion of biopolymers, especially polysaccharides with hydrocolloid properties). In all cases, the water absorption of these flours is around 60 or more percent. This, of course, has an impact on the processing (baking) properties. The water absorption did not fully correlate with the granulation of individual flours.

3.4. Solvent Retention Capacity Profile of Flours

While previous results relate to the physical impacts of the disintegration process, the SRC method is intended to outline a rough picture of the state of the microstructure of the endosperm after disintegration. SRC values of the analyzed wholemeal flours are presented in Table 4.

Table 4. SRC values of the wholemeal finely granulated flours (WM FG). Data are the means of three replicates (±standard deviation). SRC values are expressed as percent of flour weight, on a 14% moisture basis.

Flour	SRC Values (%)			
	Water	Sucrose	Lactic Acid	Sodium Carbonate
Wheat WM FG	89 ± 2 b	125 ± 2 b	101 ± 1 a	114 ± 2 c
Rye WM FG	133 ± 3 c	113 ± 2 a	117 ± 2 b	123 ± 2 c
Spelt WM FG	80 ± 2 b	106 ± 2 a	95 ± 2 a	102 ± 2 c
Buckwheat WM FG	96 ± 4 b	143 ± 3 b	103 ± 4 a	98 ± 4 b
Wheat white (T530)	59 ± 1 a	94 ± 2 a	114 ± 1 b	71 ± 2 a

Data represent the means of three replicates. Small letters in the same column denote significant differences according to Duncan's new multiple range test (MRT) ($p \leq 0.05$).

Within the bounds of possibility provided by this method, the following can be stated. The SRC values in water correspond to a higher farinographic water absorption, although the results of farinographic binding are more balanced for individual flours. The SRC values for the sucrose solution correspond to the content of pentosans (arabinoxylans) and are relatively high, which is not a problem. High values for the lactic acid solution correspond to a higher presence of glutenins proteins in whole-wheat flours (both in common wheat and in spelt); in rye, endosperm proteins have different properties than in wheat, and in buckwheat they are practically absent.

For our research, the most significant values were those of SRC in sodium carbonate solution, which correspond to the degree of starch damage. From their values, the degree of starch damage in the examined wholemeal flours appears to be higher in comparison with the values for common flours in both wheat and spelt, as well as in the case of rye. The values for buckwheat are difficult to interpret in relation to wheat or rye because its flours behave very differently when determining SRC values (there was a partial separation of the layers during determination in sodium carbonate solution). Why it occurs still needs to be investigated in future research.

3.5. Determination of Characteristics of the Saccharide–Amylase Complex Using Falling Number and Amylograph

Values of Falling Number (FN) of the analyzed wholemeal flours are listed in Table 5.

Table 5. Falling Number values of the wholemeal finely granulated flours (WM FG). Data are the means of three replicates (±standard deviation).

Flour	Falling Number (s)
Wheat WM FG	368 ± 2 c
Rye WM FG	242 ± 3 a
Spelt WM FG	256 ± 3 a
Buckwheat WM FG	* 359 ± 3 c
Wheat white (T530)	316 ± 1 b

* Sample weight 4.40 g. Data represent the means of three replicates. Small letters in the same column denote significant differences according to Duncan's new multiple range test (MRT) ($p \leq 0.05$).

This method indirectly informs about the amylolytic activity of the investigated material and the degree of damage to starch, which occurs during disintegration, affects only secondarily. Rather, if other methods indicate a higher degree of starch damage, the FN values illustrate to what extent it is enzymatic damage and to what extent physical. With examined wholemeal flours in the case of wheat, spelt, and rye, the FN values do not indicate increased amylolytic activity. Irrelevant values arise and the sample weight must be changed in order to obtain the FN values common to wheat, in the case of buckwheat under the conditions of the method for wheat or rye.

Amylographic evaluation provides a more comprehensive view of the state of the saccharide–amylase complex or starch, respectively. The values of amylolytic determination of the analyzed wholemeal flours are presented in Table 6.

Table 6. Values of amylographic measurements of wholemeal finely granulated flours (WM FG). Data are the means of duplicates (±standard deviation).

Flour	Amylograph (%)			
	Starting Temperature (°C)	Gelatinization Temperature (°C)	Gelatinization (BU)	Gelatinization Modified (BU) *
Wheat WM FG	64.0 ± 0.1 b	89.5 ± 0.1 b	746 ± 5 b	-
Rye WM FG	55.1 ± 0.1 a	71.9 ± 0.2 a	430 ± 1 a	-
Spelt WM FG	62.5 ± 0.1 b	81.2 ± 0.5 b	341 ± 4 a	-
Buckwheat WM FG	85.8 ± 0.2 c	101.9 ± 0.7 c	2370 ± 15 c	* 395 ± 5
Wheat white (T530)	60.1 ± 0.0 a	82.5 ± 0.1 b	677 ± 4 b	-

* Sample weight 40 g. Data represent the means of duplicates. Small letters in the same column denote significant differences according to Duncan's new multiple range test (MRT) ($p \leq 0.05$).

The determination of the amylographic maximum in the case of whole-wheat flours from common wheat shows rather a lower degree of starch damage even in comparison with common white wheat flour, which was also confirmed by control measurements using the RVA method (Rapid Visco Analyzer) (data unpublished). In the case of spelt or rye, on the basis of this indicator, the degree of starch damage appears to be rather higher, but not substantially so.

The problem is again in the case of determining amylographic indicators for buckwheat, similar to the determination of FN. Thus, the amylographic indicators of the examined wholemeal flours (especially in the case of sown wheat) also point to a relatively low degree of mechanical and thermal damage to starch, which does not differ significantly from the degree of damage caused by the standard grinding method.

The confrontation and the discussion with other studies was not possible because there are no available literature data on the grinding of cereals on this particular type of mill, which was not originally intended for their processing. Although similar types of grinding machines have already been used for these purposes, there are still relatively few literature data and each of the principally similar mills has its own specifics, for which comparison is difficult.

3.6. Scanning Electron Microscopy and Microstructure of Wholemeal Flours

The abovementioned methods of examining the degree of damage to the microstructure of the endosperm, especially starch granules, as well as other methods that we did not use in our work (e.g., amperometric measurement of damaged starch content according to ICC No. 172/1, on SDmatic, Chopin Technologies, Villeneuve-la-Garenne, France) always affect some aspect that occurs in the condition of starch in the given flour.

Therefore, the samples of the examined flours were subjected to scanning electron microscopy, which (at different levels of magnification) clearly shows both the condition, shape, and constitution of individual flour particles and (at higher magnification, 2500×) shape and possible deformation of individual starch granules. Scanning electron microscopy of samples of examined wholemeal flours are shown in Figure 3a–d.

The spherical particles represent starch (small and large starch grains). Some large starch grains are slightly deformed and damaged mechanically on their surface. The other particles were fragments of the disintegrated portions of the outer and inner coating layers of the cereal grains. The portion of the protein matrix was also visible.

In summary, the images show that with the relatively gentle setting of the disintegration parameters mentioned above (residence time and resulting heating of the flour particles), the grain disintegration is relatively uniform and starch granules are not significantly damaged (e.g., compared to the standard roller grinding process).

(a)

Figure 3. *Cont.*

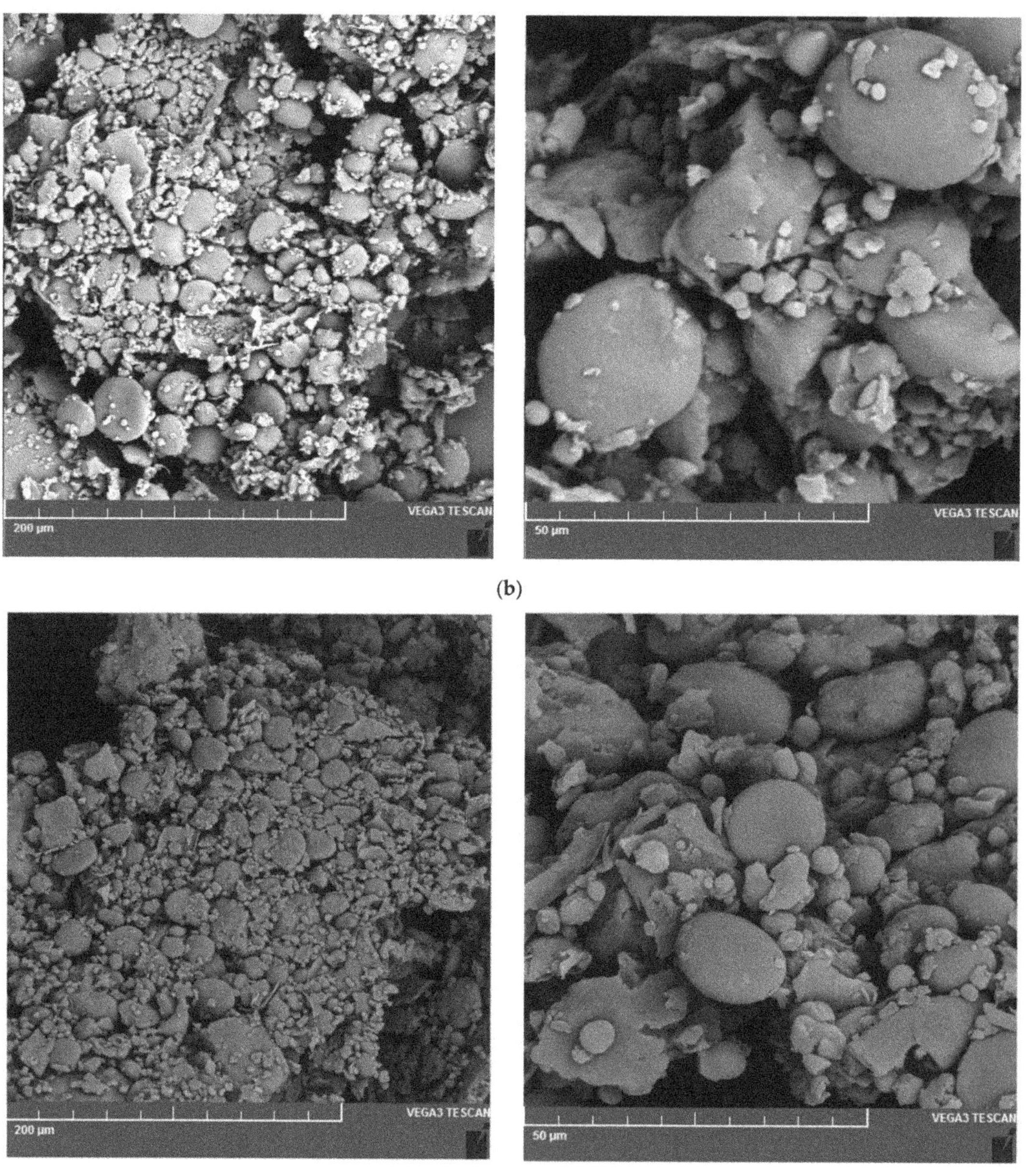

(b)

(c)

Figure 3. *Cont.*

(**d**)

Figure 3. Images of microstructure of wheat (**a**), rye (**b**), spelt (**c**), and buckwheat (**d**) wholemeal finely granulated flours from scanning electron microscopy.

4. Conclusions

The evaluation of the structure and properties of wholemeal flours produced in the Mahltechnik Görgens GmbH grinding device, depending on the disintegration conditions and the type and properties of the raw material, is currently still at the beginning. Several types of these flours were examined, depending on the source: common wheat, spelt, rye, and buckwheat flours. It turns out that when the mill was set to the abovementioned retention time of the grist in the grinding zone, there was easy and fast disintegration of rye and common wheat. The disintegration of spelt and buckwheat grains was more complicated and not sufficiently fine granulation was achieved under the given conditions.

The study results of the impact of the disintegration process on the microstructure of the endosperm, especially in the degree of damage to starch granules, show that disintegration in this mill may indeed appear promising. We can achieve fine granulation of flours, including fine granulation of bran particles for some cereals and other grains. At the same time, an undesirable degree of damage to the microstructure of the endosperm was not caused.

Further investigation of the influence of parameters of this type of disintegration process on the quality of the resulting flours will be definitely needed, as well as exploring the possibility of technological applications of these special wholemeal flours.

Author Contributions: Conceptualization, P.S.; methodology, M.S., L.J., and I.Š.; software, I.Š.; validation, I.Š. and M.S.; formal analysis, L.J. and I.Š.; investigation, P.S.; resources, P.S. and M.S.; data curation, P.S.; writing—original draft preparation, P.S.; writing—review and editing, P.S. and M.S.; visualization, P.S. and M.S.; supervision, P.S. All authors have read and agreed to the published version of the manuscript.

Funding: This work was supported by METROFOOD-CZ project MEYS Grant No: LM2018100.

Institutional Review Board Statement: Not applicable.

Informed Consent Statement: Not applicable.

Acknowledgments: The authors would like to thank Mlýn Perner Svijany, Ltd., Svijany, Czech Republic, for providing the materials (flours) used for experiments. The authors acknowledge Ivana Laknerová, Food Research Institute Prague, Czech Republic for technical support.

Conflicts of Interest: The authors declare no conflict of interest.

References

1. Sluková, M.; Skřivan, P.; Hrušková, M. *Cereální Chemie a Technologie Zpracování Obilovin—Mlýnská a Těstárenská Výroba*, 1st ed.; UCT Prague: Prague, Czech Republic, 2017; pp. 76–77, 168.
2. Godon, B.; Willm, C. *Primary Cereal Processing: A Comprehensive Sourcebook*, 1st ed.; John Wiley and Sons Ltd.: New York, NY, USA, 1994; pp. 155–158, 174, 199.
3. Poutanen, K. Rye and Rye Bread—An Important Part of the North European Bread Basket. In *Rye and Health*, 1st ed.; AACC Inc.: St. Paul, MN, USA, 2014; pp. 1–5.
4. Heiniő, R.-L.; Liukkonen, K.-H.; Katina, K.; Myllymäki, O.; Poutanen, K. Milling fractionation of rye produces different sensory profiles of both flour and bread. *LWT Food Sci. Technol.* **2003**, *36*, 577–583. [CrossRef]
5. Barrett, E.M.; Foster, S.I.; Beck, E.J. Whole grain and high-fiber grain foods: How do knowledge, perceptions and attitudes affect food choice? *Apettite* **2020**, *149*, 104630. [CrossRef]
6. Wang, L.; Yao, Y.; He, Z.; Wang, D.; Liu, A.; Zhang, Y. Determination of phenolic acid concentrations in wheat flours produced at different extraction rates. *J. Cereal Sci.* **2013**, *57*, 67–72. [CrossRef]
7. Wang, J.; Chatzidimitriou, E.; Wood, L.; Hasanalieva, G.; Markellou, E.; Iversen, P.O.; Seal, C.; Baranski, M.; Vigar, V.; Ernts, L.; et al. Effect of wheat species (*Triticum aestivum* vs *T. spelta*), farming system (organic vs conventional) and flour type (wholegrain vs white) on composition of wheat flour; results of a retail survey in the UK and Germany—2. Antioxidant activity, and phenolic and mineral content. *Food Chem.* **2020**, *X*, 100091. [CrossRef]
8. Buksa, K.; Nowotna, A.; Praznik, W.; Gambuś, H.; Ziobro, R.; Krawontka, J. The role of pentosans and starch in baking of wholemeal rye bread. *Food Res. Int.* **2010**, *43*, 2045–2051. [CrossRef]
9. Cyran, M.R.; Saulnier, L. Macromolecular structure of water-extractable arabinoxylans in endosperm and wholemeal rye breads as factor controlling their extract viscosities. *Food Chem.* **2012**, *131*, 667–676. [CrossRef]
10. Poutanen, K.; Sozer, N.; Valle, G.D. How can technology help to deliver of grain in cereal foods for a healthy diet? *J. Cereal Sci.* **2014**, *59*, 327–336. [CrossRef]
11. Kurek, M.A.; Wyrwisz, J.; Karp, S.; Brzeska, M.; Wierzbicka, A. Comparative analysis of dough rheology and quality of bread baked from fortified and high-in-fiber flours. *J. Cereal Sci.* **2017**, *74*, 210–217. [CrossRef]
12. Miś, A.; Grundas, S.; Dziki, D.; Laskowski, J. Use of farinograph measurements for predicting extensograph traits of bread dough enriched with carob fiber and oat wholemeal. *J. Food Eng.* **2012**, *108*, 1–12. [CrossRef]
13. Cardoso, R.V.C.; Fernandes, Â.; Heleno, S.A.; Rodrigues, P.; Gonzaléz-Paramás, A.M.; Barros, L.; Ferreira, I.C.F.R. Physicochemical characterization and microbiology of wheat and rye flours. *Food Chem.* **2019**, *280*, 123–129. [CrossRef] [PubMed]
14. Wang, J.; Hasanalieva, G.; Wood, L.; Markellou, E.; Iversen, P.O.; Bernhoft, A.; Seal, C.; Baranski, M.; Vigar, V.; Ernts, L.; et al. Effect of wheat species (*Triticum aestivum* vs *T. spelta*), farming system (organic vs conventional) and flour type (wholegrain vs white) on composition of wheat flour; results of a retail survey in the UK and Germany—1. Mycotoxin content. *Food Chem.* **2020**, *327*, 127011. [CrossRef] [PubMed]
15. Protonotariou, S.; Stergiou, P.; Christaki, M.; Mandal, I.G. Physical properties and sensory evaluation of bread containing micronized whole wheat flour. *Food Chem.* **2020**, *318*, 126497. [CrossRef] [PubMed]
16. Hidalgo, A.; Brandolini, A.; Gazza, L. Influence of steaming treatment on chemical and technological characteristics of einkorn (*Triticum monococcum* L. ssp. *monococcum*) wholemeal flour. *Food Chem.* **2008**, *111*, 549–555. [CrossRef]
17. AbuMweis, S.; Thandapilly, S.J.; Storsley, J.; Ames, N. Effect of barley β-glucan on postprandial glycaemic response in the helathy human population: A meta-analysis of randomized controlled trials. *J. Funct. Foods* **2016**, *27*, 329–342. [CrossRef]
18. Tucker, A.J.; Vandermey, J.S.; Robinson, L.E.; Graham, T.E.; Bakovic, M.; Duncan, A.M. Effects of breads of varying carbohydrate quality on postprandial glycaemic, incretin and lipidaemic response after first and second meals in adults with diet-controlled type 2 diabetes. *J. Funct. Foods* **2014**, *6*, 116–125. [CrossRef]
19. Brand-Miller, J.C.; Stockman, K.; Atkinson, F.; Petocz, P.; Denyer, G. Glycemic index, postprandial glycemia, and the shape of the curve in healthy subjects: Analysis of a database of more than 1000 foods. *Am. J. Clin. Nutr.* **2009**, *89*, 97–105. [CrossRef] [PubMed]
20. Venn, B.J.; Green, T.J. Glycemic index and glycemic load: Measurements issues and their effect on diet-disease relationships. *Eur. J. Clin. Nutr.* **2007**, *61*, 122–131. [CrossRef] [PubMed]

applied
sciences

Article

Buckwheat Seeds: Impact of Milling Fractions and Addition Level on Wheat Bread Dough Rheology

Ionica Coțovanu * and Silvia Mironeasa *

Faculty of Food Engineering, Stefan cel Mare University of Suceava, 13 Universitatii Street, 720229 Suceava, Romania

* Correspondence: ionica.cotovanu@usm.ro (I.C.); silviam@fia.usv.ro (S.M.);
  Tel.: +40-740-816-370 (I.C.); +40-741-985-648 (S.M.)

Abstract: Supplementation of refined wheat flour with buckwheat flour requires a good understanding of the impact of milling fractions, their functionality, and addition level on bread quality. The chemical and functional characteristics of different particle fractions (large, medium, and small) of buckwheat flour on dough Mixolab rheological properties to predict bread quality were investigated. Moisture content, proteins, ash, lipids, and carbohydrates varied irregularly depending on the particle size. The medium particle fraction is the richest in protein, lipid and ash, which are positively correlated with its water and swelling properties and negatively correlated with its volumetric density. The alpha-amylase activity increased with the particle size increase in composite flour. The Mixolab data revealed that the decrease of particle size increased water absorption, dough viscosity during the starch gelatinization and retrogradation stage, while the addition level increased the dough development time and gel stability, and decreased the rate of protein weakening. Following the optimization process and the desirability function approach, it was established that the most appropriate rheological properties are provided by buckwheat flour addition level of 10.75% for medium particle fraction. These results can be helpful for bakery producers to diversify baked products with the desired particle fraction with optimal technological and nutritional properties along with beneficial effects to consumers.

Keywords: buckwheat flour; dough rheology; particle size; optimization; wheat flour

Citation: Coțovanu, I.; Mironeasa, S. Buckwheat Seeds: Impact of Milling Fractions and Addition Level on Wheat Bread Dough Rheology. *Appl. Sci.* **2021**, *11*, 1731. https://doi.org/10.3390/app11041731

Academic Editor: Mike Boland

Received: 31 January 2021
Accepted: 12 February 2021
Published: 15 February 2021

Publisher's Note: MDPI stays neutral with regard to jurisdictional claims in published maps and institutional affiliations.

1. Introduction

Buckwheat is a pseudocereal of renewed interest that gained popularity as it often provides a distinguished nutritional and health-promoting value [1,2]. The genus *Fagopyrum* includes three species (*F. esculentum*, *F. tataricum*, and *F. cymosum*), but *F. esculentum* Moench is widely cultivated for human consumption [3,4]. Buckwheat seeds flour is considered to be a remarkable resource for functional food development and was used as an ingredient in bread, pasta, cake, and pancake [5–8] due to its nutritional quality and bioactive compounds. The major nutritional components in buckwheat seeds are represented by the carbohydrates that vary between 63.1 and 82.1%, of which from 54.5% to 54.7% was found to be starch [9] with a relatively high level of amylose (18.3–47% of total starch) [10]. Moreover, buckwheat has high levels of resistant starch (27–33.5%) which provides health benefits [11]. Buckwheat starch is concentrated in the endosperm, similarly to common cereals, while protein and lipids are located in the embryo that extends through the starchy endosperm [12]. Buckwheat grains represent an excellent source of dietary fiber (17.8%) with a low ratio of soluble–insoluble dietary fiber (0.5–0.28) [13]. The protein content of buckwheat ranges from 5.7% to 14.2%, depending on the variety and environmental conditions [11]. Buckwheat protein is characterized by a high biological value due to its excellent amino acids balance [14] and it is close to the optimum composition suggested by FAO/WHO [15], making buckwheat an essential contributor to human protein intake. This gluten-free high-quality protein is rich, especially in lysine, methionine, and tryptophan,

Appl. Sci. **2021**, *11*, 1731. https://doi.org/10.3390/app11041731

which are limiting amino acids in most cereals [14,16]. Lipid content from buckwheat (0.75–7.4%) is higher than that of wheat, being characterized by a high degree of unsaturation, a fact preferable from a nutritional point of view [2,14]. The main mineral compounds in buckwheat are potassium, magnesium, and phosphorous followed by calcium and other important minerals [2,17,18], showing variability among genotypes. Buckwheat grains contain a higher content of vitamins, especially the group of B vitamins, and buckwheat flour was highlighted by the presence of vitamins B2, B3 and B6 [1,19]. In addition to being a source of nutrients components, buckwheat contains non-nutrient components that present health benefits recognized as bioactive compounds. One group of bioactive components include the phenolic compounds that are of great interest, considering the design and the development of functional food products. Due to the high levels of phenolic compounds, buckwheat can be considered an excellent source of phenols [20,21]. The supplementation of common cereals like wheat with buckwheat possessing better nutritive value and nutraceutical characteristics would be a feasible strategy to improve the features of final products [22] and dietary diversification. In the literature, most of the study focuses on the levels of buckwheat flours added in wheat flour because the addition in high amounts led to various technological challenges related to the fiber content and due to the gluten dilution effect. As a gluten-free protein, buckwheat has negative effects on dough due to the absence of gluten proteins that form the structure, resulting in poor dough strength. To minimize such an undesirable effect, an appropriate level of buckwheat needs to be used to supplement wheat flour.

Nowadays, it is a challenge to produce flour at different particle sizes, because milling fractions varying in physical and chemical properties, and thus offering the possibility to produce flours with specific features for diverse use products [23]. The flour separated on different granularities with high concentrations of valuable components of interest such as proteins, soluble dietary fiber and polyphenols could be used to obtain bakery products with desired features [24]. The functionality of buckwheat milling fractions is essential. Functional properties describe the behavior of components during the technological process, and how they affect the final products in terms of rheological and sensorial properties [25]. Very few studies regarding the functionality of buckwheat milling fractions have been published [22,26], without studying their effects on wheat bread dough. Flour particle sizes influence dough behavior with an impact on the quality attributes of the finite products. In some previous investigations [27–29] the impact of the chemical composition and functionality of different particle size and their effect on rheological properties was highlighted, but there was no evident tendency.

Knowing the rheological behavior of composite flour is necessary for designing and developing new baked products because, in this study, the dilution of wheat gluten when buckwheat flour is added in the wheat flour dough, can impair proper dough rheological characteristics. In order to obtain information on the rheological characteristics, the Mixolab technique can be taken into account since, in only one test, it can predict the quality of the finite product, simulating dough behavior in the bread-making process. The Mixolab technique has been used to assess the impact of different milling fractions from whole buckwheat grains on the rheological properties [30], as well as the effect of amaranth and quinoa milling fractions on wheat flour dough rheology [31,32].

Response surface methodology (RSM) is recognized as a useful statistical tool that provides more information for a few numbers of experiments that require assessing the factors that influence the process and their interaction that implies mathematical models and needs small time resources [33]. In order to optimize the various composite flour formulation, multiple-response optimization, in conjunction with the desirability function, has been successfully applied [34,35].

An optimal amount of buckwheat flour that could be added in wheat flour has tried to be established by various researches, but, to the best of our knowledge, the impact of particle size of buckwheat flour on wheat flour dough rheology, assessed by applying the Mixolab technique, is yet unknown. Moreover, to our knowledge, the effect of the interaction between particle size and the level of buckwheat flour on wheat dough rheology to find the optimal values that give the best technological parameters has not yet been studied. The results will provide millers and manufacturers with scientific evidence of an optimal buckwheat flour particle size range and of an optimal supplementation level of refined wheat flour to improve the quality of baked products.

The aim of this study was to investigate the impact of buckwheat milling fractions on the chemical composition and functional properties, and to evaluate the effect of wheat flour substitution level by buckwheat flour at 0, 5, 10, 15, and 20% at three different milling fractions (large, medium, and small) on thermo-mechanical characteristics and to find the optimal particle size and level added, which inform on the bread-making potential of buckwheat—wheat composite flour.

2. Materials and Methods

2.1. Raw Materials

Commercial refined wheat flour of 650 type that has been used to formulate buckwheat-wheat composite flour was acquired from a local company S.C. Mopan S.A. (Suceava, Romania). Buckwheat seeds (*Fagopyrum esculentum* Moench) were purchased from a local market (S.C. SANOVITA S.R.L) and they were grinded with a laboratory machinery (Grain Mill, KitchenAid, Model 5KGM, Italy) to obtain buckwheat flour and sieved through a Retsch Vibratory Sieve Shaker AS 200 basic (Haan, Germany) to obtain three milling fractions: large (L > 300 μm), medium (M > 180 μm, < 300 μm) and small fractions (S < 180 μm).

2.2. Proximate Composition

The main raw material, wheat flour was investigated for its moisture (14.00%), ash (0.65%), fat (1.40%), protein (12.60%), wet gluten (30.00%), Falling number index (312 s), and gluten deformation index (6.00 mm) by the ICC methods 110/1, 104/1, 105/1, 105/2, 106/1, 107/1 and by the Romanian standard method SR 90:2007, respectively. These results make it be a suitable flour for bread making according to the Romanian standard SR 877:1996. Buckwheat grains' analytical characteristics included 13.28% moisture (SR EN ISO 665:2003), 2% ash (SR ISO 2171:2009), 3.40% fat (SR EN ISO 659:2009), and 13.26% protein (SR EN ISO 20483:2007). Total carbohydrate content was determined by difference as 100 − (moisture + ash + protein + total lipids). All measurements were taken at least in duplicate, and the results were expressed as the average ± standard deviation.

2.3. Functional Properties of Buckwheat Milling Fractions

For each buckwheat flour particle size (BL, BM, BS) the functional properties considered were water absorption and water retention capacity, swelling capacity, and volumetric density. All the analyses were determined at least in duplicate and values were calculated on a dry weight basis.

2.3.1. Water Absorption Capacity

The method described by Raghavendra et al. (2004) [36], with slight modifications, was used to determine the water absorption capacity (WAC) for the particle size of BF. From each sample, 1 g was measured and then was mixed into a 50 mL test tube with 10 mL distilled water. This mixture was left 30 min at room temperature and then placed in a centrifuge for 10 min at 2000 rpm (MPW-223e, MPW Med. Instruments, Warszaw, Poland). During this time, the following scientific phenomenon occurs: the water molecules hydrate the damaged starch, the glutenin and gliadin, and other components until molecules of starch and protein create hydrophilic interactions and hydrogen bonds with the molecules of water. When the hydration was over the supernatant was removed and the tubes were weighed.

2.3.2. Water Retention Capacity

To determine the water retention capacity (WRC), accurately weighed BF milling fractions (2.0 g) were loaded into a weighed centrifuged tube with 20 mL of water and stirred continuously on a water bath at 25 °C (Memmert Waterbath, Germany) for 1 h. After centrifugation at 1600 rpm for 25 min, the supernatant was filtrated off on a paper towel and the sample was dried at 105 °C for 2 h to obtain residue dry weight [37]. The weights obtained were specified as a % water per gram of flour.

2.3.3. Swelling Capacity

Swelling capacity (SC), also known as the swelling index (SI) was determined according to the method previously applied by Olapade et al. (2003) [38] with minor changes. This property indicates the volume in milliliter taken up by the swelling of flour (1 g) under specific conditions. One gram for each BF fractions was mixed in a centrifuge tube with 10 mL of water and then a weighed tube was heated in a water bath at 80 °C for 15 min. After centrifugation at 2000 rpm for 30 min, the supernatant was decanted and the pellet was estimated. The final volume attained by the sample was measured as a percent of swelled per gram of sample, which gave SC.

2.3.4. Volumetric Density

Volumetric density (VD), known also as bulk density, was measured following the procedure described by Ikegwu (2009) [39]. Buckwheat flour milling fractions (50 g) were placed into a dry graduated cylinder (50 mL) and it was shaken slightly until no volume differences were observed. Volumetric density was determined as a sample weight (g) per sample volume (mL).

2.4. Influence of Buckwheat Milling Fractions Addition on the Falling Number Values

The Falling number index (FN) value, determined at least in duplicate using the standard method (ICC 107/1) and FN device (Perten Instruments AB, Stockholm, Sweden), was performed. The method is a viscometer assay, based on the rapid gelatinization of a flour suspension in a boiling water bath and the measurement of its liquefaction by α-amylase. FN index value gives an estimation of the α-amylase activity from the wheat-buckwheat flour.

2.5. Dough Rheology Assessed by Mixolab

The influence of buckwheat flour milling fractions and the level added on the rheological properties of composite flours dough was assessed using a Mixolab analyzer (Chopin, Tripette et Renaud, Paris, France) and ICC-Standard Methods No. 173 (ICC, 2010) by applying the standard Chopin+ protocol, as described in previous studies [31,32]. For all wheat-buckwheat flour samples, the first stage was determined by the following mixing parameters: water absorption, WA (%), dough development time, DT (min), and dough stability, ST (min). In the following stages, the minimum torque C2 (N·m), related to protein reduction due to temperature rise; peak torque, C3 (N·m), related to starch gelatinization; minimum torque C4 (N·m), as the stability of hot-formed gel; maximum torque, C5 (N·m), as a starch retrogradation measure during dough cooling were recorded. From the Mixolab curve registered, the differences between torques C1 and C2 (C1-2), C3 and C2 torques (C2-3), C3 and C4 torques (C3-4), and C4 and C5 torques (C4-5) associated with protein weakening, starch gelatinization, cooking stability and starch retrogradation were also determined.

2.6. Factorial Design and Statistics

A full factorial design was used to study the main and interaction effect of replacing wheat flour with buckwheat milling fractions on the responses, Falling number (FN) index, water absorption of composite flour, and Mixolab properties. The studied factors were three buckwheat flour particle sizes (large, medium, and small) and the level added in wheat flour (0, 5, 10, 15, and 20%). An experimental design that consists of fifteen combinations (Table 1) was conducted. The simultaneous effect of these two factors on the responses was investigated through the response surface methodology (RSM) in conjunction with the desirability function approach. RSM is a powerful technique mainly used in the design and innovation of bakery products, and to find the optimum level of formulation factors [34,40] or the optimal processing conditions [41].

Table 1. Factors and their level in experimental design.

Trial No.	Coded and Real Values			
	A	Particle Size (μm)	B	Buckwheat Flour (%)
1	+1.00	380	0.00	10
2	−1.00	180	−0.50	5
3	0.00	280	−1.00	0
4	0.00	280	−0.50	5
5	−1.00	180	−1.00	0
6	+1.00	380	+1.00	20
7	−1.00	180	0.00	10
8	+1.00	380	−1.00	0
9	+1.00	380	−0.50	5
10	+1.00	380	+0.50	15
11	0.00	280	0.00	10
12	0.00	280	+1.00	20
13	−1.00	180	+1.00	20
14	0.00	280	+0.50	15
15	−1.00	180	+0.50	15

The quadratic polynomial regression model (Equation (1)) was proposed for all responses. In the equation, Y represents the response and the regression coefficients represented by b_0—coefficient of intercept, b_1, b_2—coefficient of linear terms, b_{11}, b_{22}—coefficient of quadratic terms, and b_{12}—coefficient of interactions between effects of A (particle size of buckwheat flour) and B (level of buckwheat flour added in wheat flour) factors.

$$Y = b_0 + b_1 \cdot A + b_2 \cdot B + b_{11} \cdot A^2 + b_{22} \cdot B^2 + b_{12} \cdot A \cdot B \tag{1}$$

The multiple linear regression analysis was applied to fit the data obtained for each response to linear, two-factor interactions, quadratic and 2FI models. The most adequately model to predict data variation for each factor was found through a sequential F-test, coefficients of determination (R^2), adjusted coefficients of determination (*Adj.-* R^2). To evaluate the significant differences ($p < 0.05$) between the samples, one-way ANOVA and Tukey's HSD post-hoc test was used to describe means with 95% confidence. All the analysis was determined in duplicate. These analyses were performed using Stat Ease Design-Expert 12.00 software (Stat-Ease, Inc., Minneapolis, MN, USA) (trial version), and the relationships between composite flour and chemical characterization and functional properties were verified using XLSTAT 2020.2 software (Addinsoft, Paris, France, 2020). To establish the optimal value of the factors, buckwheat flour particle size and addition level, the multiple responses methodology was used to adequately predict the models. For the numerical optimization applied in this study, the desired goal established for each response included: dough stability (ST) at maximum value, starch retrogradation (C5-4) was minimized and the level of the other responses which have been taken into account in this research were maintain within studied limits. The one-way analysis of variance (ANOVA) was performed by using XLSTAT 2020.2 software (Addinsoft, Paris, France, 2020) to test if the effect of the particle sizes on the physicochemical and functional properties of buckwheat flour was significant ($p < 0.05$).

3. Results

3.1. Buckwheat Milling Fractions Composition

The values of physico-chemical analyses of the buckwheat milling fractions, large (BL), medium (BM), and small (BS) are presented in Table 2. It can be seen that the moisture content ranged from 12.00% to 12.85% and the lowest value has been obtained for the medium particle size. The low moisture found for this particle size suggested its potential for higher storage stability and longer shelf life. Offia-Olua (2014) [42] mentioned that a moisture flour up to 12% indicated higher storage stability compared to the one with higher moisture. The results showed different moisture among the samples ($p < 0.05$). Similar data on the content of different buckwheat milling fractions was reported by Sciarini et al. (2020) [43] and Slukova et al. (2017) [44].

Table 2. Proximate composition and functional properties of buckwheat milling fractions.

Parameters [a]	Milling Fractions		
	BL	BM	BS
Chemical Composition			
Moisture (%)	12.85 ± 0.04 [b]	12.00 ± 0.07 [a]	12.68 ± 0.04 [b]
Ash (%)	1.01 ± 0.00 [a]	4.23 ± 0.02 [c]	1.22 ± 0.01 [b]
Protein (%)	10.19 ± 0.05 [a]	26.61 ± 0.06 [b]	10.09 ± 0.07 [a]
Fat (%)	1.93 ± 0.04 [a]	5.63 ± 0.01 [b]	2.03 ± 0.03 [a]
Carbohydrates (%)	74.02 ± 0.02 [b]	51.52 ± 0.09 [a]	73.96 ± 0.14 [b]
Functional Properties			
WAC (%)	2.52 ± 0.09 [b]	2.91 ± 0.03 [c]	2.18 ± 0.02 [a]
WRC (g/g)	4.71 ± 0.03 [b]	5.91 ± 0.09 [c]	3.81 ± 0.09 [a]
SC (mL/g)	5.02 ± 0.21 [a]	5.67 ± 0.06 [a]	5.10 ± 0.38 [a]
VD (g/mL)	0.53 ± 0.00 [b]	0.49 ± 0.00 [a]	0.61 ± 0.00 [c]

[a] Mean ± SD. WAC—water absorption capacity; WRC—water retention capacity; SC—swelling capacity; VD—volumetric density. Values followed by different lowercase letter ([a], [b], [c]) are statistically different at 95% confidence level.

The ash content for all fractions was between 1.01% and 4.23%. Similar results were found by Kasar et al. (2020) [26], Martin-Garcia et al. (2019) [45], and Bobkov S. (2016) [24]. The ash content of the studied milling fractions showed significant differences ($p < 0.05$) between all the samples. These findings can be attributed to the biochemical and morphological structure of the buckwheat seeds. In comparison with wheat flour, buckwheat is a good source of iron, zinc, copper and manganese [18,46], minerals that act as cofactors of antioxidant enzymes and serve as indirect antioxidants. A high level of magnesium, potassium, and phosphorus and a slightly low content of calcium was reported in buckwheat [18,47]. The high ash value (4.23%) observed in medium particle size (BM) can be related to the minerals such as magnesium, potassium, and phosphorus, which are stored in the embryo [48], and thus, BM was expected to have a high ash content. The supplementation of wheat flour with buckwheat flour fractions with high ash content could imply a rise in the mineral amount in the newly formulated flour.

Buckwheat flour milling fractions had a high protein content, which varied between 10.09% and 26.61% and was only different ($p < 0.05$) between fraction BM and the other two particles (BL and BS). Close results for protein content of various buckwheat milling fractions that differ from those studied in this work were reported in some studies [43,45,49]. Therefore, it can be concluded that noticeable differences in protein content can be associated with the presence of different embryo parts in the milling fractions. The protein content between BL and BS fractions did not show significant differences ($p > 0.05$), which can be possible because, during milling, the parts of the embryo where the protein is concentrated are separated as components of the BM fractions. In this case, the BM fraction, which is high in ash, may be a feasible option for use as a mineral-rich ingredient.

The fat content of the buckwheat milling fractions varied from 1.93% to 5.63% and was significantly different ($p < 0.05$) between the medium fraction and the other large and small fractions. Similar data for fat content on various buckwheat flour fractions were reported in some studies [24,50]. Most lipids in buckwheat seeds are located in the embryo and seed coat, followed by the pericarp and endosperm [51]. According to Ahmed et al. (2014) [46], the total lipid content in buckwheat grains ranged from 2.8% to 3.4%, while Dziadek et al. (2016) [19] found a significantly lower lipid content in whole common buckwheat grain samples (1.72–2.24%), a variation that can be attributed to the crop genetics and growing condition. As can be observed in Table 2, the fat content has an irregular trend with particle size variation. The highest values of fat (5.63%) were found for the medium fraction and can be related to the large amount presence of the lipidic part from buckwheat seed in this fraction.

The amount of carbohydrate in buckwheat fractions ranged from 51.52% to 74.02%, the milling fractions showing a significant ($p < 0.05$) effect on the carbohydrate content. The large particle size (BL) presented the highest carbohydrate content (74.02%), which can be related to the starchy endosperm. A decrease in carbohydrates in medium particle size was expected due to the rise in protein and ash content in this fraction.

3.2. Functional Proprieties of Buckwheat Milling Fractions

The water absorption capacity (WAC) of buckwheat milling fractions showed a significant reduction in a small fraction (BS) compared to medium (BM) and large particle size (Table 2). Data revealed that buckwheat milling fractions were statistically different ($p < 0.05$). The highest value (2.91%) observed at BM could be explained by the presence of a higher quantity of hydrophilic compounds from this fraction. The proteins and carbohydrates have a great influence on WAC due to the presence of such components as polar or charged side chains [52]. A high WAC implies high water incorporation in the dough, which can improve bread properties, especially texture and mouthfeel, with the knowledge that WAC is an index of the ability of protein and fiber to absorb and retain water. Thus, the WAC values confirmed that M particle size can absorb water well and swell for improved dough consistency, making it suitable for use in bread making. This WAC provides information on the strength of the starch inter-granular bond. A lower WAC was attributed to a close association of starch polymers in the native granule or can be attributed to the presence of the lower amounts of hydrophilic constituents in BF like carbohydrates and proteins, and influenced the cohesiveness of the food products.

Water retention capacity (WRC) for buckwheat flours fractions ranged between 3.81 and 5.91 g/g (Table 2), and are similar to the results obtained by Kasar et al. (2020) [26] for different buckwheat milling fractions. Differences were found among all the studied milling fractions ($p < 0.05$). These differences can be due to the different compositions of milling fractions and to the particle size distribution and its morphology. The increase of WRC in BM (5.91%) can be linked to a substantial content of ash and protein from this fraction.

Swelling capacity (SC) for the milling fractions of buckwheat flour studied ranged between 5.02 and 5.67 mL/g and does not present significant differences ($p > 0.05$) among the samples (Table 2). A similar trend was reported by Kasar et al. (2020) [26] when some functional properties of different buckwheat milled fractions were investigated. Buckwheat starch granules have a diameter of, on average, 5 μm and presented pores [53,54], facilitating, more or less, the penetration of water molecules into buckwheat starch granules, depending on particle size, but a regular trend was not observed. The values obtained for SC can be related to the extent of degree swelling during the heating process, and can also predict the buckwheat milling fractions behavior in further processing.

Volumetric density (VD) values for each particle size are showed in Table 2, which varied from 0.49 to 0.61 g/mL, showing a statistically difference ($p < 0.05$) for each particle size. The BS presented the highest volumetric density that can be explained, probably, by the positive association with the carbohydrates and negatively with the lipid content. Similar results (0.67 g/mL) were reported by Sindhu et Khatkar (2016) [55]. The lower VD value found for medium particle size indicated a relatively good packaging characteristic of this milling fractions compared to other particle sizes.

It can be concluded that the compositional difference of carbohydrates and proteins between the particle sizes affected the functional properties of buckwheat particle size. The hydrophilic components present in milling fractions, like polysaccharides, have a good water retention capacity, whereas the polar amino acid residues of proteins have an affinity for water molecules, imparting water binding property.

3.3. Fitting Models

The investigation was conducted following the experimental design for determining the effects of the particle size and buckwheat milling fractions level added in wheat flour on falling number (FN) value and thermo-mechanical parameters. Mixolab rheological properties during the mixing stage in terms of water absorption (WA), dough stability (ST), dough development time (DT), minimum C2 torque, the difference between torques C1 and C2 (C1-2), and in the starch behaviour process as maximum C3 torque (C3), the difference between torques C3 and C2 (C3-2), minimum C4 torque (C4), the difference between torques C3 and C4, (C3-C4), maximum C5 torque (C5) and the difference between torques C5 and C4 (C5-4) [32] were assessed. The most fitting models were highly significant for almost all responses and were used for studying the influence of factors on the responses. These adequate models represented the experimental data well, showing high values (0.66–0.92) for coefficients of determination (R^2) (Table 3). The ANOVA results, which comprise significant regression coefficients ($p < 0.05$), expressed in terms of coded value, quadratic determination coefficients (R^2), and adjusted-R^2 ($Adj.$-R^2), were calculated to assess the adequacy of quadratic models and are shown in Table 3.

Table 3. The coefficients in the predictive models for FN index and dough rheological properties during mixing and heating process of buckwheat-wheat flour dough.

Factors [a]	Parameters											
	FN (s)	WA (%)	DT (min)	ST (min)	C2 (N·m)	C1-2 (N·m)	C3 (N·m)	C3-2 (N·m)	C4 (N·m)	C3-4 (N·m)	C5 (N·m)	C5-4 (N·m)
Constant	318.82	58.28	3.09	9.10	0.42	0.66	1.73	1.31	1.61	0.12	2.34	0.73
A	−19.25 ***	−0.13 *	0.89 *	ns	ns	ns	ns	−0.04 *	ns	ns	−0.10 *	ns
B	ns	−0.43 ***	0.78 *	−0.76 *	−0.06 ***	0.04 *	−0.12 **	−0.05 *	−0.20 ***	0.08 **	−0.39 ***	−0.19 ***
A × B	−16.45 *	ns	0.83 *	ns	ns	ns	−0.06 *	−0.05 *	−0.11 *	0.04 *	ns	ns
A^2	13.95 *	ns	ns	-	ns	ns	0.10 *	0.07 *	0.13 *	ns	0.28*	0.14 *
B^2	ns	ns	ns	-	ns	ns	ns	ns	ns	ns	ns	0.14 *
R^2	0.88	0.86	0.76	0.51	0.86	0.66	0.87	0.87	0.88	0.86	0.92	0.88
$Adj.$-R^2	0.82	0.79	0.62	0.48	0.78	0.47	0.80	0.79	0.81	0.79	0.88	0.82
p-value	0.0005	0.0011	0.0130	0.0411	0.0012	0.0492	0.0009	0.0010	0.0007	0.0011	<0.0001	0.0006

Note. FN: Falling number index; WA: water absorption; DT: development time; ST: stability; C2: minimum torque during temperature increase; C1-2: protein weakening; C3: peak viscosity; C3-2: starch gelatinization; C4: cooking stability; C3-4: breakdown; C5: starch retrogradation; C5-4: setback; ns: non-significantly. [a] A: Particle size (μm); B: Addition of buckwheat flour in refined wheat flour (%); R^2, $Adj.$-R^2: is measures of fit of the model. ***, **, * indicated significance at $p < 0.0001$, $p < 0.001$, and $p < 0.05$, respectively.

3.4. Falling Number Index Estimation as Influenced by Buckwheat Milling Fractions and the Addition Level in Wheat Flour

The Falling Number (FN) index values as the parameter used for estimation of amylolytic activity of grain enzymatic complex in the buckwheat-wheat composite flours formulated ranged from 299 to 369.5 s. The model can be adequate to predict the FN index, with a high level of significance ($p < 0.001$), defining the amylolytic activity of the buckwheat-wheat flour through the FN index. As in data showed in Table 3, the milling fractions caused a significant negative influence ($p < 0.001$) on FN, while the amount of buckwheat flour (BF) added was not-significant ($p > 0.05$). The negative effect of buckwheat milling fractions addition showed that the FN index of buckwheat-wheat flour increased with decreasing particle size of BF. This consequence can be probably associated to the phenolic compounds from the lowest fractions that bind to α-amylase and changing its conformation, and thus led to a rise in the FN index. A significant negative correlation was observed between FN and the interaction effect between the factors, while the quadratic term of BF addition showed a significant positive correlation with FN (Table 3). The effect of factors, BF milling fractions, and BF addition level is presented in Figure 1a, showing that, with the increase of particle size and BF addition, the FN index decrease. This tendency suggested a rise in the α-amylase activity because an inversely correlation exists between the FN index and α-amylase activity in flour [56]. Consequently, buckwheat milling fractions can be used as an ingredient to adjust the α-amylase activity of wheat flour.

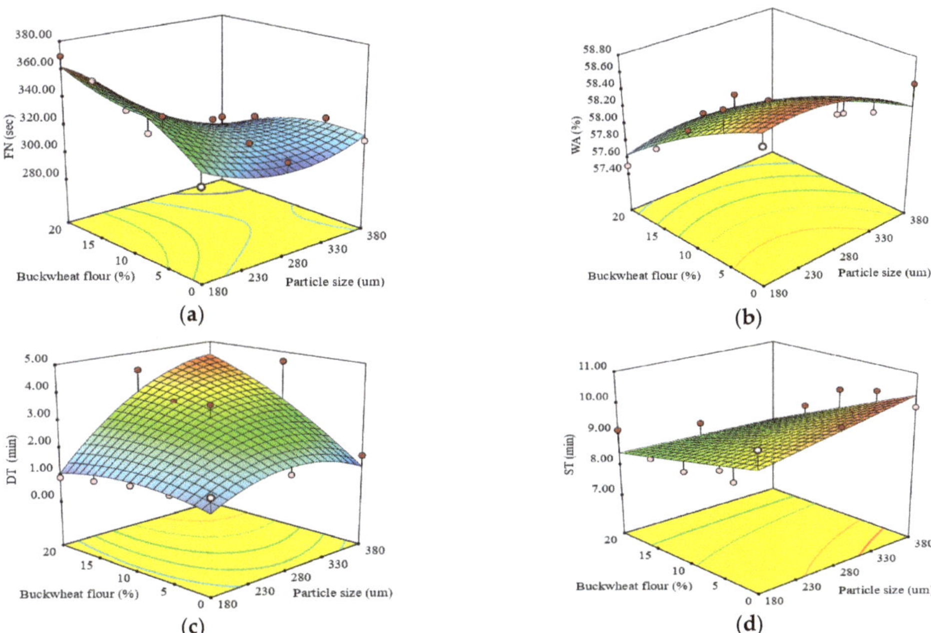

Figure 1. 3D response surface plots showing the interaction between buckwheat flour particle and addition level on (**a**) the falling number index and (**b**) water absorption (WA), (**c**) development time (DT), and (**d**) dough stability (ST) achieved during mixing.

3.5. Dough Rheological Properties Evaluation Using Mixolab Test

When some part of wheat flour is replaced with different buckwheat milling fractions, various physical and chemical changes in composite flours can occur, modifying the performance during processing. Therefore, the Mixolab test was used to observe dough rheological properties of wheat flour enriched with different buckwheat milling fractions at different levels. The Mixolab presents the advantage measure dough behavior during mixing and heating in a single test, simulating the mixing and baking processes, and offer the possibility to monitor both the protein and starch behavior during processing. The protein characteristics of dough are expressed as WA, DT, ST, C2 torque, and C1–2. The starch behavior during heating is expressed as torques C3, C4, C5, and C3-2, C3-4, C5-4, quantifying the changes in dough structure caused by temperature rise and mechanical forces of mixing.

3.5.1. Water Absorption Evaluation during Mixing

The substitution of wheat flour with BF determined changes in dough consistency and stability, which are mainly associated with the formation of the aggregates as a result of hydrogen bonding or proteins linking through disulfide or dityrosine bonds [31]. The changes in dough consistency due to added BF milling fractions in wheat flour can be related to the water quantity resulted from BF milling fractions and its addition level. Water absorption capacity has an essential role in influencing the baking process affecting the volume efficiency of the baked goods. Besides, this parameter offers an indication of the potential for the protein molecules in the sample to absorb the added water. Thus, as reported by Liu et al. (2019) [20], water absorption is considered to be an indicator of baking quality. For the formulated samples, water absorption (WA) values ranged from 57.5 to 58.7%. The predictive model for WA indicated that particle size and BF level had a significant ($p < 0.05$) effect on WA (Table 3). The negative coefficient of particle size and BF level showed that WA increased significantly ($p < 0.05$) when the BF level and particle size decreased. Figure 1b may notice that PS and the addition level of buckwheat flour have a negative effect on water absorption (WA), which indicated that WA decreased as the addition and particle size of BF increased. Nedeljković et al. (2014) [7] and Gavurníková S. (2011) [57] also reported a negative relationship between the water absorption and the BF milling fractions. In this study, the medium fraction, followed by large particles, needed more water to swell the starch than the small particle size, a finding that can be associated with WAC values (Table 2). This tendency can be related to several factors, especially fractions of chemical composition. As Sapirstein et al. (2018) [58] stated, damaged starch represents one of the major factors that contribute to the water absorption alongside the proteins and starch. At the same time, an increase in WA with a decrease of milling fractions was reported in previous studies [28,29,31,32] on the impact of substituting wheat flour with different fraction sizes of non-gluten flours, revealing that the small particle size raised the WA values.

3.5.2. Dough Development Time Evaluation

The effect of buckwheat milling fractions addition on dough development time (DT) is presented in the quadratic regression model shown in Table 3. The buckwheat flour (BF) addition level influenced the gluten network of the dough and significantly modified the dough development time depending on fraction sizes. The quadratic model was found to be adequate to predict DT, depending on the studied factors, particle size and BF addition level (Table 3). Describing the goodness of fit of the models, the quadratic determination coefficient (R^2) was 0.75, which confirms the suitability of the model, showing that the model explained 75% of data variation. DT of buckwheat-wheat flour ranged between 0.9 to 4.65 min and it was affected ($p < 0.05$) by the particle size, BF addition level, and the interaction between fraction size and BF addition level. A higher increase in the DT was observed with a particle size increase compared to the BF level increase (Table 3). Thus, DT increased notably as the particle size increased (Figure 1c), showing that it took a long time from the time the water was added to the time when the dough reached the optimal consistency (C1 torque =1.1 N·m). An increase of DT of about 2.1–2.7 times in the buckwheat-wheat flour with the largest fraction and about 2.1–2.5 times in the medium fraction as the BF addition level increased by over 10% was obtained beside the wheat dough used as control. This fact may be due to the rise in water when the BF addition was high, thus requiring mixing. Larger and medium particle sizes rich in some compounds such as dietary fibers and proteins, respectively, required a longer time for hydration, which implies a higher DT. With the addition of gluten-free flour, the gluten network is affected and leads to a rise in DT. As the DT value quantifies the dough strength, a higher value indicated stronger dough. Increased DT was also found by Sedej et al. (2011) [22] and Nikolic et al. (2011) [59] when different levels of buckwheat flour were used to supplement wheat flour, but without taking into account different particle sizes.

3.5.3. Dough Stability Evaluation

The 2FI predictive model results from the regression analysis showed that both PS and the addition level of BF had a remarkable influence ($p < 0.05$) and fitted well the experimental data for dough ST ($R^2 = 0.51$) (Table 3). From the addition level effect point of view on dough ST, a significant decrease ($p < 0.05$) was observed, while fraction size did not significantly influence ST, which varied from 7.05 to 10.17 min. The response surface plot showed that ST decreased when addition BF increased, depending on the fraction size (Figure 1d). Dough ST increased with particle size rise. This behavior may be associated with the composition of particle size fractions. Probably, at a particular ratio of fiber and gluten, gluten–fiber interactions can occur, enhancing dough stability. This result is consistent with the report by Han et al. (2013) [49], which showed that ST increased when a certain fraction of tartary buckwheat flour was added in wheat flour compared to other milling fractions due to gluten–fiber interactions that increase dough stability. However, the decrease in dough ST with the addition of BF can be explained by the gluten dilution when wheat flour was supplemented with gluten-free flour, indicating a rising degree of softening. The weakening of the gluten network was probably related to the dissimilarity between the proteins of buckwheat and the wheat proteins. Additionally, the buckwheat grain protein consists mainly of albumin and glutelins [60] and it cannot form structures such as wheat protein. Mohamed et al. (2014) [61] reported similar findings regarding the incompatibility between the protein spectrum of pulses and the semolina gluten protein. Similar changes in dough ST when whole buckwheat flour supplemented wheat flour in different amounts were reported by Stefan et al. (2018) [62] and Sedej et al. (2011) [22], revealing decreases in ST with supplementation increase.

3.5.4. Protein Weakening Evaluation during Heating

In the heating stage, the mechanical shear stress and heating constraint led to the destabilization of proteins when the dough consistency decreased and C2 torque to the minimum value. For C2, the quadratic model was found to fit well with the experimental data, showing a high R^2 value (0.86) (Table 3). The C2 torque was influenced ($p < 0.05$) by the level of BF added in wheat flour, while the particle sizes had a non-significant influence ($p > 0.05$). It was found that the BF level had a highly significant negative effect ($p < 0.01$) on C2, showing a decrease in C2 torque with the increase in the BF level added in wheat flour (Figure 2a). This behavior was probably due to the gluten dilution effect by the addition of gluten-free buckwheat flour, which caused the formation of a weaker protein network. It is well known that buckwheat flour proteins do not have structure forming ability compared to wheat flour proteins. From Figure 2a, a decrease of C2 as the fraction of BF decreased can be seen, and in the medium fractions this fact was most evident, which can be correlated with either a release of water molecules in dough or to the presence of high proteins and the BF interfering with the protein unfolding [32]. The presence of a high protein in the medium fraction compared, to large and small particles, was noticed (Table 2). C2 torque reduction for small particle sizes may be correlated with the lower water availability in the dough because of high carbohydrates contents from this particle (Table 2). Lower C2 torque denotes too weak of gluten network and lower resistance of gluten with rising temperature.

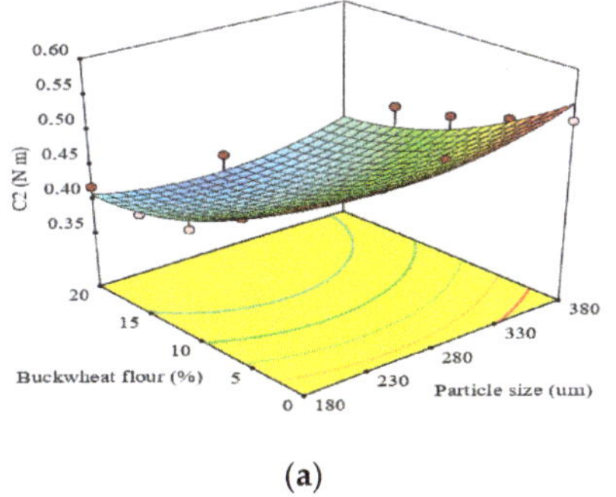

(a)

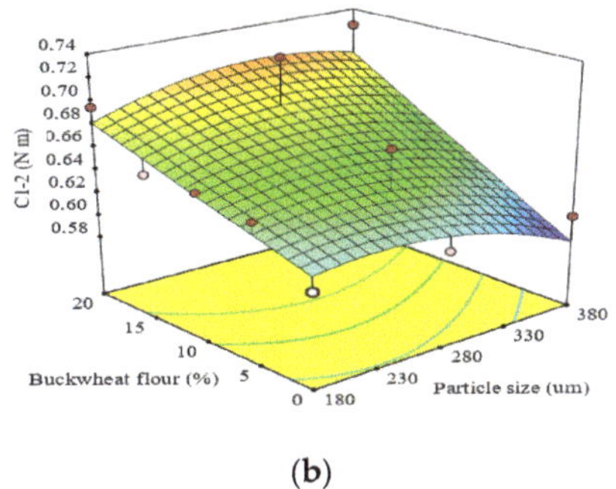

(b)

Figure 2. 3D response surface plots showing interaction between buckwheat flour particle size and addition level on (**a**) C2 torque (C2) and (**b**) difference between torques C1 and C2 (C1-2) during protein weakening.

The difference between the C1 and C2 torques (C1-2), which represents the rate of protein thermal weakening, showed that the quadratic model was adequate (Table 3) for the prediction of the effects of factors on C1-2. The BF addition level influenced ($p < 0.05$) the parameter C1-2, while the milling fractions did not affect the C1-2 torque, as is shown by the ANOVA results (Table 3). It can be observed from the response surface plot (Figure 2b) that the BF level added to obtain composite flour had a positive influence on the C1-2, revealing an increase in C1-2 as the BF level increased, whereas from the fraction size effect point of view on C1-2, a decrease was observed. This variation can be explained by the changes in protein network structure due to the kneading and temperature effects. Buckwheat presents relatively small starch granules, which can occur in aggregates and singly with the same range of diameter (3–8 μm) [53]. In this case, probably the presence of enzymatic attacking points is favored by the proteins, which become less compact in the heating, increasing in the speed of protein weakening, causing a decrease of C1-2.

3.5.5. Starch Gelatinization Evaluation

When the dough is heated above 60 °C, the starch granules are quickly broken, making them to raise the water absorption and swelling capacity, conducting to leaching of the amylose molecules that increase the viscosity and therefore the torque. The torque represented by C3 indicates peak viscosity and quantifies the degree of starch gelatinization. The effect of milling fractions and the level of BF added in wheat flour on the maximum viscosity of hot gel at 90 °C, measured by the C3 torque, showed that the quadratic model obtained was found to fit well the experimental data of C3 ($R^2 = 0.87$) (Table 3). C3 was significantly ($p < 0.05$) influenced by the linear term of the BF addition level in wheat flour, the interaction between factors, and the quadratic term of particle size. The addition of BF caused a decrease of C3 torque in composite flour and might be because buckwheat amylose forms with lipids complex compounds [54]. Thus, these complexes can decrease the swelling power and solubility of buckwheat starch, thus affecting C3 torque. Moreover, nonstarch components (lipids, proteins, and dietary fibers) present in BF milling fractions could restrict swelling and gelatinization during cooking, in addition to a diluting effect due to the interaction with starch polymers (lipids and proteins), and to the competition for water (proteins and dietary fiber) [32] interfering with starch swelling [63]. The greater negative effect on C3 was given by the BF addition level, showing a decrease of C3 as the BF amount added in wheat flour increased. The lower swelling and gelatinization of starch granules was probably explained by the morphological characteristics of buckwheat starch, which restricted water quantity in the composite flour dough could decrease the peak of C3 torque. These results were in accordance with those reported by Filipčev et al. (2015) [64]. As can be seen from Figure 3a, C3 decreased with BF supplementation, and when the BF addition in wheat flour is 20%, the C3 torque decreased at 1.60 N·m for the large particle size value. These changes, most likely, are related to the higher enzyme activity and the amylose-lipid complexes formed during the heating of starch slurries. Qian et al. (1998) [54] related that to the small granule of buckwheat-starch and their pores, and this starch is more susceptible to fungal α-amylase than wheat starch. The viscosity of the gel formed is largely influenced by granule shape and swelling power, amylopectin-amylose entanglement, and amylose and amylopectin granular interaction [65]. The differences among the assessed fraction sizes might be due to the particle microstructural and chemical features of the buckwheat-starch that governs water absorption and swelling.

The difference between C3 and C2 torques (C3–2) is significantly ($p < 0.01$) affected by the milling fractions and BF addition levels. The quadratic regression model represented well the experimental data of C3-2 and the R^2 value (0.86), confirming that this model is suitable. The increase in the amount and the BF milling fractions led to C3-2 (Figure 3b) decrease, the interaction between milling fractions and the BF amount being more pronounced (Table 3). α-amylase activity along with the composition of the BF milling fractions from the composite flour could explain these changes.

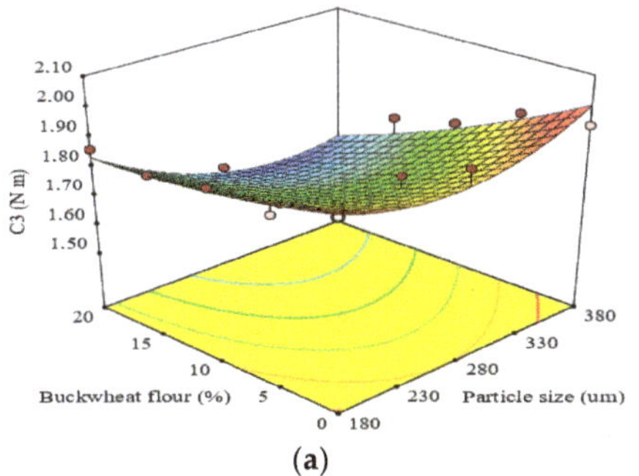
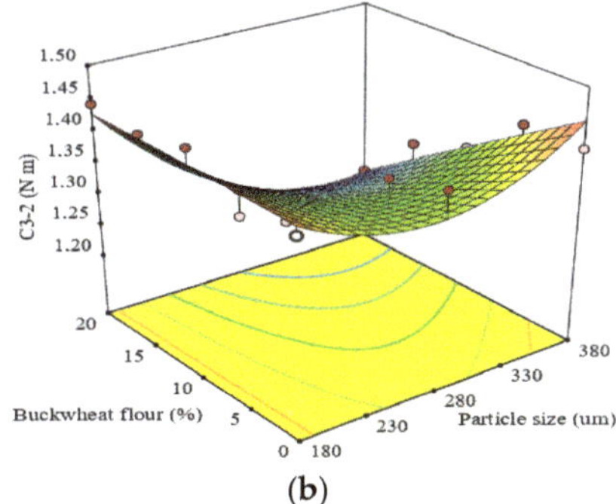

(a)

(b)

Figure 3. 3D response surface plots showing interaction between buckwheat flour particle size and addition level on (**a**) C3 torque (C3) and (**b**) difference between torques C3 and C2 (C3-2) during starch gelatinization.

3.5.6. Cooking Stability Evaluation

The cooking stability is measured by the minimum torque C4 obtained due to the rupturing of the swollen starch granules, thereby causing a decrease in the consistency of the hot-formed starch gel. The quadratic regression model for C4 torque was found to be significant and the R^2 value of 0.87 confirmed the suitability of the model (Table 3). The ANOVA results revealed that there was a significant effect ($p < 0.0001$) of the BF level added to wheat flour, and of the interaction ($p < 0.05$) between particle size and BF level on C4. The combined effect of factors on C4 torque is noticed in the response surface plot (Figure 4a). It can be observed that, with the addition of BF, the C4 torque decreased. Consequently, the cooking stability is diminished as a result of the paste resistance to disintegrate when the temperature is raised. The C4 torque indicated lower values for all samples formulated compared to the control, exhibiting a low resistance of starch against the enzymatic hydrolysis by amylase. The results obtained are in line with those reported by Filipčev et al. (2015) [64], which found a decrease in hot gel stability for buckwheat supplemented doughs. The lowest C4 value was achieved for a large fraction at a higher level of 20% BF. The C4 torque decrease when the amount of BF in composite flour increased might be correlated with the fraction composition, more likely because of the soluble fiber that can bind water by hydrogen bonds, causing a decrease in available water for the starch granules [31]. These results may be related to the protein denaturation from BF, which according to Janssen et al. (2017) [66] shows a minor and a major denaturation peak at about 80 °C and about 102 °C. The protein denaturation will change the dough network, thus determining a decrease in viscosity.

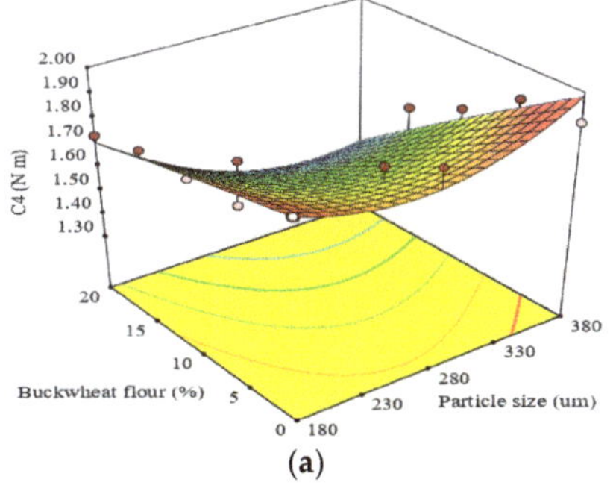
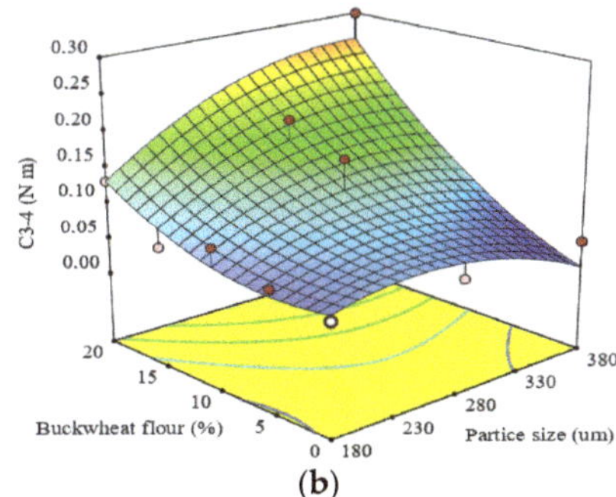

(a)

(b)

Figure 4. 3D response surface plots showing interaction between buckwheat flour particle size and addition level on (**a**) C4 torque (C4) and (**b**) difference between torques C3 and C4 (C3–4) during cooking stability.

The difference between C3 and C4 torques peak values (C3-4) reflects the hot-gel stability or cooking stability of dough. C3-4 was influenced ($p < 0.05$) by BF addition in composite flour and the interaction between fraction size and BF level. The quadratic regression model obtained fitted well to the experimental data of C3-4 with a high coefficient of determination ($R^2 = 0.86$) (Table 3). The simultaneous effect of particle size and addition level on C3-4 is shown in Figure 4b, revealing their capacity to increase the C3-4. This increase of C3-4 may be attributed to the compounds from BF fractions, which changed the α-amylase starch interaction modifying the α-amylase hydrolytic activity on starch.

3.5.7. Starch Retrogradation Evaluation

The decrease in temperature during the cooling stage to 50 °C resulted in a final viscosity associated with a higher dough resistance, and consequently with the C5 torque, which reflects starch retrogradation. This is the process in which the leached amylose chains undergo recrystallization, conditioned by the ratio of amylose and amylopectin from starch [67,68], because amylose recrystallizes with a high speed compared to the amylopectin. Evaluation of C5 torque and C5–4 is used to characterize starch retrogradation. The quadratic regression model achieved for C5 torque was statistically significant ($p < 0.001$) and was a very good one, with a high value (0.92) of R^2, which showed that only 8% of the total variation was unexplained by the model (Table 3). Both studied factors influenced ($p < 0.05$) the C5 torque. With the rise of fractions and the level of BF added in wheat flour, the C5 values decreased, indicating a low starch retrogradation and recrystallization. This result can be associated with the particular structure of starch and the granules of buckwheat presenting a low tendency toward retrogradation. Moreover, the high amounts of phenolic compounds can contribute to retrogradation lowering. The graphical representation of the effect of milling fractions and the level of buckwheat flour added on the C5 torque is presented in Figure 5a. From the response surface plot, it can be observed that, as the fraction size and the BF addition increased, the C5 torque decreased, suggesting that fraction size composition affected α-amylase activity in buckwheat-wheat flours, a lower C5 value being caused by the high activity. However, lower starch gels, which have low C5 torque, are usually linked to lower amylose content. The initial retrogradation is mostly attributed to the re-association of amylose, and to the long-term retrogradation associated with the development of gel structure and crystallinity, e.g., during staling of bread, being due to the recrystallization of amylopectin side chains [69]. As reported in some studies [31,70], the structure reorganization of starch may be favorable for the hardening of bakery goods during storage.

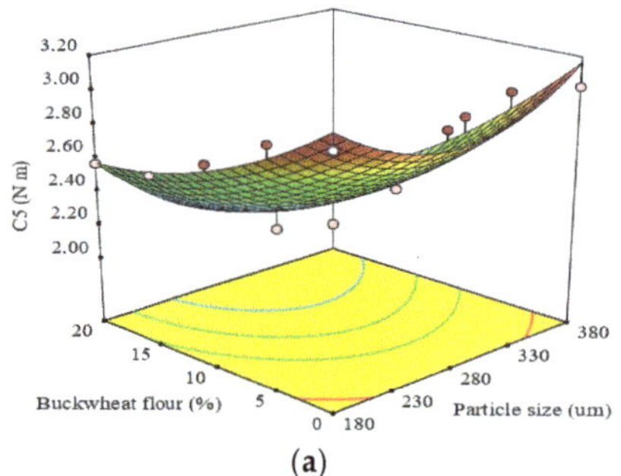
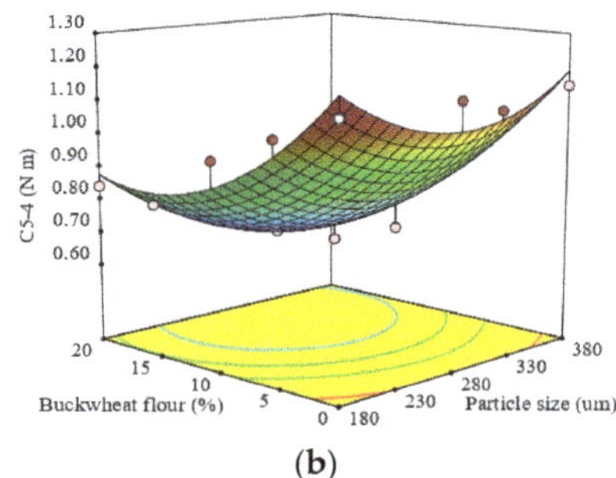

Figure 5. 3D response surface plots showing interaction between buckwheat flour particle size and addition level on (**a**) C5 torque (C5) and (**b**) difference between torques C5 and C4 (C5–4) during starch retrogradation.

The difference between C5 and C4 torques (C5–4) indicates starch retrogradation capacity, influenced by different factors such as the botanical source of the starch, amylose/amylopectin ratio and average chain length of amylose and amylopectin [71]. The quadratic model, which was obtained for the C5–4, was found to have statistical significance ($p < 0.05$) with a high R^2 (0.88), which confirms the adequacy of the model (Table 3). C5–4 was affected ($p < 0.0001$) by the addition of BF in wheat flour, but the fraction size had a non-significant influence ($p > 0.05$).

The level of BF in the composition of dough decreases the C5–4 torque in comparison to wheat flour dough (Figure 5b). This decrease could be attributed to the higher lipid content of buckwheat flour, different structure of amylose and amylopectin fraction of buckwheat starch compared to wheat starch, and lipid-amylose complex-forming ability [22]. Amylose molecules cannot reassociate and recrystallize in a free manner as in systems with low lipid content. This could have a positive effect on bread staling made from buckwheat flour, once the starch retrogradation is one of the key-bread staling factors. These results are in agreement with those reported in other studies [22,72].

3.6. Optimization of Buckwheat-Wheat Composite Flour

The models fitted in this study for dough rheological proprieties were used for simultaneous optimization by using numerical optimization and desirability function approaches. The numerical optimization allowed for the establishment of promising composite flour formulation, which was then compared to the control sample. The results highlighted that the optimal values of factors to achieve the most appropriate composite flour would be 10.75% buckwheat flour that were 280 ìm in particle size. Based on this optimal formulation, the predicted values for the evaluated responses, in terms of the FN index and Mixolab parameters, are shown in Table 4. The results revealed that the optimal values for WA, ST and C2 torque were slightly smaller than the values of the control sample, with no statistical difference.

Table 4. Features of wheat dough control and optimized buckwheat-wheat samples.

Type of Sample	Falling Number Index and Mixolab Parameters Related to Protein Behavior						Mixolab Parameters Related to Starch Behavior					
	FN (%)	WA (%)	DT (min)	ST (min)	C2 (N·m)	C1-2 (N·m)	C3 (N·m)	C3-2 (N·m)	C4 (N·m)	C3-4 (N·m)	C5 (N·m)	C5-4 (N·m)
Control	313.13	58.43	1.43	9.78	0.51	0.60	1.92	1.41	1.89	0.03	3.02	1.13
Optimized	319.04	58.24	3.14	9.04	0.41	0.66	1.72	1.30	1.59	0.12	2.31	0.71

These results suggested that, during the first stage of bread making, the dough resulting from the optimization process could retain a CO_2 similar to the wheat dough. The high rise of DT indicated an increase in the gluten network strength compared to the control, suggesting that the optimal dough can sustain the mechanical treatment for a longer period during the bread-making process, and these were findings that were in line with those previously reported [31,32].

In respect to starch behavior, a decrease in the C2, C3–2, C4, C5, and C5–4 Mixolab parameters was obtained, which highlighted that the optimal buckwheat-wheat composite flour may be appropriate for bread making. The baked goods that can be obtained from this composite flour will present a longer shelf-life in the absence of problems related to staling.

4. Conclusions

The particle size of buckwheat flour revealed variations in chemical characteristics and functional properties. The highest content of protein, lipids, and ash was found in medium particle sizes, followed by a large particle size and a small particle size, which are rich in carbohydrates. Medium particle size had more water and swelling properties, whereas the small particle sizes presented a higher volumetric density. The milling fractions and amount of buckwheat incorporated in wheat flour remarkably influenced dough rheological properties, predicted by the suitable regression models. The multi-response optimization study allowed for the proposal of the optimal milling fraction size and buckwheat flour amount added in wheat flour for obtaining composite flour with the best rheological properties. The results allowed us to confirm that the best formulation composite flour has 10.75% buckwheat flour of 280 µm particle size and 89.25% wheat flour, and was the closest for the requirement to create new products in the bakery industry, without altering the dough matrix.

Future studies are required to identify the impact of buckwheat flour milling fractions on bread and sensory characteristics that would provide further evidence and confirmation for milling fractions and bread making.

Author Contributions: Conceptualization, I.C. and S.M.; methodology, I.C.; software, I.C.; validation, I.C. and S.M.; formal analysis, I.C. and S.M.; investigation, I.C.; resources, I.C. and S.M.; data curation, I.C. and S.M.; writing—original draft preparation, I.C.; writing—review and editing, I.C. and S.M.; visualization, I.C. and S.M.; supervision, S.M. All authors have read and agreed to the published version of the manuscript.

Funding: This work was funded by Stefan cel Mare University of Suceava, Romania.

Acknowledgments: The authors thank Arcada Research laboratory, Arcada Mill, Romania for technical support.

Conflicts of Interest: The authors declare no conflict of interest.

References

1. Martínez-Villaluenga, C.; Peñas, E.; Hernández-Ledesma, B. Pseudocereal grains: Nutritional value, health benefits and current applications for the development of gluten-free foods. *Food Chem. Toxicol.* **2020**, *137*, 111178. [CrossRef]
2. Joshi, D.C.; Chaudhari, G.V.; Sood, S.; Kant, L.; Pattanayak, A.; Zhang, K.; Fan, Y.; Janovská, D.; Meglič, V.; Zhou, M. Revisiting the versatile buckwheat: Reinvigorating genetic gains through integrated breeding and genomics approach. *Planta* **2019**, *250*, 783–801. [CrossRef] [PubMed]
3. Yasui, Y.; Hirakawa, H.; Ueno, M.; Matsui, K.; Katsube-Tanaka, T.; Yang, S.J.; Aii, J.; Sato, S.; Mori, M. Assembly of the draft genome of buckwheat and its applications in identifying agronomically useful genes. *DNA Res.* **2016**, *23*, 215–224. [CrossRef] [PubMed]
4. Zhou, M.; Ivan, K.; Sun, H.W.; Nikhil, K.C.; Wieslander, G. *Molecular Breeding and Nutritional Aspects of Buckwheat*; Academic Press: Pittsburgh, PA, USA, 2016; pp. 203–207.
5. Burešová, I.; Tokár, M.; Mareček, J.; Hřivna, L.; Faměra, O.; Šottníková, V. The comparison of the effect of added amaranth, buckwheat, chickpea, corn, millet and quinoa flour on rice dough rheological characteristics, textural and sensory quality of bread. *J. Cereal Sci.* **2017**, *75*, 158–164. [CrossRef]
6. Lee, D.-G.; Woo, S.H.; Choi, J.-S. Biochemical Properties of Common and Tartary Buckwheat: Centered with Buckwheat Proteomics. In *Molecular Breeding and Nutritional Aspects of Buckwheat*; Elsevier BV: Amsterdam, The Netherlands, 2016; pp. 239–259.
7. Nedeljkovic, N.; Sakac, M.; Mandic, A.; Psodorov, D.; Jambrec, D.; Pestoric, M.; Sedej, I.; Dapcevic-Hadnadjev, T. Rheological properties and mineral content of buckwheat enriched wholegrain wheat pasta. *Chem. Ind. Chem. Eng. Q.* **2014**, *20*, 135–142. [CrossRef]
8. Campbell, C.G. Major and minor production areas. In *Buckwheat: Fagopyrum Esculentum Moench*; Bioversity International: Manitoba, MB, Canada, 1997; Volume 19, p. 52.
9. Alonso-Miravalles, L.; O'Mahony, J.A. Composition, Protein Profile and Rheological Properties of Pseudocereal-Based Protein-Rich Ingredients. *Foods* **2018**, *7*, 73. [CrossRef]
10. Repo-Carrasco-Valencia, R.; Arana, J.V. Carbohydrates of kernels. In *Pseudocereals, Chemistry and Technology*; Haros, M., Schoenlechner, R., Eds.; Wiley: Oxford, UK, 2017; pp. 28–48.
11. Zhou, Y.; Zhao, S.; Jiang, Y.; Wei, Y.; Zhou, X. Regulatory Function of Buckwheat-Resistant Starch Supplementation on Lipid Profile and Gut Microbiota in Mice Fed with a High-Fat Diet. *J. Food Sci.* **2019**, *84*, 2674–2681. [CrossRef]
12. Alvarez-Jubete, L.; Wijngaard, H.; Arendt, E.; Gallagher, E. Polyphenol composition and in vitro antioxidant activity of amaranth, quinoa buckwheat and wheat as affected by sprouting and baking. *Food Chem.* **2010**, *119*, 770–778. [CrossRef]
13. Dziedzic, K.; Górecka, D.; Kucharska, M.; Przybylska, B. Influence of technological process during buckwheat groats production on dietary fibre content and sorption of bile acids. *Food Res. Int.* **2012**, *47*, 279–283. [CrossRef]
14. Shukla, A.; Srivastava, N.; Suneja, P.; Yadav, S.K.; Hussain, Z.; Rana, J.C.; Yadav, S. Genetic diversity analysis in Buckwheat germplasm for nutritional traits. *Indian J. Exp. Biol.* **2018**, *56*, 827–837.
15. FAO/WHO. *Protein Quality Evaluation. Report of a Joint FAO/WHO Expert Consulation*; Food and Agriculture Organiza-tion/World Health Organization of United Nations: Rome, Italy, 1990.

16. Harasym, J.; Satta, E.; Kaim, U. Ultrasound Treatment of Buckwheat Grains Impacts Important Functional Properties of Resulting Flour. *Molecules* **2020**, *25*, 3012. [CrossRef]
17. Alvarez-Jubete, L.; Arendt, E.K.; Gallagher, E. Nutritive value and chemical composition of pseudocereals as gluten-free ingredients. *Int. J. Food Sci. Nutr.* **2009**, *60*, 240–257. [CrossRef]
18. Bonafaccia, G.; Marocchini, M.; Kreft, I. Composition and technological properties of the flour and bran from common and tartary buckwheat. *Food Chem.* **2003**, *80*, 9–15. [CrossRef]
19. Dziadek, K.; Kopeć, A.; Pastucha, E.; Piątkowska, E.; Leszczyńska, T.; Pisulewska, E.; Witkowicz, R.; Francik, R. Basic chemical composition and bioactive compounds content in selected cultivars of buckwheat whole seeds, dehulled seeds and hulls. *J. Cereal Sci.* **2016**, *69*, 1–8. [CrossRef]
20. Liu, Y.; Cai, C.; Yao, Y.; Xu, B. Alteration of phenolic profiles and antioxidant capacities of common buckwheat and tartary buckwheat produced in China upon thermal processing. *J. Sci. Food Agric.* **2019**, *99*, 5565–5576. [CrossRef] [PubMed]
21. Rocchetti, G.; Lucini, L.; Rodriguez, J.M.L.; Barba, F.J.; Giuberti, G. Gluten-free flours from cereals, pseudocereals and legumes: Phenolic fingerprints and in vitro antioxidant properties. *Food Chem.* **2019**, *271*, 157–164. [CrossRef] [PubMed]
22. Sedej, I.; Sakač, M.; Mandić, A.; Mišan, A.; Tumbas, V.T.; Hadnađev, M. Assessment of antioxidant activity and rheological properties of wheat and buckwheat milling fractions. *J. Cereal Sci.* **2011**, *54*, 347–353. [CrossRef]
23. Sakhare, S.D.; Inamdar, A.A.; Soumya, C.; Indrani, D.; Rao, G.V. Effect of flour particle size on microstructural, rheological and physico-sensory characteristics of bread and south Indian parotta. *J. Food Sci. Technol.* **2014**, *51*, 4108–4113. [CrossRef]
24. Bobkov, S. Biochemical and Technological Properties of Buckwheat Grains. In *Molecular Breeding and Nutritional Aspects of Buckwheat*; Elsevier BV: Amsterdam, The Netherlands, 2016; pp. 423–440.
25. Awuchi, C.G.; Igwe, V.S.; Echeta, C.K. The functional properties of foods and flours. *Int. J. Curr. Adv. Res.* **2019**, *5*, 139–160.
26. Kasar, C.; Thanushree, M.P.; Gupta, S.; Inamdar, A.A. Milled fractions of common buckwheat (Fagopyrum esculentum) from the Himalayan regions: Grain characteristics, functional properties and nutrient composition. *J. Food Sci. Technol.* **2020**, *2020*, 1–11. [CrossRef]
27. Ahmed, J.; Thomas, L.; Arfat, Y.A. Functional, rheological, microstructural and antioxidant properties of quinoa flour in dispersions as influenced by particle size. *Food Res. Int.* **2019**, *116*, 302–311. [CrossRef]
28. Mironeasa, S.; Iuga, M.; Zaharia, D.; Mironeasa, C. Rheological Analysis of Wheat Flour Dough as Influenced by Grape Peels of Different Particle Sizes and Addition Levels. *Food Bioprocess Technol.* **2018**, *12*, 228–245. [CrossRef]
29. Iuga, M.; Mironeasa, C.; Mironeasa, S. Oscillatory Rheology and Creep-Recovery Behaviour of Grape Seed-Wheat Flour Dough: Effect of Grape Seed Particle Size, Variety and Addition Level. *Bull. Univ. Agric. Sci. Veter. Med. Cluj Napoca. Food Sci. Technol.* **2019**, *76*, 40–51. [CrossRef]
30. Bressiani, J.; Oro, T.; Da Silva, P.; Montenegro, F.; Bertolin, T.; Gutkoski, L.; Gularte, M. Influence of milling whole wheat grains and particle size on thermo-mechanical properties of flour using Mixolab. *Czech J. Food Sci.* **2019**, *37*, 276–284. [CrossRef]
31. Coțovanu, I.; Stoenescu, G.; Mironeasa, S. Amaranth influence on wheat flour dough rheology: Optimal particle size and amount of flour replacement. *J. Microbiol. Biotechnol. Food Sci.* **2020**, *10*, 336–373. [CrossRef]
32. Coțovanu, I.; Batariuc, A.; Mironeasa, S. Characterization of Quinoa Seeds Milling Fractions and Their Effect on the Rheological Properties of Wheat Flour Dough. *Appl. Sci.* **2020**, *10*, 7225. [CrossRef]
33. Mironeasa, S.; Mironeasa, C. Dough bread from refined wheat flour partially replaced by grape peels: Optimizing the rheological properties. *J. Food Process. Eng.* **2019**, *42*, 1–14. [CrossRef]
34. Iuga, M.; Mironeasa, S. Grape seeds effect on refined wheat flour dough rheology: Optimal amount and particle size. *Ukr. Food J.* **2019**, *8*, 799–814. [CrossRef]
35. Mironeasa, S.; Iuga, M.; Zaharia, D.; Mironeasa, C. Optimization of grape peels particle size and flour substitution in white wheat flour dough. *Sci. Study Res. Chem. Eng. Biotechnol. Food Ind.* **2019**, *20*, 29–42.
36. Raghavendra, S.N.; Rastogi, N.K.; Raghavarao, K.S.M.S.; Tharanathan, R.N. Dietary fiber from coconut residue: Effects of different treatments and particle size on the hydration properties. *Eur. Food Res. Technol.* **2004**, *218*, 563–567. [CrossRef]
37. Onipe, O.O.; Beswa, D.; Jideani, A.I.O. Effect of size reduction on colour, hydration and rheological properties of wheat bran. *Food Sci. Technol.* **2017**, *37*, 389–396. [CrossRef]
38. Olapade, A.A.; Adeyemo, M.A. Evaluation of cookies produced from blends of wheat, cassava and cowpea flours. *Int. J. Food Stud.* **2014**, *3*, 175–185. [CrossRef]
39. Ikegwu, O.J.; Nwobasi, V.N.; Odoh, M.O.; Oledinma, N.U. Evaluation of the pasting and some functional properties of starch isolated from some improved cassava varieties in Nigeria. *Afr. J. Biotechnol.* **2009**, *8*, 2310–2315.
40. Mironeasa, S.; Iuga, M.; Zaharia, D.; Mironeasa, C. Optimization of White Wheat Flour Dough Rheological Properties with Different Levels of Grape Peels Flour Addition. *Bull. Univ. Agric. Sci. Veter. Med. Cluj Napoca Food Sci. Technol.* **2019**, *76*, 27–39. [CrossRef]
41. Peluola-Adeyemi, O.A.; Bolaji, O.T.; Abegunde, T.A.; Ogungbesan, V.T. Optimization of Processing Conditions on Some Selected Mechanical Properties of Cookies from Wheat-Cocoyam Flour Using Response Surface Methodology (RSM). *Am. J. Food Sci. Technol.* **2019**, *7*, 22–26. [CrossRef]
42. Offia-Olua, B.I. Chemical, Functional and Pasting Properties of Wheat (Triticumspp)-Walnut (Juglansregia) Flour. *Food Nutr. Sci.* **2014**, *5*, 1591–1604. [CrossRef]

43. Sciarini, L.S.E.; Steffolani, M.; Fernández, A.; Paesani, C.; Pérez, G.T. Gluten-free breadmaking affected by the particle size and chemical composition of quinoa and buckwheat flour fractions. *Food Sci. Technol. Int.* **2020**, *26*, 321–332. [CrossRef]

44. Marcela, S.; Julie, L.; Alena, M.; Šárka, H.; Pavel, S. Effect of the dough mixing process on the quality of wheat and buckwheat proteins. *Czech J. Food Sci.* **2017**, *35*, 522–531. [CrossRef]

45. Martín-García, B.; Pasini, F.; Verardo, V.; Gómez-Caravaca, A.M.; Marconi, E.; Caboni, M.F.; García, M.; Caravaca, G. Use of Sieving as a Valuable Technology to Produce Enriched Buckwheat Flours: A Preliminary Study. *Antioxidants* **2019**, *8*, 583. [CrossRef]

46. Ahmed, A.; Khalid, N.; Ahmad, A.; Abbasi, N.A.; Latif, M.S.Z.; Randhawa, M.A. Phytochemicals and biofunctional properties of buckwheat: A review. *J. Agric. Sci.* **2013**, *152*, 349–369. [CrossRef]

47. Ikeda, S.; Yamashita, Y.; Kusumoto, K.; Kreft, I. Nutritional characteristics of minerals in various byckwheat groats. *Fagopyrum* **2005**, *22*, 71–75.

48. Li, S.-Q.; Zhang, Q.H. Advances in the Development of Functional Foods from Buckwheat. *Crit. Rev. Food Sci. Nutr.* **2001**, *41*, 451–464. [CrossRef] [PubMed]

49. Han, L.; Zhou, Y.; Tatsumi, E.; Shen, Q.; Cheng, Y.; Li, L. Thermomechanical Properties of Dough and Quality of Noodles Made from Wheat Flour Supplemented with Different Grades of Tartary Buckwheat (Fagopyrum tataricum Gaertn.) Flour. *Food Bioprocess Technol.* **2012**, *6*, 1953–1962. [CrossRef]

50. Dziedzic, K.; Kurek, S.; Mildner–Szkudlarz, S.; Kreft, I.; Walkowiak, J. Fatty acids profile, sterols, tocopherol and squalene content in Fagopyrum tataricum seed milling fractions. *J. Cereal Sci.* **2020**, *96*, 103118. [CrossRef]

51. Golijan, J.; Milinčić, D.D.; Petronijević, R.; Pešić, M.B.; Barać, M.B.; Sečanski, M.; Lekić, S.; Kostić, A.Ž. The fatty acid and triacylglycerol profiles of conventionally and organically produced grains of maize, spelt and buckwheat. *J. Cereal Sci.* **2019**, *90*, 102845. [CrossRef]

52. Kinsella, J.E. Functional properties of soy protein. *J. Am. Oil Chem. Soc.* **1979**, *56*, 254–264. [CrossRef]

53. Wolter, A.; Hager, A.-S.; Zannini, E.; Arendt, E.K. In vitro starch digestibility and predicted glycaemic indexes of buckwheat, oat, quinoa, sorghum, teff and commercial gluten-free bread. *J. Cereal Sci.* **2013**, *58*, 431–436. [CrossRef]

54. Qian, J.; Rayas-Duarte, P.; Grant, L. Partial Characterization of Buckwheat (Fagopyrum esculentum) Starch. *Cereal Chem. J.* **1998**, *75*, 365–373. [CrossRef]

55. Sindhu, R.; Khatkar, B.S. Composition and Functional Properties of Common Buckwheat (Fagopyrum Esculentum Moench) Flour and Starch. *IJIRAS* **2016**, *3*, 154–159.

56. Struyf, N.; Verspreet, J.; Courtin, C.M. The effect of amylolytic activity and substrate availability on sugar release in non-yeasted dough. *J. Cereal Sci.* **2016**, *69*, 111–118. [CrossRef]

57. Gavurníková, S.; Havrlentová, M.; Mendel, Ľ.; Čičová, I.; Bieliková, M.; Kraic, J. Parameters of Wheat Flour, Dough, and Bread Fortified by Buckwheat and Millet Flours. *Agriculture* **2011**, *57*, 144–153. [CrossRef]

58. Sapirstein, H.; Wu, Y.; Koksel, F.; Graf, R. A study of factors influencing the water absorption capacity of Canadian hard red winter wheat flour. *J. Cereal Sci.* **2018**, *81*, 52–59. [CrossRef]

59. Nikolic, N.; Sakač, M.; Mastilović, J. Effect of buckwheat flour addition to wheat flour on acylglycerols and fatty acids composition and rheology properties. *LWT* **2011**, *44*, 650–655. [CrossRef]

60. Joshi, D.C.; Zhang, K.; Wang, C.; Chandora, R.; Khurshid, M.; Li, J.; He, M.; Georgiev, M.I.; Zhou, M. Strategic enhancement of genetic gain for nutraceutical development in buckwheat: A genomics-driven perspective. *Biotechnol. Adv.* **2020**, *39*, 107479. [CrossRef]

61. Mohammed, I.; Ahmed, A.R.; Senge, B. Effects of chickpea flour on wheat pasting properties and bread making quality. *J. Food Sci. Technol.* **2014**, *51*, 1902–1910. [CrossRef]

62. Stefan, E.-M.; Voicu, G.; Constantin, G.-A.; Ipate, G.; Munteanu, M. Effect of whole buckwheat flour on technological properties of wheat flour and dough. In *Proceedings of the 17th International Scientific Conference Engineering for Rural Development, Jelgava, Latvia, 23–25 May 2018*; Latvia University of Life Sciences and Technologies: Jelgava, Latvia, 2018.

63. Collar, C. Significance of heat-moisture treatment conditions on the pasting and gelling behaviour of various starch-rich cereal and pseudocereal flours. *Food Sci. Technol. Int.* **2017**, *23*, 623–636. [CrossRef]

64. Filipčev, B.; Šimurina, O.; Bodroža-Solarov, M. Impact of buckwheat flour granulation and supplementation level on the quality of composite wheat/buckwheat ginger-nut-type biscuits. *Italic J. Food Sci.* **2015**, *27*, 495–504.

65. Kumar, R.; Khatkar, B.S. Thermal, pasting and morphological properties of starch granules of wheat (*Triticum aestivum* L.) varieties. *J. Food Sci. Technol.* **2017**, *54*, 2403–2410. [CrossRef]

66. Janssen, F.; Pauly, A.; Rombouts, I.; Jansens, K.J.; Deleu, L.J.; Delcour, J.A. Proteins of Amaranth (Amaranthusspp.), Buckwheat (Fagopyrumspp.), and Quinoa (Chenopodiumspp.): A Food Science and Technology Perspective. *Compr. Rev. Food Sci. Food Saf.* **2017**, *16*, 39–58. [CrossRef] [PubMed]

67. BeMiller, J.N. Starches: Molecular and granular structures and properties. In *Carbohydrate Chemistry for Food Scientists*, 3rd ed.; Elsevier Inc.: Amsterdam, The Netherlands, 2019; pp. 159–189.

68. Gujral, H.S.; Sharma, B.; Singh, K. Rheological characterization of wheat flour as modified by adding barley glucagel (a β-glucan isolate) under thermo-mechanical stress using Mixolab. *J. Food Meas. Charact.* **2021**, *15*, 228–236. [CrossRef]

69. Wang, S.; Li, C.; Copeland, L.; Niu, Q.; Wang, S. Starch Retrogradation: A Comprehensive Review. *Compr. Rev. Food Sci. Food Saf.* **2015**, *14*, 568–585. [CrossRef]

70. Cai, L.; Choi, I.; Hyun, J.-N.; Jeong, Y.-K.; Baik, B.-K. Influence of Bran Particle Size on Bread-Baking Quality of Whole Grain Wheat Flour and Starch Retrogradation. *Cereal Chem. J.* **2014**, *91*, 65–71. [CrossRef]
71. Ozturk, S.; Kahraman, K.; Tiftik, B.; Koksel, H. Predicting the cookie quality of flours by using Mixolab®. *Eur. Food Res. Technol.* **2008**, *227*, 1549–1554. [CrossRef]
72. Hadnađev, T.D.; Torbica, A.; Hadnađev, M. Rheological properties of wheat flour substitutes/alternative crops assessed by Mixolab. *Procedia Food Sci.* **2011**, *1*, 328–334. [CrossRef]

Review

Strategies for Reducing Sodium Intake in Bakery Products, a Review

Georgiana Gabriela Codină *, Andreea Voinea and Adriana Dabija

Faculty of Food Engineering, Stefan cel Mare University of Suceava, 720229 Suceava, Romania;
andreea.musu@fia.usv.ro (A.V.); adriana.dabija@fia.usv.ro (A.D.)
* Correspondence: codina@fia.usv.ro or codinageorgiana@yahoo.com; Tel.: +40-745-460-727

Abstract: Nowadays, the dietary sodium chloride intake is higher than the daily recommended levels, especially due to its prominent presence in food products. This may cause an increase of high blood pressure leading to cardiovascular diseases. Cereal products, and in particular bread, are the main source of salt in human diet. However, salt is a critical ingredient in bread making, and its reduction can have a negative impact on bread quality. This review focuses on physiological role of sodium chloride, its effect on the human body and legislative recommendations on its consumption. Moreover, it presents sodium chloride effects on the bread making from the technological and sensory point of view and presents different options for salt reduction in foods focusing on bakery products. It may be concluded that salt reduction in bread making while maintaining dough rheological properties, yeast fermentation rate, bread quality through its loaf volume, color, textural properties, sensory characteristics is difficult to be achieved due to sodium chloride's multifunctional role in the bread-making process. Several strategies have been discussed, focusing on sodium chloride replacement with other type of salts, dry sourdough and flavor enhancers.

Keywords: salt reduction; legislative recommendations; bread making; salt replacement; bread quality

Citation: Codină, G.G.; Voinea, A.; Dabija, A. Strategies for Reducing Sodium Intake in Bakery Products, a Review. *Appl. Sci.* **2021**, *11*, 3093. https://doi.org/10.3390/app11073093

Academic Editor: Monica Gallo

Received: 1 March 2021
Accepted: 27 March 2021
Published: 31 March 2021

Publisher's Note: MDPI stays neutral with regard to jurisdictional claims in published maps and institutional affiliations.

1. Introduction

Bread is a part of the foodstuffs that are the basis of many people's diets due to its nutritional value and the low price that is reflected from the flour from which it is obtained, the auxiliary materials used, and the technology applied. Food experts define bread as a staple food at the top of the food pyramid due to its rich content in carbohydrates, fiber, protein, B vitamins and mineral salts [1]. In most European countries bread is the most important sources of salt, its contribution to salt intake ranging between 19.1% in Spain to 28% in France [2].

According to the World Health Organization, processed foods such as bakery products are the main daily source of sodium in consumers' diet for developed countries, with an average of about 75–85% of the total sodium intake, while 5–10% are naturally provided from foods consumption that are part of the daily diet and the remaining part of 10–15% of sodium are provided from sodium chloride addition during cooking or eating [3]. However, in developing countries, salt addition during meal preparation presents a much more important role [4]. Globally, the sodium intake from processed foods is much higher than the intake of unprocessed, naturally consumed foods [5].

Although sodium is a normal constituent of the human body, distributed in the extracellular compartments, performing many functions with beneficial effects on the body, excessive sodium intake is associated with cardiovascular diseases caused by increased hypertension [6].

Epidemiological studies on hypertension have shown that many people from countries where salt consumption is high also presented high hypertension values [7]. Often, the salt consumption covers 35–50 times the renal needs and it can be concluded that one of the fundamental characteristics of contemporary diet is an excessive salt intake. It is considered

95

that two main factors have contributed to this situation: a behavioral factor in humans salt intake which is not dependent on real needs but on the taste for salt artificially created and which is part of a hedonic behavior that develops since childhood; the second factor is the urbanization that has led to a consumption of industrialized foods in which sodium chloride is used as a flavor additive and preservative. As a result, increased consumption of salt is a relatively recent food habit, which greatly demands the body's ability to adapt [8].

Nowadays, consumers' concerns about excessive sodium intake and its associated effects have increased, and that is why some food companies have changed their product portfolios to reduce sodium intake and to promote healthier diets. An example of this is Nestle, which eliminated almost 7500 tons of sodium from their products starting 2005 [5].

Although consumers are now becoming aware of the negative effects of salt excess consumption on their health, they do not have much information on the salt connections with sodium consumption. In developed countries, consumer awareness of proper nutrition and nutritionally healthy behavior is increasing nowadays especially through education [9,10]. Unfortunately, in underdeveloped countries the level of education regarding proper nutrition behavior is very low and therefore the population awareness on the negative effect of excessive sodium consumption on health is not very high [10].

From the consumer behavior point of view, sodium chloride increases the acceptability of many foods by intensifying the salty taste and flavor and by trans-modal interactions which increase the taste of other aromatic compounds and decrease or eliminate the bitter taste [3]. It seems that the sensitivity to salt varies in the same individual from one moment of life to another: depending on age, blood pressure level, obesity, pregnancy, various diseases, drugs consumed, and even race etc. risk [11].

Sodium chloride is one of the raw materials in the bakery industry, which is used to make all bakery products except salt-free dietary products. It has an important role on the sensory characteristics of bakery products but also on the technological characteristics of bread making such as dough rheological properties, enzymatic and microbiological dough activity, and bread quality [11]. Nowadays, bread is considered one of the most important sources of salt in the diet, contributing 25% of the amount of salt consumed by the population. Therefore, for a reduction in salt intake it is necessary a reduction of salt in the bakery products. In general salt is used as a food ingredient, as a preservative, to improve moisture retention and to increase food sensory characteristics. Although in some cases it is impossible to reduce the salt content from foods, in many others it is possible to obtain processed foods with lower sodium content. This is also the case for bakery products which are the largest contributor to dietary sodium intake in Great Britain and the USA [12].

Nowadays, almost every EU country has different strategies which includes recommendation for salt reduction via food reformulation to reduce the salt content from food products including bakery ones. For example, different programs are developed in EU such as "STOP SALT!" in Hungary, "Gaining health: making healthy choices easier" in Italy which encourages in especially salt reduction in bread, in Bulgaria the National Food and Nutrition Action Plan 2012–2017 promotes salt reduction, etc. The bakery products reformulation strategies for salt reduction are continuing in EU countries, some examples in this regard being the following: in Austria the salt reduction has been established of 15% up to 2015 by the Federal Ministry for Health, in Italy of 10% up to 2012 by the Ministry of Health, in Spain of 20% up to 2014 by the Ministry of Health and Social Policy, in Hungary, the Hungarian Bakery Association recommended reducing salt in bread to reach, after December 2018, a maximum level of 2.35%, etc. [13].

Numerous strategies have been proposed to reduce sodium chloride in foodstuffs including bakery products, in order to improve the health of the population. The challenge of these strategies is to solve the technological and sensory problems caused by sodium chloride removal from bakery products recipe. From the technological point of view, a salt reduction up to 0.6% may conduct to bakery products without a significant negative effect on dough rheological properties. However, its effect on the sensory properties of bakery products may be a problem due to the fact that the salty taste is difficult to be achieved [1]. There are a number of combinations of substances proposed as salt substitutes, which will be discussed in a more detailed way in point 4 of this review, but they still may offer to the bakery products an unpleasant taste. Different combinations based on flavor enhancers, other type of salts, only partial substitution of sodium chloride from bakery recipe may offer some solutions for sodium chloride reduction. More, to reduce the negative effects of lack of salt on technological properties of bakery products it is recommended the use of another type of salt with similar effects on dough rheology as the one produced by sodium chloride. The proposed strategies for salt reduction are shown in Figure 1, methods that as we mentioned may be combined in order to increase bakery products flavor and salty taste.

Figure 1. Strategies of salt reduction in bakery products.

All these strategies along with sodium chloride effect on bread making, its physiological role on the human body and legislative norms regarding the international recommendation of daily sodium intake will be discussed further.

2. The Physiological Role of Sodium Chloride and Legislative Recommendations on Its Consumption

Sodium is among the top six elements in the earth's crust, comprising 2.83% sodium in all its forms. The most important sodium salts found in nature are sodium chloride, sodium carbonate, sodium borate, sodium nitrate and sodium sulfate [14]. To obtain sodium chloride, the weight of sodium must be multiplied by 2.54. Regulating the levels of sodium and chloride in the human body is an important biological process, maintained by multiple mechanisms that work to control them. In the human body, sodium and chloride are the major constituents of the extracellular fluid, participating in maintaining the electrochemical gradient between the extracellular space and the cytoplasm. In the absence of this gradient, life is not possible. Their presence in the human body maintains the membrane potential that is absolutely necessary in the transmission of nerve impulses, muscle contraction and therefore cardiac function. Absorption of sodium in the small intestine promotes the assimilation of amino acids, glucose and water [1].

Sodium is a colorless crystalline compound found naturally in many foodstuffs and its most widely used form is sodium chloride, also named table salt. Salt has an important role in the human history. The ancient populations used salt to preserve food, salt being for them so precious that it could be exchanged for gold. Until the end of antiquity, salt was, along with amber and tin, one of the main currencies. In American history, salt has been a vital element for survival and during the Civil War it was used not only for foodstuffs but also for tanning the skin, dyeing clothes and preserving rations [15]. The need for permanent salt intake is one of the physiological needs to which man is exposed. Blood sodium concentration is an important homeostatic parameter that controls extracellular fluid volume and blood tonicity [11].

Nowadays salt is used as a spice and preservative in processed foods, being used in various forms such as sodium nitrite, sodium benzoate, monosodium glutamate and baking soda. It is present in processed foods such as soy sauce, canned meat and vegetables, soups, processed meat and almost any food with a long shelf life [15].

2.1. Aspects Regarding the Effect of Sodium Chloride Consumption on the Human Body

Sodium (Na^+) is the dominant cation in the extracellular fluid of the human body. The functions of sodium consist in its participation in the control of the volume and systemic distribution of the total water in the body allowing cellular absorption of dissolved substances and generating, by interactions with potassium, of the trans membrane electrochemical potential [16]. Sodium ions are involved in transmitting electrochemical impulses along cell membranes to maintain normal nerve and muscle susceptibility. They contribute to the swelling of colloids in the tissues and thus cause the retention of bound water in the body. At the same time, sodium takes an active part in neutralizing the acids that form in the body. It is an element present in all organs, tissues and biological fluids, which plays an important role in intercellular processes and interstitial metabolism [17].

In the body of an adult with a weight of about 70 kg, sodium is found in the body in an amount of 92 g and is distributed differently from person to person but in identical concentrations in all people, regardless of sex, age and physical or intellectual effort [18].

Sodium has multiple roles in the human body, including:

- maintaining a normal excitability of the muscular and nervous system;
- activation of enzymatic systems;
- maintaining a normal pH in the stomach, intestines and blood plasma;
- regulation of water absorption and retention [19].

Sodium is essential for cellular homeostasis and physiological function. Claude Bernard was the first to highlight the "internal environment". Walter Cannon defined homeostasis more clearly when he referred to the "fluid matrix" of the body and emphasized the role of sodium [20]. In the last few decades, there has been an increasing amount of work exploring sodium and dietary health. The amount of sodium needed to maintain homeostasis in adults is extremely low (<500 mg) compared to the average intake of most Americans (>3200 mg) [21].

Dietary sodium deficiency is rare in healthy European populations. Sodium chloride and other sodium salts are daily used in the diet and there are adaptive physiological mechanisms that reduce the loss of sodium from urine, feces and sweat to a low sodium intake. Sodium chloride addition during industrial food processing or food preservation is the main source of dietary sodium in Western diets. Other sources of sodium include inherent native sources and sodium-containing food additives, in which sodium may be associated with anions other than chloride.

In healthy people, almost all dietary sodium is absorbed, even at a high sodium intake. After absorption, sodium ions are distributed through portal and systemic circulations, where their concentrations are maintained in a limited range. Up to 95% of the body's sodium content is in the extracellular fluid, including a large amount in bone, skin and muscle. Sodium excretion and retention (homeostasis) is performed by an integrated neurohormonal control from the centers located in the hypothalamus. The kidney is the main organ that mediates the excretion and retention of sodium. It effectively excretes sodium in response to high sodium intakes from food and stores sodium when dietary intake is deficient. In contrast, the excretion of sodium in the feces is relatively stable and usually limited to a few mmol/day. The amount of sodium excreted through perspiration can vary greatly, depending on environmental conditions or levels of physical activity [16].

Excess sodium in the diet has been linked to high blood pressure. The sensitivity of blood pressure to salt varies greatly, but certain subgroups tend to be more sensitive to salt. The mechanisms underlying sodium-induced increases in blood pressure are not fully understood, but may involve changes in renal function, fluid volume, fluid-regulating hormones, vasculature, cardiac function, and autonomic nervous system. It was established that in hypertension cases excess salt is not only an aggravating factor, but can also be one of the triggers. It appears that there is a functional abnormality of red blood cell membranes in hypertensive population, which it is reported to the ratio between the net flows of Na^+ leaving and the net flow of K^+ entering. This abnormality is genetically transmitted and allows sodium to enter into the cells in excess. Many arguments support the role of salt as a factor in promoting atherosclerosis. Recent preclinical and clinical data suggest that, even in the absence of an increase in blood pressure, excess dietary sodium can adversely affect target organs, including blood vessels, heart, kidneys, and brain [22]. In normal individuals, the increase in excess salt has no short-term effects. In the long term, if the kidney does not have the ability to regulate the concentration of sodium, as a result of excess of NaCl consumption, there is an increase in blood pressure in the peripheral vessels. To compensate the excess of Na^+ and to prevent high blood pressure, a higher amount of natriuretic hormone is secreted. As a result, the sodium pump is injected from the erythrocytes and the smooth muscles of the vessels, which causes the membrane depolarization and the internal accumulation of calcium. Gradually, hypertonia is reached [8].

2.2. Legislative Norms on Recommended Daily Sodium Intake

Globally, institutions such as the European Union are actively promoting the reduction of salt content in food. The European Parliament and the EU Council approved in 2006 a regulation on nutrition and healthy food (EC) No. 1924/2006 (European Commission (EC), 2006), this document allowing, among other things, the use of nutritional claims regarding the sodium/sodium chloride content of foods. These regulations allow consumers to focus on the salt content of food with the help of inscriptions found on product packaging [1]. Article 8 of EU Regulation 1924/2006 lists the following restrictions for sodium/sodium chloride claims it's shown in Table 1.

Table 1. Nutrition claims regarding salt/sodium content—per Article 8 of EU Regulation 1924/2006 [1,5].

Salt Content, g/(100 g/100 mL)	Sodium Content, g/(100 g/100 mL)	Nutrition Claims
-	-	Low content (sodium chloride/sodium) certifies that the sodium or equivalent salt has been reduced by at least 25% compared to a similar product.
0.30	0.12	Low sodium chloride/sodium content certify that the product does not contain more than 0.12 g of sodium or the equivalent value for sodium chloride, 0.3 g per 100 g or per 100 mL.

Table 1. *Cont.*

Salt Content, g/(100 g/100 mL)	Sodium Content, g/(100 g/100 mL)	Nutrition Claims
0.10	0.04	A very low sodium/sodium chloride content certifies that the product does not contain more than 0.04 g of sodium or the equivalent value for sodium chloride, 0.1 g per 100 g or per 100 mL.
0.013	0.005	Sodium-free or sodium chloride-free certifies that the product does not contain more than 0.005 g of sodium or the equivalent value for sodium chloride, 0.013 g per 100 g.

2.3. Sodium Chloride Content on Bakery Products and Their Degree of Consumption

The main source of salt in food in most European countries is generally foodstuffs such as bakery products, followed by meat and meat products, cheese and dairy products [23–25]. Bakery products have a major contribution to the daily intake of sodium in the diet, along with other products obtained from cereals such as biscuits, cakes, breakfast cereals, pastries, noodles, cereal bars, etc. The origin of sodium in these products is caused by the addition of sodium chloride to obtain them but also other raw materials and ingredients used. For example, a source of sodium is also ingredients such as baking powder or leavening agents [26]. It was found that the daily consumption of 150 g of bread containing 20 g of salt/kg of flour contributes with 25% of the average amount of salt consumed, which is 10 g/day/person. Bread is thus the main food contributor to the sodium chloride intake in the diet. Per capita, bread consumption varied between EU countries. The highest bread consumption was reported in Turkey (104 kg per year) and Bulgaria (95 kg) while the lowest one was reported in Great Britain (32 kg). It seems that the average consumption of bread by European people are 59 kg per year. Bread consumption is generally stable but in some countries such as Netherlands, Belgium, UK, Poland has been reported a slightly decrease. In 2015, in Poland, the bread consumption was of 145 g of bread daily. Thus, consumption of 145 g of wheat baguettes provides 4.4 g of salt per day, covering 87% of the maximum daily salt requirement given by the WHO. However, other types of bread consumption provide 1.6–2.2 g of salt, covering 32–44% of the demand [2]. According to the studies made on several European countries it was concluded that in Ireland bread accounts for 25.9% of total salt intake, in Turkey 25.5%, in Belgium 24.8%, in France 24.2%, in Spain 19.1% and in the UK 19% [12]. Compared to other European countries, in Romania, bread represents 30% of the total salt intake, which indicates a higher value compared to other European countries [27]. According to consumer taste tests, it was reported that the optimal salt content for white wheat flour bread was between 1.29% and 1.43%. In contrast, a similar study in Argentina reported a value of 1.74%, which was much higher than expected and could be due to some geographical preferences. There are major differences in the salt content of different types of bread. In France, for example, bread contains an average of 1.7% NaCl. This amount is found in bread, country style, French baguette whereas in croissant, puff pastry and sweet bakery products the amount of salt is lower of an average of 1.3 and 1% respectively [1]. In the UK bread has only 1.0% salt [12] whether it is white or brown whereas in the fruit buns, plain cake and fruited cake the amount of salt is lower between 0.32–0.72%. In Ireland, the amount of salt in bread is 1.10% in white bread and 1.09% in brown bread and in Germany the amount of salt in all types of bread and rolls is between 1.0–2.9% [1]. In Italy, the salt content of bread varies from 0.7 to 2.3% for artisanal bread and from 1.1 to 2.2% for bread produced to the industrial level [28] and in Spain all types of bread have an amount of 1.63% salt [1]. Thus, the bread contribution to salt consumption highly differs depending on the type and location. In Romania, the salt content of bread varies depending on its type from 0.17% for non-salt bread and 1.79% for potato bread. An important factor in the salt variation of bread is its type. Thus, for example, baguette bread has a lower salt content and potato bread a higher salt content.

The most consumed bread in Romania is the one obtained from refined wheat flour, with a low selling price. This type of bread has an average salt content of 1.25% [27].

3. The Technological Effect of Sodium Chloride in Bread Making

Bread is one of the oldest foods consumed in the world and is obtained by baking a dough, prepared from wheat or rye flour, possibly mixed with other legumes or potatoes flours [29]. In bakery products, the salt is introduced into the dough phase in the form of saturated or concentrated solutions, but also in an undissolved state. This has a major impact on the dough rheological properties but also on the finished product quality [24].

The dough is a heterogeneous system which contains carbohydrates, proteins, lipids, mineral salts, water, and air in different proportions, with different degrees of homogenization, being from a rheological point of view a viscoelastic mass and from a technological point a view a homogeneous semi-finished product. It is obtained by mixing flour with water, salt, with or without baking yeast and other auxiliary materials [29]. The dough rheological properties present an important role in the bread making process, being closely correlated with the quality characteristics of the finished product [24,30].

The most important stages in the bread making process are dough making, its processing and dough baking to obtain the finished bakery product [31]. The NaCl effect in bread making process is shown in Figure 2.

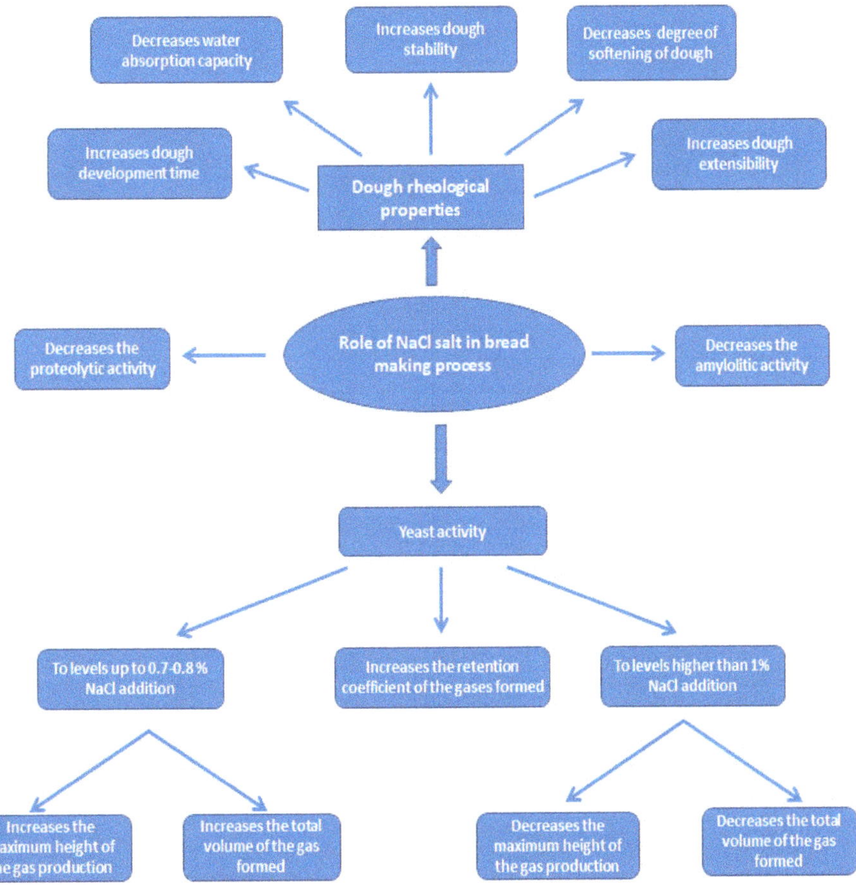

Figure 2. The role of sodium chloride in bread making process.

As it may be seen, salt, due to its ionic nature has significant influences in the dough making such as dough development, water absorption capacity, mixing time, its intensity, etc. [32–35].

Many parameters such as temperature, mixing time and the ratio between the amount of water and flour are essential in order to obtain optimal dough rheological properties [36]. The addition of salt decreases the flour water absorption capacity and increases the dough development time and its stability. The increase in dough development time shows that salt delays the formation of gluten during mixing, therefore extending the mixing time [37]. Consequently, the more salt is added, the higher the dough development time and the lower the dough softening is. Studies have shown that the dough consistency is lower as the salt addition is higher if the amount of water added to the dough making remains constant. Many researchers concluded that the effect of salt in the dough system is primarily related to the change in the gluten proteins hydration which changes the ratio between free water and bound water, in the sense of increasing the amount of free water. The addition of 2% salt decreases the hydration capacity of gluten by 8% without changing the hydration capacity of starch [38]. This was attributed to the conformational changes from gluten proteins which occur in the presence of salt. The conformational changes may occur due to the salt ions interaction with the electrically charged groups from the protein molecules. As a result, the intermolecular and intramolecular electrostatic repulsion forces between protein molecules are reduced, these becoming more compact. It was suggested that salt, for a wheat dough pH value of 5.6–6, decreases the intermolecular repulsion to a higher extent than the intra-molecular one. The results of these conformational changes of the protein molecules may lead to a decrease of the proteins ability to bind water, as they become more compact and less penetrable to water molecules. Moreover, the fewer hydrophilic groups are more available to interact with water. Additionally, some hydrophobic groups may be greater exposed, leading to more hydrophobic intermolecular interactions between protein molecules which become more compact for water and more resistant to the action of enzymes.

The increase in the amount of free water can also be attributed to the increase in inter-micellar osmotic pressure (external to the protein micelle) following the salt dissolution in the free water from the dough system. As a result, there is a difference between internal and external osmotic pressure and in order to balance the osmotic pressure some of the initial water bounded to the gluten proteins diffuses outside, making them more compact and more resistant [39].

Salt increases the dough extensibility and its strength. This increase depends on the type of cation it includes [38]. Reducing the salt content from the bread recipe mainly affects dough elasticity, without impacting viscosity. Sodium and chloride ions compete for water from the dough system affecting the hydration of the wheat flour proteins. They cannot hold water for a longer period of time, which causes an increase in the amount of free water, changing dough rheological properties [40]. During fermentation, the salt may have an inhibitory effect depending on the yeast strain and its concentration in the dough system. Thus, for concentrations below 1.5% in relation to wheat flour, the inhibitory effect is less, but it will increase with higher salt concentration, due to the rise of osmotic pressure from the dough system. Studies have shown that salt mostly inhibits the maltose fermentation and, in a lower extent, the glucose, fructose and sucrose fermentation.

Some studies reported that the surge of salt addition in dough system led to a significant reduction in the maximum height of the dough, which causes an increase in the total volume of the gas released. Through salt addition the gluten network becomes stronger. Therefore, the gas capacity to retain gases grows as the amount of salt used in the dough recipe increases [41].

On the finished bakery products, the addition of sodium chloride has an important role on their technological and sensory properties as it may be seen from Figure 3. From a technological point of view, it was concluded that the bread with 0.3% and 0.6% sodium chloride addition in wheat flour does not present significant differences compared to the bread recipe with 1.2% sodium chloride regarding its loaf volume, moisture and losses during baking. However, the lack of sodium chloride causes significant changes in the structure of the bread crumb and the shelf life after five days of storage. Regarding the flavor, crust formation and shelf life of low-sodium bread, important changes have been reported [35,42,43].

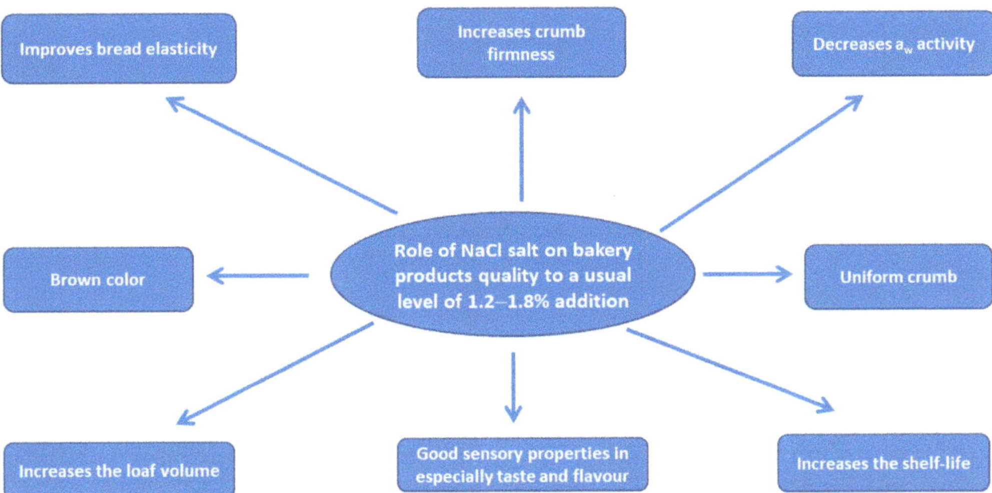

Figure 3. The role of sodium chloride on bakery products quality.

The reduction of sodium chloride content in bread influences the textural properties (due to the change in the dough rheological properties), its color and flavor (due to the more intense fermentation activity leading to a reduction in the amount of free reducing sugars resulted from the Maillard reaction) [24].

In bakery products, salt, in addition to its role on the flavor of the finished products, also acts as a preservative. The shelf life of the bread is closely related to its moisture content because it has been found that the migration of water from the crumb to the crust is related to the amount of salt present in the bread [35]. The shelf-life process of bread is due to the starch retrogradation [44], a process which means the tendency of its macromolecular components, amylose and amylopectin, to aggregate, to associate, resulting into a more insoluble form. Degradation is accompanied by the reappearance of crystalline areas into the starch structure partially destroyed during baking, which causes a more rigid crumb. Proteins also contribute to this structure, losing their elasticity and hydration. The water content and its condition play an important role in this process. Thus, products with a moisture content below 16% do not stale. The optimum humidity for staling is between 16–37%. In staling bread, bound water reaches 70% of the total water. The free water from the gel passes into crystalline structures, being strongly bound. It disappears from the bonding layer between the gluten and the partially gelatinized starch granules, causing stronger gluten-starch bonds, which reduces the elasticity of the structure [45]. Salt prevents water migration from crumb to crust and keeps the product fresh for a longer period of time compared to bread without salt addition [46]. The main responsible for foodstuffs' deterioration with high humidity content are microorganisms. Bread is known as a product with high humidity, with a_w values between 0.96–0.98 [47]. In general, the addition of salt increases the gelatinization temperature of starch and delays its gelatinization process [39]. Through its action on starch and gluten, but also on the activity of water from the dough system, salt has a major contribution on starch staling [48]. For example, reducing NaCl from 1.2% to 0.3% decreases the shelf life of the bakery products with almost two days [24].

On the bread color, the reduction of the sodium chloride content in bread recipe or its lack thereof leads to products with a light-colored crust [31]. The brown color of the crust resulting from the bread baking process is due to the Maillard reaction, as an interaction between reducing sugars and amino acids. Therefore, a certain amount of free reducing sugars and amino acids is necessary for this reaction to take place. The lack of salt addition leads to a more intense fermentation which induces a high consumption of carbohydrates during the initial stages of the bread making process. Therefore, during baking the dough which presents less reduced sugars to form more melanoidins, will yield a bread sample with a light color. Many factors such as pH, protein/amino acid ratio, water, sugar amount from the dough system, temperature and baking time have a major influence on crust formation. The intake of sodium chloride produces a plasticizing effect during the baking process of bakery products, this favoring the Maillard reaction and therefore producing a darker crust [49]. Salt has the ability to control the yeast fermentation from the dough system and therefore, reducing the amount of salt will increase the activity of yeast. Fermentable carbohydrates are needed to the yeast fermentation process [50] and therefore, yeast fermentation will reduce the availability of free sugars involved in the Maillard reaction [24].

Sodium chloride is also responsible for the flavor of bakery products, which also intensifies the sweet taste of the finished product. Sodium salts such as sodium chloride or monosodium glutamate influences the food flavor in a positive way, improving it. However, Girgis et al. [51] concluded that consumers did not observe a gradual reduction of 5% sodium per week over six weeks (corresponding to a final reduction of 25% salt). A 20% reduction of sodium in bread did not affect the taste perception of 60 participants in this study. Instead, Lynch et al. [35] showed that a 50% reduction in salt changed the bread flavor. Salt-free bread has been described by them as sour and having the taste of yeast [12]. Research has shown that sodium, and to a lesser extent lithium, are the only salted cations with good effects on sensory food properties while calcium and potassium ions have a bitter or metallic taste, which is undesirable for consumers [52].

This effect of sodium chloride on sensory perceptions are due to the fact that free Na^+ ions in solution are the mainly responsible for the perception of salty taste on the tongue while the contribution of Cl^- ion in terms of taste is not yet fully clarified. Due to the interaction between sodium chloride and wheat proteins during dough formation, sodium ions interconnect with dough components due to their ionic interactions with negatively charged amino acids [1]. Pflaum et al. [53] reported that Na^+ and Cl^- ions are not irreversibly bounded in bread, sodium ions being released by chewing. The speed with which sodium is released in the first seconds when chewing takes place is quite important for the perception of salty taste by the consumers. The rate of release of the salty taste during chewing depends on the structure of the bread crumb and the amount of NaCl in the product. Bread with low amounts of NaCl has been described as having a more acidic or yeasty taste. Moreover, significant changes in crumb structure have been reported when different amounts of NaCl were added [12].

It has been shown that the use of sodium as compound in different ingredients significantly reduces the perception of the bitter taste [54]. It seems that the positive effects of sodium chloride on the aroma are due to its effect on the water activity. Reducing the amount of free water by adding salt changes the ratio between free water and the bounded water, which influences the volatility of flavor compounds [55]. Moreover, an essential contribution in the formation of flavor compounds is due to the Maillard reaction [56]. As mentioned above, melanoidins are also responsible for the golden-brown color of the bread crust during baking and therefore unsalted or low-salt bread will have a light crust, a weaker aroma, and a bland taste [24]. Obtaining bread with a low salt content (up to 0.3%) is technically easy to be done via a few changes in the bread making process, with a finished product maintaining the same qualities, except for taste. An interesting technique in this respect, although not one that can be applied to all production systems, involves the inhomogeneous distribution of salt in the bread, which allows a 28% reduction of the salt content, while maintaining the intensity of the salty taste [12].

The textural properties of bread are very important in terms of their acceptance and choice by consumers. They can be evaluated both sensorially and instrumentally. Salt, as mentioned before, has a major impact on the gluten network development during the dough mixing process. The addition of sodium chloride leads to a strengthening effect on the gluten network, stabilizes the fermentation rate of the yeast, increases the dough mixing time, and improves the aroma and porosity of the bread crumb [57].

During the dough making and baking, the appearance of gas bubbles leads to the dough expansion and finally, the loaf volume and texture of the finished product. Gas bubbles have a limited expansion correlated with their stability and possible gases released from the dough system. Therefore, it is very important that the walls of the gas bubbles to be stable during baking but also during other technological stages. Moreover, the dough must present a good gas retention capacity for porosity, elasticity and a specific loaf volume in order to be accepted by consumers [1].

As the salt content of the dough decreases, the activity of the yeast increases leading to a finished product with a higher loaf volume. At low salt amounts levels of 0.7–0.8% the yeast cells multiplication increases, but above this concentration the yeast multiplication process is slowed down due to the plasmolysis process from the yeast cells. Moreover, the addition of salt in the dough reduces the activity of enzymes, both proteolytic and amylolytic, due to the action of salt on the protein part of enzymes [45]. However, less salt can induce a weakening effect on the protein network and to a lower gas retention capacity which will lead to bread with irregular porosity and low loaf volume [31].

From a textural point of view, a comparative study between a bread without salt and a bread with a salt content of 4% reported that there was a high difference between bread samples of about 40% in the firmness value of the bread crumb after a 24 h storage period. This shows that the firmness of the bread crumb decreases with the reduction of the amount of salt from bread and is closely related to the increase of the loaf volume of the finished product [58]. The increase in firmness with the presence of salt in bread can be explained by the conformational changes of the gluten proteins in the salt presence [45].

4. Methods for Sodium Content Reduction in Bakery Products

With the growing interest of consumers in a healthy diet, their concern for low-salt products has also increased. This is also due to the media regularly sending messages about the diseases caused by excessive consumption of sodium chloride and its impact on the human body. An example of this is the "Salt Kills" campaign, in which the UK government regularly reports on the harmful effects of excessive salt consumption [59].

According to Powles et al. [60], the average global level of sodium consumption in 2010 was approximately 3.95 g/day, with a specific regional intake of 2.18 to 5.51 g/day. This value is almost twice the limit of 2 g sodium/day (equivalent to 5 g/day of salt intake) recommended by the World Health Organization (WHO) [61].

Salt consumption among adults in most European countries varies between 7 and 13 g per day, according to the European Commission. Germany, Cyprus, Bulgaria and Latvia reported the lowest salt intake (6.3–7.3 g/day), while the Czech Republic, Slovenia, Hungary and Portugal reported the highest salt intake (12.3–13.6 g/day). It seems that are different levels of salt intake in Europe, the lowest values being noticed in Denmark, the Netherlands and Belgium (8.3–8.8 g/day) and the highest in Hungary, Slovenia, Slovakia, Portugal and Italy (10.7–11.2 g/day) [5].

It was concluded that from all the risk factors related to our diet, a diet high in sodium accounted for the highest number of deaths from all the analyzed causes. Most people consume more salt than the World Health Organization's maximum daily salt recommendation of <5 g per day, with an average overall salt intake of about 10 g/day. The daily intake of salt in most countries varies between 9–12 g, which is far above the recommendations of the World Health Organization. It is estimated that the overall reduction in sodium intake to the recommendation levels would prevent approximately 2.5 million deaths annually, which has led WHO Member States to agree to reduce the sodium intake of the population by 30% until 2025 [62]. A decrease in salt intake by 3 g/day would reduce systolic blood pressure by 5 mm Hg at the age of 60, hence a decrease in stroke and myocardial infarction. Many studies in Germany, Canada and France have shown a correlation between left cardiac ventricular mass and salt intake (assessed by urinary sodium excretion). These negative effects of salt appear to be independent of blood pressure. In addition, the role of sodium in osteoporosis has been demonstrated since the 1970s [11]. Since 2013, World Health Organization member states have committed to reduce the population's salt intake by 30% until 2025 [63]. Many countries have initiated salt reduction interventions, and a number of reviews have shown some progress in reducing salt intake among the population. To further support countries in implementing salt reduction interventions and achieving a reduction in salt intake among the population, the World Health Organization published the SHAKE technical package for salt reduction in 2016, consisting of implementation strategies in five key areas, including surveillance, product reformulation through lower salt intake, labeling, consumer education, etc. Continuous monitoring and evaluation of these efforts is essential to strengthen policies and actions to reduce salt consumption.

The main sources of salt are bread, cheese, meat and meat products, snacks, sauces, soups and pastries. Redeveloping these types of foods to reduce salt content (or improving/reformulating foods) is considered a promising strategy for reducing the dietary intake of salt. In order to reduce the salt intake of a certain population, large-scale structural efforts are needed to reduce the salt content of foodstuffs at the time of production, as well as to initiate behavioral changes. The World Health Organization encourages a multisectoral approach, including public-private partnerships to improve the composition of the food supply [12].

However, traditional culinary habits and consumers' preference for salty taste creates difficulty to the initiative to reduce the consumption of salt in food products. That is why research is underway for different solutions to reduce the content of sodium chloride in food products to be accepted by a large mass of consumers [64]. The gradual reduction in sodium levels by various food industry producers has been somewhat successful, which has been reported by the Food and Drink Federation and the British Retail Consortium. The strategy entitled "reduction in small steps" aims to slowly and gradually to reduce the salt content in manufacturing recipes without notifying the consumers [12].

Bread is one of the main sources of salt in the world and that is why some countries are trying to reduce the sodium content of the general consumption of the population by reducing it in bread. For example, in New Zealand, there has been a 7% reduction in salt content in bread products for four years. In Ireland, through a salt reduction program, the bakery industry set 0.45 g of salt per 100 g of bread as the average value for white and dark bread. In 2007, as part of the second national nutrition and health program, a third of bakeries in France reported a reduction in the amount of salt addition in bread. Similarly, in 2005, the Spanish Confederation of Bakeries agreed to reduce the salt content of bread from 22 to 18 g to 100 g, reducing, on average, 1 g/kg of flour per year. Thus, evidence from various countries shows that salt reduction programs in bakery products are efficient, as part of a cost-effective strategy for improving public health [65].

Studies have shown that a lot of effort has been made to reduce sodium in bread. The salt content of bread in the UK decreased by about 20% from 2001 to 2011 (from 1.23 to 0.98 g/100 g). By 2011, 71% of all bakeries met FSA targets of ≤ 1 g/100 g, while in 2001 only 25% of all bakeries met this limit [5]. In terms of feasibility and marketing, it seems that salt reduction can be achieved in most bakery products [12]. Different strategies of salt reduction in bakery products are shown it Table 2.

Table 2. Strategies of salt reduction in bakery products.

Strategies	Effects	References
Sodium chloride replacement with different types of salts	Partial replacement of NaCl with KCl up to 20–30% in bakery products without negative effects on its metallic or bitter taste. This is one of the best solutions reported so far to reduce the sodium chloride from bread making	[24,66,67]
	Partial replacement of NaCl with a mix of KCl, $MgSO_4$, $MgCl_2$ up to 32% in dark bread conducted to finished products with a similar flavor and texture with that of the control sample	[68]
	KCl addition in bread making presented similar effects on dough rheological properties compared to NaCl salt	[69–73]
	NaCl replacement with potassium citrate up to 30% conducted to good results from bread quality point of view	[74]
	A mix of KCl and potassium bicarbonate of 1:1 conducted to good results for bread quality and also to an improvement of potassium availability for the human body	[74]
	NaCl replacement with $CaCl_2$ up to 25% conducted to good results from the rhelogical point of view with a slightly decrease to water absorption capacity value	[69]
	NaCl replacement with $CaCl_2$ changed in a differently way dough rheological properties by decreasing dough development time and by increasing water absorption capacity	[75]
	The use of sea salt (a natural mix of NaCl, KCl, $MgCl_2$) with a low sodium content conducted to good results from the rheological, technological and sensorial point of view and may lead to a sodium reduction. For the best bread quality results, it is necessary a low amount of sea salt addition of only 0.5%.	[75,76]
	The use of calcium salts as gluconate and lactate as replacements for sodium chloride conducted to similar data for the rheological point of view compared with those obtained when NaCl was incorporated in bread making. Their addition conducted to a strengthening effect (more in the case of gluconate than lactate) an increase of dough stability, a decrease of in degree of softening, a positive effect on yeast activity leading to an increase of the gas retention coefficient. However, contrary to the NaCl effect, the calcium salts improved the alpha amylase activity of the dough system.	[77]
	The use of magnesium salts as gluconate and lactate as replacers for sodium chloride conducted to dough mixing properties to similar behavior for gluconate salt and a different one for lactate salt (a more weakening effect on dough) compared with those obtained when NaCl was incorporated in bread making. From the yeast activity point of view, it seems that both magnesium salts had a similar effect to that obtained by the sodium chloride addition in wheat flour.	[78,79]
Gradual reduction of sodium chloride in bakery products	The sodium chloride reduction up to 25% in six weeks or even up to 52% in four weeks did not affect the consumer's acceptability or bakery products consumption. However, if the reduction of salt in food is done too quickly without necessary adjustments in food taste, the opposite effect can be reached, namely it can lead consumers to other foods rich in salt or to add it during the preparation of meals or to supplement the lack of salt in that food	[51,80–82]
Inhomogeneous distribution of salt in the bread recipe	May lead to a reduction up to 20–30% NaCl content from the bakery products with good results from the sensory point of view but may presents negative effects on yeast activity	[83–86]

Table 2. *Cont.*

Strategies	Effects	References
Use of encapsulated salt in the bread recipe	The sodium chloride level may be half reduced in the bakery products with similar effects on salty taste on consumers, but this high reduction may present negative effects from the technological point of view on dough development time, on yeast fermentation rate, water activity, bread loaf volume and textural properties.	[87]
The use of flavor enhancers in the bread recipe	Intensifies the salty taste perception by using them in combination with NaCl (glutamates, yeast extracts, hydrolyzed vegetable proteins nucleotides, amino acids, and fermented sugars). The use of fermented sugars in addition with sea salt and dry sourdough it may reduce the sodium chloride content up to 0.02%.	[83,88,89]
The use of dry sourdough in bread recipe	It improves the salty taste perception, flavor, shelf life and textural properties of the bakery products by using it in combination with NaCl. It may reduce sodium chloride content in bread from 1.5% up 1% with good results from the bakery products quality point of view and in addition with sea salt up to 1.39% for the optimum rheological properties reducing therefore the sodium content from bread by 22%.	[24,90]
The use of different spices in the bread recipe	The use of different spices in combination with NaCl presents the disadvantages that may alter the taste and aroma of bakery products and may not be appreciated by everyone. Moreover, a high level of specific spices used might lead to toxicity. In general, this solution is the easiest to be applied in restaurants or when preparing meals at home	[10,88,91–94]
The use of B$_4$ vitamin in the bread recipe	It improves the salty taste perception of bakery products and may replace the NaCl up to a level of 25% without affecting the sensory characteristics of the bakery products.	[95,96]

The difficulties of sodium replacement in bakery products are due in especially to the salty taste produced by it. This taste in produced by the specificity of the Na+ ions on the epithelial channel known as ENaC (epithelial sodium channel). The primary process by which salty taste is detected is due to epithelial sodium channel receptors that respond almost exclusively to sodium ions (Na$^+$). Therefore, the central gustatory system and mesolimbic structures are needed to process the taste signal and hedonic responses to food. A consequence of stimulating the reward pathways of the brain is palatability, which suggests that an individual's preference may be associated with the hedonic neuronal properties of salty foods [97]. Further, are presented in a detailed way the best solutions reported so far to reduce sodium chloride from bread making.

4.1. The Use of Different Types of Mineral Salts as Substitutes for Sodium Chloride in Bakery Products

Today, all kinds of combinations of different ingredients are proposed as salt substitutes that do not give an unpleasant taste to the finished product. A natural combination of different salts is the Dead Sea salt (contains NaCl, KCl, MgCl$_2$) which can be successfully used as a substitute for sodium chloride [75]. Natural salt extracted from the Dead Sea has a much lower sodium content compared to normal salt used in food (maximum 7% sodium in the form of sodium chloride) and even more it contains a number of minerals useful to the human body such as magnesium and potassium [98]. Research on the use of low-sodium salt from the Dead Sea in bakery products have been less conducted. This substitution is considered interesting, has potential and may be useful especially when it is used with other strategies, but more studies are needed, given that some changes in bread quality parameters have been reported [75,76]. The most used strategy to reduce the sodium chloride content is its partial replacement with potassium chloride in a proportion level of 20–30% and not higher because it would give the products a bitter and metallic taste. Thus, a combination of NaCl and KCl may be a viable alternative for reducing the sodium content in the food industry [74]. Other studies have shown that a series of mixtures of various salts represented by chlorides of K^+, Mg^{2+}, NH_4^+ and carbonates can provide a taste similar to that of sodium chloride. Charlton et al. [68] also replaced 32% NaCl with a mixture of potassium chloride, magnesium sulfate and magnesium chloride in dark bread, the finished product obtained having a texture and flavor similar to a bread with only NaCl addition. [12].

However, currently the most widely used method of replacing sodium chloride is with other types of salt. This is due to the fact that sodium chloride also influences bakery products quality from the technological point of view. The lack of salt influenced in a negatively way the dough rheological properties, especially on the final leavening phase, when the dough gas retention capacity decreases. As a result, the crumb elasticity is reduced, and the porosity of salt-free products is not uniform and insufficiently developed. The bread is obtained with a pale and light crust. Technologically, a reduction of the salt up to 0.6–0.3% would be possible without a significant deterioration of the rheological properties or the performance of the yeasts during the bread making process. However, its effect on the sensory properties of bread is still a critical factor in consumer acceptance [12].

Replacing sodium chloride with other types of salt can technologically improve the quality of bakery products. Several types of salt have been used successfully as substitutes for sodium chloride. The basic principle is to replace sodium cations with others or replace chloride anions with anions such as glutamate and phosphate, as a way to give a salty taste. Currently, the easiest option for bread making is to replace NaCl with mineral salts or with other cations such as calcium, magnesium and potassium. The effect of different cations on the rheological properties of the dough is related to their position in the lyotropic series, also known as the Hofmeister series, which classifies ions based on their ability to cause aggregation or dissociation of proteins. Within the series, both anions and cations are classified in the order of the most stabilizing to destabilizing. Stabilizing ions lead to less hydration, more structure and a decrease in protein solubility, while destabilizing ions lead to more hydration and increased protein solubility, thus affecting both hydrophobic interactions and hydrogen binding.

The ranking, starting from the least stabilizing cations is: $NH4^+ > Cs^+ > Rb^+ > K^+ = Na^+ > H^+ > Ca_2^+ > Mg_2^+ > Al_3^+$. Therefore, with the use of a stabilizing cation, it would be expected that the protein-protein interaction to increase and to promote the formation of a stronger gluten network and therefore of a non-sticky dough. Studies have shown that K^+ is the best option for maintaining the rheological properties similar to dough with sodium chloride, because K^+ is equivalent to Na^+ in the lyotropic series. However, this replacement leads to significant challenge of a metallic/bitter taste [99].

4.1.1. The Use of Calcium Chloride as a Substitute for Sodium Chloride in Bakery Products and the Benefits of Its Use on the Human Body

Calcium chloride is used in the bakery industry as an anti-caking agent but also as a substitute for sodium chloride. It is a solid inorganic compound at room temperature, soluble in water and its anhydrous salt is hygroscopic. Moreover, calcium chloride can also be a source of calcium for the human body. Given that billions of people suffer from osteopenia and osteoporosis, the addition of calcium chloride as a substitute for sodium chloride in bakery products can be considered appropriate. In the last decade, interest in the effects of calcium on the human body has increased and studies have expanded to include the entire life cycle. A whole range of foods and supplements in which calcium is added are widely used today. Calcium is distributed throughout the body in small amounts and is involved in the processes of vascular contraction, vasodilation, transmission of nerve signals, transmission of intracellular signals and hormonal secretion.

There are three major categories of people at risk for dietary calcium deficiency. These include women (amenorrhea, post-menopause), people with milk allergy or lactose intolerance, and risk groups for poor food intake (adolescents and the elderly) [100].

Calcium is an important element for the health of the human body and has vital functions inside cells by transmitting signals between the plasma membrane and intracellular mechanisms. Extracellular calcium is an essential cofactor in the proper formation of bones. Decreased calcium content in the bones can cause conditions such as osteoporosis, high blood pressure, arteriosclerosis, neurodegenerative diseases, malignancy, degenerative joint diseases. Increased calcium is recommended in colorectal cancer prevention treatments to lower blood pressure and blood cholesterol levels [101]. The uses of calcium chloride in the food industry are multiple as: anti-caking agent, antimicrobial agent, hardening agent, flavor enhancer, moisturizer, nutritional supplement, pH control agent, stabilizer and thickening agent, to improve textural properties, and so on.

Various studies in the bakery products on the total substitution of NaCl with calcium chloride ($CaCl_2$) have shown that this has led to a significant increase in water absorption capacity and a higher degree of dough softening due to the fact that a certain amount of water remains unabsorbed from the dough system. However, it was found that at lower levels (~25%) of NaCl substitution, the water absorption capacity decreased slightly, which a positive fact is because more protein-protein interactions occurred, leading to a stronger gluten network and stronger dough through cohesive forces [69,102].

4.1.2. The Use of Magnesium Chloride as a Substitute for Sodium Chloride in Bakery Products and the Benefits of Its Use on the Human Body

Magnesium chloride ($MgCl_2$) is one of the most common water-soluble natural magnesium compounds. It is found naturally in the salt waters of lakes or in sea water. Magnesium chloride is obtained from mineral salt deposits after the extraction of potassium chloride but also by direct extraction. Among the main sources of raw materials for the production of magnesium chloride, the most important is sea water. Magnesium can be directly precipitated from sea water in the form of magnesium hydroxide and converted to magnesium chloride [103]. Among the uses of magnesium chloride in the food industries are: coagulant; tofu production from soy milk; formula milk for babies, etc. The advantage of using magnesium chloride in food industry is that it can be a source of magnesium for the human body where it is one of the main minerals present quantitatively after potassium, calcium, phosphorus and sodium participating in carbohydrate, lipid metabolism, growth and cellular permeability. It is a catalytic element and also a plastic one and is a growth factor that helps regulate the balance of calcium in the human body having an anti-aging, anti-anaphylaxis and anti-atherosclerotic role [79]. To replace the nutrient losses in flour that occur during wheat processing and to reduce the risk of a deficiency in the body, the bakery industry has enriched white bread with various nutrients such as iron, thiamine, riboflavin and niacin [104]. Magnesium is considered to be one of the deficient elements in our diet although it is involved in the enzymatic reactions of carbohydrates, proteins, energy metabolism and maintaining the structural and functional integrity of human body tissues [105]. Along the benefits of magnesium consumptions are: fighting constipation, hypomagnesaemia, preventing convulsions in eclampsia/preeclampsia, preventing acute nephritis (pediatric and adolescent patients), cardiac arrhythmias secondary to hypomagnesaemia, etc. [106].

Salovaara [70] found that the addition of magnesium chloride in the dough reduces it development time and increases the water absorption capacity. Compared to the dough samples in which sodium chloride and potassium chloride were added in the bread recipe, it was found that there are significant differences from the rheological point of view only for the samples with magnesium chloride in dough recipe. According to this study, it was recommended the use of KCl as a substitute for NaCl even at high doses of 25–50% because this salt maintains the dough rheological properties, leading at the same time to a significant reduction of sodium in the finished products. This similar behavior between NaCl and KCl on the dough rheological properties is due to the fact that K^+ was classified equivalent to Na^+ in the lyotropic series which means that it would have similar abilities to cause protein aggregation and fortify the gluten network. It is considered a stabilizing cation that causes less protein hydration and stronger development of protein structure [99].

4.1.3. The Use of Potassium Chloride as a Substitute for Sodium Chloride in Bakery Products and the Benefits of Its Use on the Human Body

Potassium chloride (KCl) is a natural mineral salt obtained from salts from rocks and seawater, its extraction being similar to that of sodium chloride. Taking into account the dietary intake of potassium, it is clear that it has opposite effects to sodium consumption, namely a low risk of hypertension. While sodium intake is significantly increasing, the average overall potassium intake is below the WHO recommendations of at least 3510 mg of potassium per day.

Potassium is the most abundant cation in intracellular fluids and is an essential nutrient in maintaining cell functions, especially in excitable cells such as muscles and nerves. Being a major intracellular ion, it is manly found in foodstuffs that are obtained from living tissues. A higher amount of potassium was found in fruits and vegetables and less into meat products and cereals. In Western diets, food practices are not based on the fruits and vegetables consumption but rather on cereals and processed foods with a low content of nutrients which led to diets low in potassium and higher in sodium [107]. The best sources of potassium are fruits, vegetables, meat, fish, dairy and nuts. In starchy foods, potassium is found in higher amounts in whole meal flour and brown rice compared to rice and white flour. Milk, coffee, tea and other non-alcoholic beverages are among the main sources of potassium in the American adult's diets. According to the National Institutes of Health (NIH), the US estimates that the body absorbs about 85–90% of dietary potassium and different forms of potassium from fruits and vegetables including potassium phosphate, sulfate, citrate and others, but not potassium chloride (used in salt substitutes and some dietary supplements). Globally, mixtures of salts with potassium chloride are widely used for the partial or total substitution of sodium chloride. The use of potassium chloride can thus reduce the intake of sodium in food in a short period of time and to increase the potassium intake. There is an antagonism between the metabolism of potassium and sodium. Increasing the concentration of potassium in the human body leads to a decrease of the sodium concentration and to an increase in its elimination. At the same time, fluids are eliminated from the body. Diets high in potassium can help us to eliminate sodium from the body. Of all the types of salts used, potassium chloride is one of the most widely used as a substitute for sodium chloride because it has the best ability to transmit the salty perception of taste in food.

Therefore, the potassium chloride may be an interesting substitute for sodium chloride from the perspective of consumers, processors from the food industry but also from the point of view of consumer health. However, potassium chloride cannot be used in unlimited quantities because at high levels it loses its ability to give a salty taste in food and can often lead to a bitter, chemical and metallic taste.

Depending on the category of foodstuff in which it was introduced, potassium chloride has been used in different percentages to replace sodium chloride without affecting in a negative way the sensory characteristics of foods. For example, in aqueous solutions the taste of potassium chloride is perceived at a concentration of 20%, in food products such as pizza type good sensory properties was achieved up to a sodium chloride replacement with potassium chloride of a 25% level, in white and dark bread with 30%, in cheddar cheese with 46% and in feta cheese even up to 50% [66]. Studies have shown that this ingredient can have approximately the same technological functions in the dough making process and bread quality as sodium chloride leading to an improvement in the texture and shelf life of bread [67]. Doyle [52] suggested that the influence of KCl on dough rheological properties is similar to those obtained by using NaCl. Gengjun et al. [71] concluded that KCl could adjust the growth rate of the yeast cells, allowing the incorporation of more gas bubbles in the gluten network, which may have a positive effect on the rheological properties of wheat flour dough.

Consumption of potassium chloride and its use as a substitute for sodium chloride is safe for consumers health and is supported by the presence of potassium in a natural way in different foods. Therefore, the addition of potassium chloride to food has gained regulatory acceptance in the US and the European Union. Experts recommend increasing potassium consumption by its addition in foods for the population because it has a low risk in terms of its adverse effects on consumers.

There is currently no upper limit on potassium intake at the global level, but based on estimates of current consumption in Europe, the European Food Safety Authority (EFSA) states that the risk of adverse effects on potassium intake from food sources per 5000–6000 mg/day is considered low for the clinically healthy population. Moreover, the long-term intake of potassium from supplements at a level of 3000 mg/day in addition to the consumption of foods containing potassium is also considered safe for the clinically healthy adult population [108].

Potassium chloride is one of the most common substitutes for sodium chloride in bread due to its ability to lead to a salty taste perception. It can replace sodium chloride up to a level of 30–40% without changing the characteristics of the finished product [5]. Although KCl is a possible option for reducing sodium in bakery products, a significant disadvantage of it is, as we mentioned before, the metallic taste conferred by this compound at high levels. That's way various studies have assessed the threshold of sensory acceptability. For example, Wyatt and Ronan [109] did not find significant differences between a control sample (with 100% NaCl) and other bread samples with 50%/50% (NaCl/KCl) bread, the highest scores in terms of acceptability having the NaCl ratio/KCl, 75%/25%. In contrast, Salovaara (1982) [70] found significant differences with a mixture of 60/40, but not with a mixture of 80/20. Replacement is not as critical factor for dark bread, and the use of a salt mixture in which the sodium content of the bread obtained is reduced by 32.3% and the K content is increased by 34.8% showed good results in terms of quality and taste of the bread obtained. Finally, in a systematic study involving the replacement of NaCl with K^+ salts, Braschi et al. [74] concluded that the best results, other than those for the control sample, were obtained with a 70/30 ratio of K-citrate or a 1:1 mixture of KCl and potassium bicarbonate; they also concluded the complete bioavailability of the incorporated potassium using these salts. Other possibilities that have developed include mixtures of commercial salt and salt from sources of low sodium and high potassium, calcium, or magnesium. The use of KCl in combination with Na glutamate or ribonucleotides may mask the bitter aftertaste. This is another interesting alternative, although this taste tends to be used only in the substitution of salt in meat products [110]. Potassium chloride (KCl) is usually the main choice and can be used to replace from 10% to 20% without major technological and sensory problems. Total KCl replacement is not recommended due to the unpleasant bitter and metallic taste, which limits consumer acceptability. For this reason, for a higher reduction of sodium in bakery products it is recommended to combine it with other food ingredients [12].

Therefore, the advantage of using KCl in the bread making recipe is a technological and healthy one due to the fact that it increases the potassium intake from the diet which is associated with a very low risk of hypertension, an opposite effect to sodium consumption [66]. A recent study estimated that potassium intake for the United States, Mexico, France, and the United Kingdom was 80%, 95%, 77%, and respectively 95%, under the recommendations of the World Health Organization on potassium intake [111].

4.2. The Use of Dry Sourdough as a Substitute for Sodium Chloride in Bakery Products

Another possibility of partial substitution of sodium chloride in bakery products is the use of sourdough. Sourdough is defined as a fermented semi-finished product obtained from flour and water in the presence of its own, natural microbiota and then dried in conditions to keep lactic bacteria in a viable state [112]. The industrial production of dry sourdough has over 40 years old and was initially used for obtaining products with a high acidity. Further, its main uses were to obtain bakery products with specific taste and aroma [113,114]. The advantage of using sourdough in baking is that it eliminates the leaven phase by shortening the time for bread making process. It allows obtaining the bread dough in a single phase leading to bakery products of a very high quality from the technological and sensory point of view, similar to those obtained through dough making in a double or a triple phase. Nowadays, a wide range of sourdoughs are available on the market, which differs according to the flour used, as well as the specific flavor that exist on each sourdough. Sourdough in the form of an ingredient for bakery products corresponds to the new trend of clean labels, natural products, including a reduced use of additives [115]. The advantages of using dry sourdough in low-salt products to create healthy foods are evident. Belz et al. [24] suggested that the dry sourdough may compensate the effect of salt reduction on bread flavors and may lead to the good sensory characteristics of the finished products, such as crumb texture. Moreover, they also reported that the addition of sourdough in bread making, fermented with *Lactobacillus amylovorus*, for obtaining a low-salt bread, extended the shelf life compared to a control sample. Bread containing lactic acid bacteria (LAB) from fermented wheat germ had a saltier taste compared to a control bread. The salty taste was thought to be a combined effect of acidification and proteolysis. Due to the addition of a sourdough from rye malt fermented with glutamate that accumulated bacteria of the *Lactobacillus reuteri* species, it may be possible to reduce the salt content of bread from 1.5 to 1% (compared to flour), maintaining the taste and other characteristics of a standing quality. Sourdough improves the perception of salty taste and brings an additional intake of aromatic compounds. This is a useful functional ingredient for low-salt bread. Moreover, the use of dry sourdough is not limited to bread. One possibility is to incorporate sourdough into pastries or croissants to improve their flavor, texture and therefore palatability [12].

4.3. The Use of Flavor Enhancers as Substitutes for Sodium Chloride in Bakery Products

Flavor enhancers are compounds that do not have a salty taste, but they have the ability to intensify the salty taste of NaCl by activating receptors in the oral cavity. Some of them also mask the unpleasant taste of KCl. Yeast extracts are natural flavor enhancers, which the food industry commonly uses as substitutes for monosodium glutamate (MSG) and other artificial flavor enhancers. However, the effectiveness of MSG (as a salt substitute) is only partial, as it also contains sodium. Moreover, another problem is that the safety of MSG use is controversial because it has been associated with health problems (such as headaches, hyperactivity, and metabolic disorders) [13]. Glutamic acid-based combinations for salt replacement by flavor enhancement and intensification lead to good results [116]. As for yeast extract, although it is considered a natural and healthy alternative, it often contains MSG, which has been used in the food industry since the 1950s and has changed significantly in terms of quality, taste and functionality. Several low-sodium natural yeast extracts have been developed and can be used in a variety of salty foods. Yeast extracts are obtained from the water-soluble content of the cells, which contains concentrated amino acids, peptides, carbohydrates and mineral salts. In general, two types of extracts can be produced to the industrial level, using two different methods namely autolysis and hydrolyses. Autolysis of yeast extract is produced by processing yeast used in the bakery or beer industry. Cell walls are broken down by the use of heat or salt. This allows the enzymes present in the cell to break down the proteins and other cells compounds. The soluble compounds are then separated from the insoluble compounds and concentrated and pasteurized before being used in the food products [117]. The production of hydrolyzed yeast extracts implies the use of an acid, which starts the peptide bonds hydrolysis, releasing glutamic acid. Yeast extracts can be added to any salty food and are commonly used in sauces, spices and culinary products. Strong fleshy notes are also used to mask any unpleasant bitter taste resulting from the incorporation of potassium chloride into foods. This is a positive fact because the sensation of bitter taste induced by potassium chloride limits the amount in which it could be incorporated into food. This allows an increase in the amount of potassium chloride that could be added to foods containing yeast extracts in the bread making recipe [118]. It should also be mentioned that these ingredients do not have the same profile, in terms of NaCl functionality, on dough rheology, yeast fermentation rate, control of water activity and inhibition of microbial growth, which can create difficulties in bakery industry. Therefore, even it presents a high potential for flavor improvement, yeast extracts are not well received by the consumers, which limits the acceptability of the food products where they are incorporated. To balance the overall flavor and overall characteristics of bread quality, the combination of salt substitutes with flavor enhancers is recommended to be used [12].

5. Perspectives

The strategies of bakery products reformulation in order to reduce salt are nowadays of a great interest to an international level, supported by the authorities and research community to benefit consumer health. It represents a great challenge for the bakery industry, since sodium chloride is one of the raw materials for bakery products with important effects on their technological and sensory properties. Due to the important role of NaCl it is difficult to reduce or to eliminate it completely from bakery products. However, baking industry undertakes efforts to reduce the salt content from baked products.

In the salt reduction strategies will be taken into account the economic and technical criteria, but also the acceptability of the bakery products with a low salt content by consumers.

Gradual reduction of salt from bakery products may be one of the best solutions for the industry and consumers in the future without significant costs. Moreover, the use of different salts with similar effects from the technological point of view (KCl, $MgCl_2$, $CaCl_2$, calcium lactate, calcium gluconate, etc.) to those of NaCl addition may play an important role of salt reduction in baked products but their uses are limited due to unpleasant taste. Some taste enhancers (amino acids, yeast extracts, glutamates, hydrolyzed vegetable proteins nucleotides, fermented sugars, etc.), B4 vitamin may be used in the future to the industrial level in combination with other salts to improve the salty taste of baked products.

To obtain bakery products with a low salt content the NaCl may be totally replaced with sea salt (a natural mix of NaCl, KCl, $MgCl_2$) with a low sodium content but it use may be limited due to it higher cost which may be problem in poor and developing countries in which bakery products are also low cost ones.

In order to produce bakery products with low salt amount, with technological and sensory characteristics comparable to bakery assortments obtained with a normal salt content, in the future it is recommended to continue researches to combine different ingredients with similar technological effects as those of NaCl which can improve each other their salty taste, to optimize production recipes and to implement different industrial processes such as encapsulation, inhomogeneous distribution of salt in the bakery process, etc.

Therefore, further researches are needed to obtain bakery products assortments with a low salt amounts, with good technological characteristics, by complying with WHO recommendation and consumers demand.

6. Conclusions

Salt is a minor component, which influences all phases from the bread making technological process, as well as the sensory properties of the bakery products obtained. From the technological point of view, salt presents a significant effect on dough mixing, fermentation and baking. Its affects dough development time, it presents a strengthening effect on the gluten network, makes it more extensible, it inhibits yeast activity leading to a decrease of the gas formed during the fermentation process, it extends the shelf life of the bakery products and it improves the bread sensory characteristics, especially taste, flavor, and the color of the crust. Due to the fact that World Health Organization recommends a daily sodium intake of 2 g equivalent to 5 g/day of salt intake food processors are concerned to reduce the salt content from their products. Bread is the most consumed food worldwide and that way it is the mainly contributor to the sodium daily intake.

The research made so far presents different strategies to reduce sodium from bakery products content (gradual reduction of salt levels from foods, uses of different types of mineral salts, of flavor enhancers, different ingredients with flavor compounds, etc.). Any of these strategies of salt reduction are efficient in some ways. The easiest solution in salt reduction is it gradual reduction over time up to 25% from bakery products recipe. This does not imply any additional costs and technological changes. If this reduction is made under a longer period of time it does not affect the consumers acceptability. Salt replacers (KCl, MgCl$_2$, CaCl$_2$, calcium and magnesium salts as gluconate and lactate) may substitute in a limited way NaCl from bakery products recipe due to their unpleasant flavor. However, they have technological advantages since their effect are similar to that obtained by NaCl addition in bakery products. KCl presents the most similar technological effects in baked products being the most common used salt for NaCl substitution. Up to 20–30% of NaCl substitution, KCl may lead to good bakery products quality without any negative effects on its metallic and bitter taste. The use of sea salt with low sodium content it is in an increasing trend nowadays for it use in bakery products due to the fact that may conduct to good bakery products quality with similar characteristics with those obtained through NaCl addition. Some taste enhancers (glutamates, yeast extracts, hydrolyzed vegetable proteins nucleotides, amino acids, fermented sugars, etc.) can also be used to intensify the salty taste perception in combination with different salts. These present an insignificant effect from the technological point of view and that way their combinations with other types of salts are necessary. The use of dry sourdough in bread recipe may also improve bakery products flavor and may be used in combination with different type of salts or taste enhancers to reduce the sodium content from bakery products.

The results indicate that different strategies may be used in obtaining bakery products with low sodium content of a good quality from a technological and sensory point of view, with benefic effects on consumer health and in agreement with WHO sodium recommendations values.

Author Contributions: A.V., A.D. and G.G.C. contributed equally to the study design, collection of data, development of the sampling, analyses, interpretation of results and preparation of the paper. All authors have read and agreed to the published version of the manuscript.

Funding: This research received no external funding.

Conflicts of Interest: The authors declare no conflict of interest.

References

1. Silow, C.; Axel, C.; Zannini, E.; Arendt, E.K. Current status of salt reduction in bread and bakery products—A review. *J. Cereal Sci.* **2016**, *72*, 135–145. [CrossRef]
2. Gębski, J.; Jezewska-Zychowicz, M.; Szlachciuk, J.; Kosicka-Gębska, M. Impact of nutritional claims on consumer preferences for bread with varied. *Food Qual. Prefer.* **2019**, *76*, 91–99. [CrossRef]
3. Dötsch, M.; Busch, J.; Batenburg, M.; Liem, G.; Tareilus, E.; Müller, R.; Meijer, G. Strategies to reduce sodium consumption: A food industry perspective. *Crit. Rev. Food Sci. Nutr.* **2009**, *49*, 841–851. [CrossRef] [PubMed]
4. He, F.J.; Campbell, N.R.C.; MacGregor, G.A. Reducing salt intake to prevent hypertension and cardiovascular disease. *Rev. Panam. Salud Publica.* **2012**, *32*, 293–300. [CrossRef] [PubMed]
5. Kloss, L.; Meyer, J.D.; Graeve, L.; Vetter, W. Sodium intake and its reduction by food reformulation in the European Union-A review. *NFS J.* **2015**, *1*, 9–19. [CrossRef]
6. Strazzullo, P.; D'Elia, L.; Kandala, N.B.; Cappuccio, F.P. Salt intake, stroke, and cardiovascular disease: Meta-analysis of prospective studies. *BMJ* **2009**, *339*, 1–9. [CrossRef]
7. He, F.J.; MacGregor, G.A. Reducing population salt intake worldwide: From evidence to implementation. *Prog. Cardiovasc. Dis.* **2010**, *52*, 363–382. [CrossRef]
8. Dumitrescu, C.; Segal, B.; Segal, R. Cytoprotection and nutrition. In *Sodium Chloride and Hypertension*; Medical Publishing House: Bucharest, Romania, 1991; pp. 340–342. ISBN 973-39-0101-6.
9. Rizvi, S.A.F. Cost effectiveness of health expenditures: A macro level study for developing and developed countries. *Int. J. Humanit. Appl. Soc. Sci.* **2020**, 26–42. [CrossRef]
10. Kovac, B.; Blaznik, U. Systematic Reduction of Excessive Salt Intake. In *Salt in the Earth*; Cengiz Çinku, M., Karabulut, C., Eds.; IntechOpen: London, UK, 2019; pp. 1–16. ISBN 978-1-78984-635-5.
11. Raffo, A.; Carcea, M.; Moneta, E.; Narducci, V.; Nicoli, S.; Peparaio, M.; Sinesio, F.; Turfani, V. Influence of different levels of sodium chloride and of a reduced-sodium salt substitute on volatiles formation and sensory quality of wheat bread. *J. Cereal Sci.* **2018**, *79*, 518–526. [CrossRef]

12. Dabija, A. Unconventional raw materials for the bakery industry. In *Salt, a Benefic Ingredient in Bakery Products?* Performantica Publishing House: Iași, Romania, 2020; pp. 173–200. ISBN 978-606-685-731-4.
13. Belc, N.; Smeu, I.; Macri, A.; Vallauri, D.; Flynn, K. Reformulating foods to meet current scientific knowledge about salt, sugar and fats. *Trends Food Sci Tehnol.* **2019**, *84*, 25–28. [CrossRef]
14. Munteanu, C.; Iliuta, A. The role of sodium in the body. *Balneo Res. J.* **2011**, *2*, 70–74. [CrossRef]
15. Bloch, M.R. Salt in human history. *Interdiscip. Sci. Rev.* **1976**, *1*, 336–352. [CrossRef]
16. EFSA Panel on Nutrition, Novel Foods and Food Allergens (NDA); Turck, D.; Castenmiller, J.; de Henauw, S.; Hirsch-Ernst, K.I.; Kearney, J.; Knutsen, H.K.; Maciuk, A.; Mangelsdorf, I.; McArdle, H.J.; et al. Scientific Opinion on the dietary reference values for sodium. *EFSA J.* **2019**, *17*, e0577.
17. Grillo, A.; Salvi, L.; Coruzzi, P.; Salvi, P.; Parati, G. Sodium Intake and Hypertension. *Nutrients* **2019**, *11*, 1970. [CrossRef]
18. Chrysant, S.G. There are not racial, age, sex or weight differences in the sensitive hypertensive patients. *Arch. Intern. Med.* **1997**, *157*, 2489–2494. [CrossRef]
19. Whelton, P.K.; He, J. Health effects of sodium and potassium in humans. *Curr. Opin. Lipidol.* **2014**, *25*, 75–79. [CrossRef]
20. Cannon, W.B. Organization for physiological homeostasis. *Physiol. Rev.* **1929**, *9*, 399–431. [CrossRef]
21. Bernstein, A.M.; Willett, W.C. Trends in 24-h urinary sodium excretion in the United States, 1957–2003: A systematic review. *Am. J. Clin. Nutr.* **2010**, *92*, 1172–1180. [CrossRef]
22. Farquhar, W.B.; Edwards, D.G.; Jurkovitz, C.T.; Weintraub, W.S. Dietary sodium and health: More than just blood pressure. *J. Am. Coll. Cardiol.* **2015**, *65*, 1042–1050. [CrossRef]
23. Anderson, C.A.M.; Appel, L.J.; Okuda, N.; Brown, I.J.; Chan, Q.; Zhao, L.; Ueshima, H.; Kesteloot, H.; Miura, K.; Curb, J.D.; et al. Dietary sources of sodium in China, Japan, the United Kingdom, and the United States, women and men aged 40 to 59 years: The INTERMAP study. *J. Am. Diet. Assoc.* **2010**, *110*, 736–745. [CrossRef]
24. Belz, M.C.E.; Ryan, L.A.M.; Arendt, E.K. The Impact of Salt Reduction in Bread: A Review. *Crit. Rev. Food Sci. Nutr.* **2012**, *52*, 514–524. [CrossRef]
25. Guallar-Castillón, P.; Muñoz-Pareja, M.; Aguilera, M.T.; León-Muñoz, L.M.; Rodríguez-Artalejo, F. Food sources of sodium, saturated fat and added sugar in the Spanish hypertensive and diabetic population. *Atherosclerosis* **2013**, *229*, 198–205. [CrossRef]
26. Taylor, C.; Doyle, M.; Webb, D. "The safety of sodium reduction in the food supply: A cross-discipline balancing act"-Workshop proceedings. *Crit. Rev. Food Sci. Nutr.* **2018**, *58*, 1650–1659. [CrossRef] [PubMed]
27. Soptica, F.; Zugravu, C.; Tarcea, M.P.D.; Pantea, S.A. Salt in fast food products and ready-to-eat sauces: Should it be a matter of concern for public health. *Bull. UASVM Agric.* **2012**, *21*, 454–460.
28. Carcea, M.; Narducci, V.; Turfani, V.; Sinesio, F.; Raffo, A.; Bianchi, P.G.; Losi, M.; Zecchinelli, R.; Delogu, C.; Andreani, L. Pane buono con poco sale. *Molini D'italia* **2016**, *12*, 36–42.
29. Mondal, A.; Datta, A.K. Bread baking-a review. *J. Food Eng.* **2008**, *86*, 465–474. [CrossRef]
30. Popa, N.C.; Tamba-Berehoiu, R.; Popescu, S.; Varga, M.; Codină, G.G. Predictive model of the alveografic parameters in flours obtained from Romanian grains. *Rom. Biotechnol. Lett.* **2009**, *14*, 4234–4242.
31. Nahar, N.; Madzuki, I.N.; Izzah, N.B.; Karim, S.B.; Ghazali, H.M.; Karim, R. Bakery Science of Bread and the Effect of Salt Reduction on Quality: A Review. *BJRST* **2019**, *1*, 9–14.
32. Preston, K.R. Effects of neutral salts upon wheat gluten protein properties. I. Relationship between the hydrophobic properties of gluten proteins and their extractability and turbidity in neutral salts. *Cereal Chem.* **1981**, *58*, 317–324.
33. Wehrle, K.; Grau, H.; Arendt, E.K. Effects of lactic acid, acetic acid, and table salt on fundamental rheological properties of wheat dough. *Cereal Chem.* **1997**, *74*, 739–744. [CrossRef]
34. Larsson, H. Effect of pH and sodium chloride on wheat flour dough properties: Ultracentrifugation and rheological measurements. *Cereal Chem.* **2002**, *79*, 544–545. [CrossRef]
35. Lynch, E.J.; Dal Bello, F.; Sheehan, E.M.; Cashman, K.D.; Arendt, E.K. Fundamental studies on the reduction of salt on dough and bread characteristics. *Int. Food Res. J.* **2009**, *42*, 885–891. [CrossRef]
36. Codină, G.G.; Mironeasa, S. Influence of mixing speed on dough microstructure and rheology. *Food Technol. Biotechnol.* **2013**, *51*, 509–519.
37. Carcea, M.; Narducci, V.; Turfani, V.; Mellara, F. A Comprehensive Study on the Influence of Sodium Chloride on the Technological Quality Parameters of Soft Wheat Dough. *Foods* **2020**, *9*, 952. [CrossRef] [PubMed]
38. Butow, B.J.; Gras, P.W.; Haraszi, R.; Bekes, F. Effects of Different Salts on Mixing and Extension Parameters on a Diverse Group of Wheat Cultivars Using 2-g Mixograph and Extensigraph Methods. *Cereal Chem.* **2002**, *79*, 826–833. [CrossRef]
39. Codină, G.G. Rheological properties of wheat flour dough. In *Effects on Salt Addition on Dough Rheological Properties of Wheat Flour*; AGIR Publishing House: Bucharest, Romania, 2010; pp. 129–134. ISBN 978-973-720-335-9.
40. Jekle, M.; Becker, T. Dough microstructure: Novel analysis by quantification using confocal laser scanning microscopy. *Food Res. Int.* **2011**, *44*, 984–991. [CrossRef]
41. McCann, T.H.; Day, L. Effect of sodium chloride on gluten network formation, dough microstructure and rheology in relation to breadmaking. *J. Cereal Sci.* **2013**, *57*, 444–452. [CrossRef]
42. Arena, E.; Muccilli, S.; Mazzaglia, A.; Giannone, V.; Brighina, S.; Rapisarda, P.; Fallico, B.; Allegra, M.; Spina, A. Development of Durum Wheat Breads Low in Sodium Using a Natural Low-Sodium Sea Salt. *Foods* **2020**, *9*, 752. [CrossRef]

43. Pasqualone, A.; Caponio, F.; Pagani, M.A.; Summo, V.; Paradiso, M. Effect of salt reduction on quality and acceptability of durum wheat bread. *Food Chem.* **2019**, *15*, 575–581. [CrossRef] [PubMed]

44. Mironeasa, S.; Codină, G.G.; Mironeasa, C. Optimization of wheat-grape seed composite flour to improve alpha-amylase activity and dough rheological behavior. *Int. J. Food Prop.* **2016**, *19*, 859–872. [CrossRef]

45. Bordei, D. Modern technology of bread making. In *Salt*; AGIR Publishing House: Bucharest, Romania, 2005; pp. 60–63. ISBN 973-720-012-8.

46. He, H.; Hoseney, R. Changes in bread firmness and moisture during long-term storage. *Cereal Chem.* **1990**, *67*, 603–605.

47. Smith, J.P.; Daifas, D.P.; El-Khoury, W.; Koukoutsis, J.; El-Khoury, A. Shelf life and safety concerns of bakery products: A review. *Crit. Rev. Food Sci. Nutr.* **2004**, *44*, 19–55. [CrossRef]

48. Gray, J.A.; Bemiller, J.N. Bread Staling: Molecular Basis and Control. *Compr. Rev. Food Sci. Food Saf.* **2003**, *2*, 1–21. [CrossRef]

49. Moreau, L.; Lagrange, J.; Bindzus, W.; Hill, S. Influence of sodium chloride on colour, residual volatiles and acrylamide formation in model systems and breakfast cereals. *Int. J. Food Sci. Technol.* **2009**, *44*, 2407–2416. [CrossRef]

50. Codină, G.G.; Voica, D. The influence of different forms of bakery yeast Saccharomyces cerevisie type strain on the concentration of individual sugars and their utilization during fermentation. *Rom. Biotechnol. Lett.* **2010**, *15*, 5417–5422.

51. Girgis, S.; Neal, B.; Prescott, J.; Prendergast, J.; Dumbrell, S.; Turner, C. A one-quarter reduction in the salt content of bread can be made without detection. *Eur. J. Clin. Nutr.* **2003**, *57*, 616–620. [CrossRef] [PubMed]

52. Doyle, M.E.; Glass, K.A. Sodium reduction and its effect on food safety, food quality, and human health. *Compr. Rev. Food Sci. Food Saf.* **2010**, *9*, 44–56. [CrossRef]

53. Pflaum, T.; Konitzer, K.; Hofmann, T.; Koehler, P. Analytical and sensory studies on the release of sodium from wheat bread crumb. *J. Agric. Food Chem.* **2013**, *61*, 6485–6494. [CrossRef] [PubMed]

54. Breslin, P.A.S.; Beauchamp, G.K. Suppression of bitterness by sodium variation among bitter taste stimuli. *Chem. Senses* **1995**, *20*, 609–623. [CrossRef]

55. Costa-Corredor, A.; Serra, X.; Arnau, J.; Gou, P. Reduction of NaCl content in restructured dry-cured hams: Post-resting temperature and drying level effects on physicochemical and sensory parameters. *Int. J. Meat Sci.* **2009**, *83*, 390–397. [CrossRef]

56. Paravisini, L.; Sneddon, K.A.; Peterson, D.G. Comparison of the aroma profiles of intermediate wheatgrass and wheat bread crusts. *Molecules* **2019**, *24*, 2484. [CrossRef]

57. Fanari, F.; Desogus, F.; Scano, E.A.; Carboni, G.; Grosso, M. The Effect of the relative amount of ingredients on the rheological properties of semolina doughs. *Sustainability* **2020**, *12*, 2705. [CrossRef]

58. Beck, M.; Jekle, M.; Becker, T. Impact of sodium chloride on wheat flour dough for yeast-leavened products. II. Baking quality parameters and their relationship. *J. Sci. Food Agric.* **2012**, *92*, 299–306. [CrossRef]

59. Kenten, C.; Boulay, A.; Rowe, G. Salt. UK consumers' perceptions and consumption patterns. *Appetite* **2013**, *70*, 104–111. [CrossRef] [PubMed]

60. Powles, J.; Fahimi, S.; Micha, R.; Khatibzadeh, S.; Shi, P.; Ezzati, M.; Engell, R.E.; Lim, S.S.; Danaei, G.; Mozaffarian, D. Global Burden of Diseases Nutrition and Chronic Diseases Expert Group (NutriCoDE). Global, regional and national sodium intakes in 1990 and 2010: A systematic analysis of 24 h urinary sodium excretion and dietary surveys worldwide. *BMJ Open* **2013**, *23*, e003733. [CrossRef] [PubMed]

61. Lopes, M.; Cavaleiro, C.; Ramos, F. Sodium Reduction in Bread: A Role for Glasswort (*Salicornia ramosissima J. Woods*). *Compr. Rev. Food Sci. Food Saf.* **2017**, *16*, 1056–1071. [CrossRef] [PubMed]

62. Pravst, I.; Lavriša, Ž.; Kušar, A.; Miklavec, K.; Žmitek, K. Changes in average sodium content of prepacked foods in Slovenia during 2011–2015. *Nutrients* **2017**, *9*, 952. [CrossRef]

63. Dougkas, A.; Vannereux, M.; Giboreau, A. The Impact of Herbs and Spices on Increasing the Appreciation and Intake of Low-Salt Legume-Based Meals. *Nutrients* **2019**, *11*, 2901. [CrossRef]

64. Li, Q.; Cui, Y.; Jin, R.; Lang, H.; Yu, H.; Sun, F.; He, C.; Ma, T.; Li, Y.; Zhou, X.; et al. Enjoyment of spicy flavor enhances central salty-taste perception and reduces salt intake and blood pressure. *Hypertension* **2017**, *70*, 1291–1299. [CrossRef]

65. Saavedra-Garcia, L.; Sosa-Zevallos, V.; Diez-Canseco, F.; Miranda, J.J.; Bernabe-Ortiz, A. Reducing salt in bread: A quasi-experimental feasibility study in a bakery in Lima, Peru. *Public Health Nutr.* **2016**, *19*, 976–982. [CrossRef]

66. Van Buren, L.; Dötsch-Klerk, M.; Seewi, G.; Newson, R.S. Dietary Impact of Adding Potassium Chloride to Foods as a Sodium Reduction Technique. *Nutrients* **2016**, *8*, 235. [CrossRef]

67. Tahreem, I.; Allah, R.; Muhammad, S.; Summer, R.; Aamir, S. Salt reduction in baked products: Strategies and constraints. *Trends Food Sci. Technol.* **2016**, *51*, 98–105.

68. Charlton, K.; MacGregor, E.; Vorster, N.; Levitt, N.; Steyn, K. Partial replacement of NaCl can be achieved with potassium, magnesium and calcium salts in brown bread. *Int. J. Food Sci. Nutr.* **2007**, *58*, 508–521. [CrossRef] [PubMed]

69. Kaur, H.; Datt, M.; Ekka, M.K.; Mittal, M.; Singh, A.K.; Kumaran, S. Cys-Gly specific dipeptidase Dug1p from S. cerevisiae binds promiscuously to di-, tri-, and tetra-peptides: Peptide-protein interaction, homology modeling, and activity studies reveal a latent promiscuity in substrate recognition. *Biochimie* **2011**, *93*, 175–186. [CrossRef] [PubMed]

70. Salovaara, H. Sensory limitations to replacement of sodium with potassium and magnesium in bread. *Cereal Chem.* **1982**, *59*, 427–430.

71. Chen, G.; Hu, R.; Li, Y. Potassium chloride affects gluten microstructures and dough characteristics similarly as sodium chloride. *J. Cereal Sci.* **2018**, *82*, 155–163.

72. Voinea, A.; Stroe, S.-G.; Codină, G.G. Use of Response Surface Methodology to Investigate the Effects of Sodium Chloride Substitution with Potassium Chloride on Dough's Rheological Properties. *Appl. Sci.* **2020**, *10*, 4039. [CrossRef]

73. Reißner, A.M.; Wendt, J.; Zahn, S.; Rohm, H. Sodium-chloride reduction by substitution with potassium, calcium and magnesium salts in wheat bread. *LWT* **2019**, *108*, 153–159. [CrossRef]

74. Braschi, A.; Gill, L.; Naismith, D.J. Partial substitution of sodium with potassium in white bread: Feasibility and bioavailability. *Int. J. Food Sci. Nutr.* **2009**, *60*, 507–521. [CrossRef]

75. Simsek, S.; Martinez, M.O. Quality of dough and bread prepared with sea salt or sodium chloride. *J. Food Process Eng.* **2016**, *39*, 44–52. [CrossRef]

76. Miller, R.A.; Jeong, J. Sodium reduction in bread using low-sodium sea salt. *Cereal Chem.* **2014**, *91*, 41–44. [CrossRef]

77. Codină, G.G.; Zaharia, D.; Stroe, S.G.; Ropciuc, S. Influence of calcium ions addition from gluconate and lactate salts on refined wheat flour dough rheological properties. *CyTA J. Food* **2018**, *16*, 884–889. [CrossRef]

78. Codină, G.G.; Zaharia, D.; Mironeasa, S.; Ropciuc, S. Evaluation of wheat flour dough rheological properties by magnesium lactate salt addition. *Bull. Univ. Agric. Sci. Vet. Med. Food Sci. Tehnol.* **2018**, *75*, 21–26. [CrossRef]

79. Codină, G.G.; Zaharia, D.; Ropciuc, S.; Dabija, A. Influence of magnesium gluconate salt addition on mixing, pasting and fermentation properties of dough. *EuroBiotech J.* **2017**, *3*, 222–225. [CrossRef]

80. Zandstra, E.H.; Lion, R.; Newson, R.S. Salt reduction: Moving from consumer awareness to action. *Food Qual. Prefer* **2016**, *48*, 376–381. [CrossRef]

81. El Ati, J.; Doggui, R.; El Ati-Hellal, M. A Successful Pilot Experiment of Salt Reduction in Tunisian Bread: 35% Gradual Decrease of Salt Content without Detection by Consumers. *Int. J. Environ. Res. Public Health* **2021**, *18*, 1590. [CrossRef]

82. Diler, G.; Le-Bail, A.; Chevalier, S. Salt reduction in sheeted dough: A successful technological approach. *Food Res. Int.* **2016**, *88*, 10–15. [CrossRef] [PubMed]

83. Noort, M.W.J.; Bult, J.H.F.; Stieger, M.; Hamer, R.J. Saltiness enhancement in bread by inhomogeneous spatial distribution of sodium chloride. *J. Cereal Sci.* **2010**, *55*, 378–386. [CrossRef]

84. Konitzer, K.; Pflaum, T.; Oliveira, P.; Arendt, E.; Koehler, P.; Hofmann, T. Kinetics of sodium release from wheat bread crumb as affected by sodium distribution. *J. Agric. Food Chem.* **2013**, *61*, 10659–10669. [CrossRef] [PubMed]

85. Guilloux, M.; Prost, C.; Courcoux, P.; Le Bail, A.; Lethuaut, L. How Inhomogeneous salt distribution can affect the sensory properties of salt-reduced multi-component food: Contribution of a mixture experimental design approach applied to pizza. *J. Sens. Stud.* **2015**, *30*, 484–498. [CrossRef]

86. Stieger, M.A.; van de Velde, F. Microstructure, texture and oral processing: New ways to reduce sugar and salt in foods. *Curr. Opin. Colloid Interface Sci.* **2013**, *18*, 334–348. [CrossRef]

87. Noort, M.W.J.; Bult, J.H.F.; Stieger, M. Saltiness enhancement by taste contrast in bread prepared with encapsulated salt. *J. Cereal Sci.* **2012**, *55*, 218–225. [CrossRef]

88. Ponzo, V.; Pellegrini, M.; Costelli, P.; Vázquez-Araújo, L.; Gayoso, L.; D'Eusebio, C.; Ghigo, E.; Bo, S. Strategies for reducing salt and sugar intakes in individuals at increased cardiometabolic risk. *Nutrients* **2021**, *13*, 279. [CrossRef]

89. Voinea, A.; Stroe, S.-G.; Codină, G.G. The Effect of Sea Salt, Dry Sourdough and Fermented Sugar as Sodium Chloride Replacers on Rheological Behavior of Wheat Flour Dough. *Foods* **2020**, *9*, 1465. [CrossRef]

90. Voinea, A.; Stroe, S.-G.; Codină, G.G. The Effect of Sodium Reduction by Sea Salt and Dry Sourdough Addition on the Wheat Flour Dough Rheological Properties. *Foods* **2020**, *9*, 610. [CrossRef]

91. Anderson, C.A.; Cobb, L.K.; Miller, E.R.; Woodward, M.; Hottenstein, A.; Chang, A.R.; Mongraw-Chaffin, M.; White, K.; Charleston, J.; Tanaka, T.; et al. Effects of a behavioral intervention that emphasizes spices and herbs on adherence to recommended sodium intake: Results of the SPICE randomized clinical trial. *Am. J. Clin. Nutr.* **2015**, *102*, 671–679. [CrossRef]

92. Wang, C.; Lee, Y.; Lee, S.-Y. Consumer acceptance of model soup system with varying levels of herbs and salt. *J. Food Sci.* **2014**, *79*, S2098–S2106. [CrossRef]

93. Bo, S.; Fadda, M.; Fedele, D.; Pellegrini, M.; Ghigo, E.; Pellegrini, N. A critical review on the role of food and nutrition in the energy balance. *Nutrients* **2020**, *12*, 1161. [CrossRef]

94. Bhattacharya, S. "Taste modification"—A new strategy for reduction of salt intake among Indian population. *Epidemiol. Int.* **2017**, *1*, 10–14.

95. Crucean, D.; Debucquet, G.; Rannou, C.; le-Bail, A. Vitamin B4 as a salt substitute in bread: A challenging and successful new strategy. Sensory perception and acceptability by French consumers. *Appetite* **2019**, *134*, 17–25. [CrossRef] [PubMed]

96. Locke, K.W.; Fielding, S. Enhancement of salt intake by choline chloride. *Physiol. Behav.* **1994**, *55*, 1039–1046. [CrossRef]

97. Oliveira-Maia, A.J.; Roberts, C.D.; Simon, S.A.; Nicolelis, M.A.L. Gustatory and reward brain circuits in the control of food intake. In *Advances and Technical Standards in Neurosurgery*; Pickard, J.D., Akalan, N., Benes, V., Jr., Rocco, C.D., Dolenc, V.V., Antunes, J.L., Schramm, J., Sindou, M., Eds.; Springer: Vienna, Austria, 2011; Volume 36, pp. 31–59.

98. Barbarisi, C.; De Vito, V.; Pellicano, M.P.; Boscaino, F.; Balsomo, S.; Laurino, C.; Sorrentino, G.; Volpe, M.G. Bread chemical and nutritional characteristics as influenced by food grade sea water. *Int. J. Food Prop.* **2019**, *22*, 280–289. [CrossRef]

99. Tuhumury, H.C.D.; Small, D.M.; Day, L. Effects of Hofmeister salt series on gluten network formation: Part, I. Cation series. *Food Chem.* **2016**, *212*, 789–797. [CrossRef]

100. Beto, J.A. The role of calcium in human aging. *Clin. Nutr. Res.* **2015**, *4*, 1–8. [CrossRef] [PubMed]

101. Krupa-Kozak, U.; Altamirano-Fortoul, R.; Wronkowska, M.; Rosell, C.M. Breadmaking performance and technological characteristic of gluten-free bread with inulin supplemented with calcium salts. *Eur. Food Res. Technol.* **2012**, *235*, 545–554. [CrossRef]
102. Bassett, M.N.; Pérez-Palacios, T.; Cipriano, I.; Cardoso, P.; Ferreira, I.M.; Samman, N.; Pinho, O. Bread NaCl reduction and Ca fortification. *J. Food Qual.* **2014**, *37*, 107–116. [CrossRef]
103. Zohdy, K.; Kareem, M.A.; Abdel-Aal, H. Separation of magnesium chloride from sea water by preferential salt separation (pss). *Int. J. Bioassays* **2013**, *2*, 376–378.
104. Codină, G.G.; Dabija, A.; Stroe, S.G.; Ropciuc, S. Optimization of iron–oligofructose formulation on wheat flour dough rheological properties. *J. Food Process Preserv.* **2019**, *43*, e13857. [CrossRef]
105. Ranhotra, G.S.; Loewe, R.J.; Lehmann, T.A.; Hepburn, F.N. Effect of various magnesium sources on breadmaking characteristics of wheat flour. *J. Food Sci.* **2008**, *41*, 952–954. [CrossRef]
106. Barbagallo, M.; Veronese, N.; Dominguez, L.J. Magnesium in aging, health and diseases. *Nutrients* **2021**, *13*, 463. [CrossRef]
107. Stone, M.S.; Martyn, L.; Weaver, C.M. Potassium intake, bioavailability, hypertension, and glucose control. *Nutrients* **2016**, *8*, 444. [CrossRef]
108. European Food Safety Authority. Opinion of the Scientific Panel on Dietetic Products, Nutrition and Allergies on a request from the Commission related to the Tolerable Upper Intake Level of Potassium. *EFSA J.* **2005**, *193*, 1–19.
109. Wyatt, C.J.; Ronan, K. Evaluation of potassium chloride as a salt substitute in bread. *J. Food Sci.* **1982**, *47*, 672–673. [CrossRef]
110. Quilez, J.; Salas-Salvado, J. Salt in bread in Europe: Potential benefits of reduction. *Nutr. Rev.* **2012**, *70*, 666–678. [CrossRef] [PubMed]
111. Drewnowski, A.; Rehm, C.D.; Maillot, M.; Mendoza, A.; Monsivais, P. The feasibility of meeting the WHO guidelines for sodium and potassium: A cross-national comparison study. *BMJ Open* **2015**, *5*, e006625. [CrossRef] [PubMed]
112. Siepmann, F.B.; Ripari, V.; Waszczynskyj, N.; Spier, M.R. Overview of sourdough technology: From production to marketing. *Food Bioproc. Technol.* **2018**, 242–270. [CrossRef]
113. Arendt, E.K.; Liam, A.M.R.; Bello, F.D. Impact of sourdough on the texture of bread. *Food Microbiol.* **2007**, *24*, 165–174. [CrossRef]
114. Yu, Y.; Wang, L.; Qian, H.; Zhang, H.; Qi, X. Contribution of spontaneously fermented sourdoughs with pear and navel orange for the bread-making. *LWT* **2018**, *89*, 336–343. [CrossRef]
115. Decock, P.; Cappele, S. Bread technology and sourdough technology. *Trends Food Sci. Technol.* **2005**, *16*, 113–120. [CrossRef]
116. Maluly, H.D.B.; Arisseto-Bragotto, A.P.; Reyes, F.G.R. Monosodium glutamate as a tool to reduce sodium in foodstuffs: Technological and safety aspects. *Food Sci. Nutr.* **2017**, *5*, 1039–1048. [CrossRef]
117. Tanguler, H.; Erten, H. Utilisation of spent brewer's yeast for yeast extract production by autolysis: The effect of temperature. *Food Bioprod. Process.* **2008**, *86*, 317–321. [CrossRef]
118. Spina, A.; Brighina, S.; Muccilli, S.; Mazzaglia, A.; Rapisarda, P.; Fallico, B.; Arena, E. Partial Replacement of NaCl in Bread from Durum Wheat (*Triticum turgidum L subsp. durum Desf.*) with KCl and Yeast Extract: Evaluation of Quality Parameters During Long Storage. *Food Bioprocess Technol.* **2013**, *8*, 1089–1101. [CrossRef]

Article

Assessing the Performance of Different Grains in Gluten-Free Bread Applications

Iuliana Banu and Iuliana Aprodu *

Faculty of Food Science and Engineering, Dunarea de Jos University of Galati, 111 Domneasca Str.,
800201 Galati, Romania; iuliana.banu@ugal.ro
* Correspondence: iuliana.aprodu@ugal.ro

Received: 19 November 2020; Accepted: 7 December 2020; Published: 8 December 2020

Abstract: A comparative analysis of quinoa, sorghum, millet and rice flours and breads in terms of proximate composition, resistant starch, antioxidant activity and total phenolic content was realized in this study. Quinoa whole flour had the highest content of proteins, fat, ash and total dietary fiber, followed by millet and sorghum flours. Quinoa and rice breads had higher specific volume (192.22 and 181.04 cm^3/100 g, respectively) and lower crumb firmness (10.81 and 13.74 N, respectively) compared to sorghum and millet breads. The highest total phenol content was obtained in the case of bread prepared with quinoa flour (398.42 mg ferulic acid equiv/100 g d.w.), while the lowest content was obtained for the rice flour bread (70.34 mg ferulic acid equiv/100 g d.w). The antioxidant activity of gluten-free breads decreased in the following order: sorghum > quinoa > millet > rice. Quinoa bread had the highest resistant starch content of 3.28% d.w., while the rice bread had the highest digestible starch content of 81.48% d.w. The slowly digestible starch varied from 15.5% d.w. for quinoa bread, to 6.51% d.w. for millet bread. These results revealed the huge potential of quinoa, sorghum and millet to be used for developing functional gluten-free bread.

Keywords: quinoa; millet; sorghum; rice; gluten-free bread

---

1. Introduction

Coeliac disease is a lifelong autoimmune enteropathy induced by gluten consumption, affecting persons which are genetically susceptible. Due to the increasing prevalence worldwide, the coeliac disease started to be considered a major public health problem [1]. Taking into account that no efficient cure was reported yet for this condition, avoidance of gluten-containing products in the diet was recommend as the main solution. Therefore, identifying ingredients and tools for obtaining high-quality gluten-free products is highly desired. Rice flour is considered the most important ingredient in gluten-free bread making, but in the last period the minor grains like millet, sorghum and quinoa became increasingly important for producing gluten-free baking products [2].

Originating from South America but with a great adaptability to different growing conditions, quinoa has become more and more popular due to the well-balanced nutritional composition [3–5]. Quinoa is recommended primarily by the content and quality of its proteins, being rich in amino acids which are deficient in cereals. In fact, the amino acids composition of quinoa proteins is close to the ideal protein recommended by FAO [3,6–8]. In addition, quinoa contains high levels of unsaturated fatty acids [9–11], vitamins [9], minerals [12], dietary fiber and polyphenols [13].

Sorghum and millet are two cereals which are insufficiently used in bakery products, even if they have nutritional compositions comparable to other cereals [14], and could be used if not as the based flour, at least as functional ingredients, mainly due to phenolic compounds [2,15,16] and resistant starch contents [14,17]. Moreover, the two cereals can be grown on semi-arid regions, where other cereals do not give consistent productions.

Several studies pointed out that quinoa, sorghum and millet have huge potential to be used for developing functional foods, due to the high level of phenolic compounds, antioxidant capacity and low starch digestibility [2,3,17].

The objective of this study was to compare quinoa, sorghum, millet and rice in terms of composition, bread-making potential and physical properties, resistant starch, antioxidant activity and total phenolic content of breads. The rice was included in this study due the large applications in gluten-free bread making.

2. Materials and Methods

2.1. Materials

The quinoa seeds (Titicaca variety cultivated through conventional farming in Galati, Romania, harvest, 2017) used in the experiments were preliminary treated to eliminate the bitter taste and toxic saponins using the procedure described by Nascimento et al. [18]. Quinoa was washed with tap water for 20 min, then the seeds were dried at 45 °C for 12 h. After drying the seeds were ground using a blade mill grinder (Bosch MKM6003, Gerlingen, Germany). The quinoa whole flour was then stored at 4 °C until further analyses.

The sorghum whole flour (origin Hungary, distributed by Adams Vision SRL Tg Mures, Romania), brown millet flour (distributed by La Finestra sul Cielo Vilareggia Italia) and wholegrain rice flour (Solaris Plant SRL, Bucharest, Romania) were purchased from the local market (Galati, Romania).

Other ingredients used for bread making, such as salt, sugar, lecithin and compressed yeast, were purchased from local market (Galati, Romania).

2.2. Proximate Compositions

The proximate composition of quinoa, sorghum, millet and rice whole flours was determined as follows: moisture content with SR ISO 712:2005 [19], protein content through semimicro-Kjeldahl method (Raypa Trade, R Espinar, SL, Barcelona, Spain) using a nitrogen-to-protein conversion factor of 6.25 (AACC method 46–11.02 [20]), fat content through Soxhlet extraction method (SER-148; VELP Scientifica, Usmate Velate (MB), Italy) with ether, total, insoluble and soluble dietary fiber contents with a combination of enzymatic and gravimetric methods [21] (Merck KGaA, Darmstadt, Germany) and ash contents with SR ISO 2171/2002 [19]. The starch contents were afterwards determined by subtracting from one hundred the total percentage of the components experimentally assayed.

Amylose content was determined using the Amylose/Amylopectin Assay Kit and the procedure recommended by the manufacturer (Megazyne International Ireland Ltd. Wicklow, Ireland), which is based on the method of Gibson et al. [22].

The starch damage was determined though AACC Method 76–31.01 [20], using the dedicated kit from Megazyme International Ireland Ltd. (Wicklow, Ireland).

2.3. Physical Properties

Fineness module was determined using the method of Godon and Willm [23], the flour being sieved through 500, 400, 315, 160 and 125 µm mesh.

The brightness value (L*), redness value (a*) and yellowness value (b*) were measured using the Chroma Meter CR-410 (Konica Minolta Business Solutions Europe GmbH). Furthermore, based on a* and b* values, the chroma (C*) and hue angle (h°) were calculated.

2.4. Bread-Making Procedure

The doughs were prepared through the one stage method described by Banu et al. [24], using the following formula on a 100 g whole flour: 1.5% salt, 2% sugar, 6% lecithin, 3% compressed baker's yeast, and water (78 mL). The dough was fermented for 150 min at 30 °C in a laboratory proofer, was divided in two pieces, molded and placed in baking trays. After the additional leavening of 30 min at 30 °C, the samples were baked for 30 min at 230 °C

2.5. Breads Characterization

Prior to characterization, the breads were stored for 60 min at room temperature.

The specific volume of the bread samples was measured through the rapeseed displacement method [19].

Crumb firmness was determined using the MLFTA apparatus (Guss, Strand, South Africa) and a probe with diameter of 7.9 mm. Three distinct measurements were performed on two bread slices originating from the center of the bread samples. The following parameters were considered when measuring the crumb firmness: test speed of 5 mm/s, trigger threshold force of 1.96 N and bread slices penetration wide of 25 mm.

In order to determine the total phenolic content, 2,2-diphenyl-1-picrylhydrazyl (DPPH) radical scavenging activity, trolox equivalent antioxidant capacity and iron reducing antioxidant power, the extraction procedure described by Aprodu and Banu [25] was used.

The total phenolic content, the DPPH-radical scavenging activity, the Trolox equivalent antioxidant capacity (TEAC) and the ferric reducing antioxidant power (FRAP) were determined using the procedures described by Aprodu and Banu [25]. The total phenolic content was reported as mg ferulic acid equivalent (FAE) per 100 g d.w. bread sample, the TEAC was expressed as μmols Trolox/g d.w., while results of the FRAP were expressed in μmol Fe^{2+}/g d.w.

The content of resistant starch was determined through the AACC Method 32–40.01 [20], using the Megazyne assay kit (Megazyne International Ireland Ltd. Wicklow, Ireland).

Rapidly and slowly digestible starch contents were determined using the procedure described by Miao et al. [26]. The glucose content of the samples was determined using the glucose oxidase/peroxidase (GOPOD) assay kits (Wicklow, Ireland). The amount of hydrolyzed starch was determined by multiplying the assayed glucose content by a factor of 0.9. Finally, the rapidly and slowly digestible starch values were obtained by considering the contents of glucose released after 20 min and 120 min (G20 and G120, respectively) and free glucose (FG).

2.6. Statistical Analysis

The results are reported as average values of three replicates together with standard deviation. Analysis of variance was carried out with Microsoft Excel Soft to detect significant differences among results. Pearson's correlation coefficients were calculated to identify eventual statistical relationships.

3. Results and Discussion

3.1. Proximate Composition of Flours

The proximate composition of quinoa, millet, sorghum and rice whole flours is shown in Table 1. The highest proteins content of 14.05% was registered for quinoa flour. According to Basile et al. [4], quinoa has a similar protein content with the highest strains of wheat, with lower amounts of glutinous proteins. In addition, quinoa was reported as a source of complete protein, having a well-balanced amino acids composition needed for human diet, having high contents of methionine and lysine [6]. Moreover, Vega-Galvez et al. [6] noted that the value of quinoa protein is similar to casein in milk. According to Srichuwong et al. [8] the percentage of lysine in the quinoa proteins is 6.9%, much higher than in sorghum (2.2%), millet (3.1%) or wheat (2.9%). Regarding the methionine content of the total proteins from quinoa, millet and sorghum, the percentages are in the 2.5–3.6% range, significantly higher compared to wheat proteins which have 1.8% [8]. Fairbanks et al. [27] appreciated that the high level of lysine from quinoa proteins is due to the high amount of albumins and globulins which are rich in this amino acid.

Table 1. Proximate compositions of whole quinoa, sorghum, millet and rice flours (% d.w.).

Component	Quinoa Flour	Sorghum Flour	Millet Flour	Rice Flour
Protein, %	14.05 ± 0.07	10.29 ± 0.03	11.01 ± 0.06	6.18 ± 0.03
Fat, %	5.29 ± 0.06	3.17 ± 0.03	3.91 ± 0.05	2.16 ± 0.03
Ash, %	2.39 ± 0.01	1.61 ± 0.01	2.70 ± 0.01	1.53 ± 0.01
Total dietary fiber, %	9.11 ± 0.08	7.42 ± 0.06	8.57 ± 0.08	4.69 ± 0.06
Insoluble dietary fiber, %	6.74 ± 0.05	6.52 ± 0.04	7.46 ± 0.05	3.72 ± 0.03
Soluble dietary fiber, %	2.36 ± 0.02	0.90 ± 0.01	1.10 ± 0.01	0.97 ± 0.01
Starch, %	58.84 ± 0.07	69.63 ± 0.06	63.92 ± 0.07	74.09 ± 0.05
Amylose, % starch	17.76 ± 0.08	23.23 ± 0.10	20.36 ± 0.11	19.92 ± 0.09

The proteins content of sorghum whole flour was 10.29%, close to the results reported by Khan et al. [14] for red sorghum whole flour (10.05%) and white sorghum whole flour (11.77%). Mokrane et al. [28] analyzed the sorghum protein quality, and observed the high amino acid scores that varied between 0.9 and 2.6, except for lysine, methionine and cysteine. The sorghum proteins include albumins, globulins, kafirins, cross-linked kafirins and glutelins, among which the kafirins represent about 80% of decorticated flour protein [29].

In case of millet, the proteins content was 11.01%, comparable to the value of 10.75% reported by USDA [30]. According to Devi et al. [31], different millet species present large variations of proteins content, ranging from 7.3–8.3% in case of finger and kodo millet, to 14.5% in case of pearl millet; the proso and foxtail millet have protein contents of 11–11.3%. Kalinova and Moudry [32] reported for proso millet proteins an essential amino acid index of 0.51, and the amino acids scores of 0.47 for lysine, of 0.78 for tyrosine and of 0.75 of cysteine and methionine.

Among the investigated whole protein flours, the smallest proteins content of 6.18% was registered for rice flour (Table 1), lower compared to the results of 7.23% reported by USDA [30], or 7.5% by Devi et al. [31].

The results shown in Table 1 indicated that the fat content varied from 5.29%, for quinoa flour, to 2.16%, for rice flour. Higher fat contents of 6.0, 6.4 and 6.8% were reported by Pereira et al. [10] for whole quinoa white grain, red grain and black grain, respectively. On the other hand, Vidueiros et al. [11] reported lower fat content of 4.7–5.3% for different varieties of quinoa. Sorghum and millet flours presented fat contents of 3.17 and 3.91%, respectively, while Srichuwong et al. [8] reported values of 5 and 3.7%, in case of white and red sorghum, and 4.1% for millet flour. The fat content of rice whole flour was 2.16%.

The lipids are divided into free and bound fractions, the free lipids being the most present fractions [6]. Ragaee et al. [16] reported the presence of free lipid fractions of 2.0–4.1 and 5.6%, and of 0.1–0.56 and 0.6–0.9% for bound fractions in sorghum and millet, respectively. In case of quinoa flour, Collar and Angioloni [6] reported for free and bound lipids fractions values of 3.23 and 0.28 g/100 g flour, respectively.

The lipid composition of quinoa stands out in the relation to the rest of cereals analyzed, through the fatty acid profile that is comparable to that of corn and soybeans [9]. The major saturated fatty acid is palmitic acid (about 10%), while the oleic, linoleic and alpha-linolenic acids represents about 19.7–29.5, 49–56.4 and 8.7–11.7%, respectively [8–10]. According to Rooney [33], when compared to quinoa, the sorghum and millet grains contain higher levels of palmitic acid (of 12 and 20%, respectively), oleic acid (34 and 26%, respectively), and lower levels of linoleic acid (of 50 and 45%, respectively), and linolenic acid (of 3 and 4%, respectively).

The ash content of millet (2.7%) and quinoa (2.39%) flours is higher compared to rice and sorghum flours that had values of 1.53 and 1.61%, respectively (Table 1). The ash content of quinoa flour used in the present study is lower than that reported by other authors [11], probably due the intensity of the washing process of grains. On the other hand, in case of rice flour the ash content was higher than that reported by other authors [34]. Instead, in case of sorghum and millet our results were between the values reported by other authors, namely 2–3.6% [31] and 1.52–2.57% [14], respectively.

The highest amounts of total dietary fibers were registered in case of quinoa and millet flours, of 9.11 and 8.57%, respectively, whereas the rice flour had the lowest total dietary fiber content (4.69%) among the investigated grains. The soluble fiber content ranged between 2.36, in quinoa flour, and 0.90% in sorghum flour. The insoluble fiber content decreased in the following order: millet (7.46%), quinoa (6.74%), sorghum (6.52%) and rice (3.72%) flours. The results reported by different authors present a large variation, even if the same method used for investigations. For instance, Srichuwong et al. [8] reported much lower value for total fiber content for quinoa and millet (9.5% and 8.4–10%, respectively), compared to the Kurek et al. [35] (16.43% and 11.71%, respectively), even if the authors used the same method. These results highlight the great variability in terms of chemical composition among grain varieties.

Regarding fiber compositions, Lamothe et al. [36] reported that, in case of quinoa, the soluble and insoluble dietary fiber were composed mainly of pectic polysaccharides and xyloglucans, unlike cereals that contain mostly arabinoxylans. Lai et al. [37] reported that the total dietary fiber extracted from nonwaxy brown rice contained important amount of pectic substances, whereas the waxy counterparts were richer in hemicellulose or cellulose.

Due to the high contents of protein, fiber, fat and ash from quinoa, sorghum and millet flours, the amount of starch present in the samples was lower compared to the rice flour (Table 1). The lowest starch content was registered in case of quinoa flour (58.84%), followed by millet and sorghum flours, while the highest value was obtained for rice flour (74.09%). The amylose content in starch varied from 17.76 for quinoa to 23.23% for sorghum flours. Navruz-Varli and Sanlier [9] showed that the amylose content in the quinoa starch can vary from 3 to 22%, being lower than in wheat. The amylose content of rice starch depends by type of rice. Thus, Chung et al. [38] reported value of 27.2% of the amylose content for the rice starch isolated from long grains, and significantly lower values of 15.4 and 18.8% for the amylose contents for the rice starches isolated from medium and round grains, respectively. Srichuwong et al. [8] reported for millet and sorghum starches values of amylose about of 24 and 24.6–25.8%, respectively.

3.2. Physical Properties of Flours

The granularity of the flour samples is an important physical parameter because of the high influence on the quality of the bakery products. The granularity of the flours was estimated by determining the tailing and sieve fractions on a set of sieves with mesh aperture ranging from 500 to 125 μm, and the fineness modulus.

The sorghum and millet flours had highest percentage of particles with size between 125 and 315 μm, of 80.7 and 74.9%, respectively (Table 2). In case of rice flour highest percentage of particles had larger size, ranging between 315 and 500 μm, while in case of quinoa flour the particles size was most homogenous, even if the percentage of particles with size between 125 and 315 μm was about two times higher than of those between 315 and 500 μm, and three times higher than of those with size less than 125 μm. The higher percentage of larger particles in case of rice flour resulted in higher fineness module (of 2.85) compared to the rest of flours. Sorghum flour had the lowest percentage of particles higher than 315 μm and the lowest fineness module of 1.63.

The fineness module is a measure of the distribution of fine and coarse particles in the analyzed sample, and can be influenced on one hand by grains related factors such as compactness of the endosperm, composition and texture of the pericarp and embryo, and by the milling method on the other hand. Taking into account that sorghum, millet and rice flours are commercial flours, it was not possible to factor the effect of grinding on the particle size distribution. Anyway, the difference between the values of the fineness modulus of the investigated flour might be due to the differences in terms of compactness of the rice endosperm.

Table 2. Physical properties of whole quinoa, sorghum, millet and rice flours.

Properties		Quinoa Flour	Sorghum Flour	Millet Flour	Rice Flour
Fineness module		2.31 ± 0.03	1.63 ± 0.02	1.90 ± 0.02	2.85 ± 0.03
- Particles size 500–315 µm, %		35.9 ± 0.11	10.0 ± 0.10	18.7 ± 0.10	59.5 ± 0.11
- Particles size 315–125 µm, %		46.0 ± 0.10	80.7 ± 0.12	74.9 ± 0.12	32.7 ± 0.10
- Particles size < 125 µm, %		18.1 ± 0.09	9.3 ± 0.08	7.0 ± 0.07	7.8 ± 0.05
Damaged starch, %		4.44 ± 0.05	8.17 ± 0.06	5.27 ± 0.05	3.85 ± 0.05
Color values	L*	81.46 ± 0.16	81.62 ± 0.16	83.79 ± 0.29	83.89 ± 0.62
	a*	1.32 ± 0.05	4.49 ± 0.02	2.62 ± 0.01	0.59 ± 0.01
	b*	15.28 ± 0.12	12.45 ± 0.39	15.18 ± 1.12	15.33 ± 0.06
	C*	15.34 ± 0.43	13.23 ± 0.01	15.40 ± 0.02	15.34 ± 0.02
	h°	56 ± 0.00	51 ± 0.00	54 ± 0.01	57 ± 0.00

The starch damaged of sorghum flour (8.17%) was higher than other flours (3.85–4.44%) (Table 2). A positive correlation of 0.87 was registered between the fineness module and damaged starch ($p < 0.05$).

The color parameters of the investigated flours are presented in Table 2. The lowest lightness values (L*) were obtained in case of quinoa and sorghum flours. Sorghum flour presented the highest value of redness (a*) and lowest values of yellowness (b*), chroma (C*) and hue angle (h°), and most probably these values might be explained by the presence of colored polyphenolics such as tannins and anthocyanins [17]. On the other hand, the higher values of yellowness (b*) registered in case of quinoa, millet and rice flour might be explained by the presence of carotenoids in these samples [39].

3.3. Breads Characterization

Sorghum and millet breads are characterized by lower specific volumes and higher crumb firmness compared to quinoa and rice flour breads (Table 3). Marston et al. [40] reported an improvement of the specific volume and crumb firmness of gluten-free breads based on heat-treated sorghum flour compared to the unheated flour. The authors noted that these results are possible due the modification of sorghum proteins by oxidizing the free sulfhydryl groups. Moreover, Taylor et al. [2] suggested that the lack of glyco- and phospholipids in sorghum flour might be a reason for the lower volume of the bread prepared with sorghum flour compared to the wheat. In addition to starch, the properties of the lipids influence to high extent the crumb firmness.

Table 3. Physical properties of quinoa, sorghum, millet and rice breads.

Physical Properties	Breads Prepared With			
	Quinoa Flour	Sorghum Flour	Millet Flour	Rice Flour
Specific volume, cm³/100 g	192.22 ± 0.12	152.35 ± 0.10	164.39 ± 0.11	181.04 ± 0.12
Crumb firmness, N	10.81 ± 0.14	21.47 ± 0.32	25.70 ± 0.15	13.74 ± 0.23

The quinoa bread had the highest specific volume and lowest crumb firmness. Elgeti et al. [1] reported enhanced specific volume of the bread made with quinoa flour compared to the bread made with a blend of rice and corn flours. The authors explained these results through the better ability of the dough prepared with quinoa flour to retain a high amount of gas. It is well known that the proteins from rice flour do not have great gas retention properties [41]. The high gas volume resulting throughout fermentation was stabilized within the network during baking, resulting in a nice structure with homogeneously distributed fine pores. It was considered that different surface-active components present in quinoa flour, such as peptides or polar lipids, might contributed to the stabilization of the gas bubbles. In addition, the viscoelastic properties of the dough might be optimized such as to allow easier and more efficient gas inclusion and stabilization. Furthermore, Elgeti et al. [1] noted that specific volume of bread increased when quinoa flour was used to replace the quinoa whole flour. They considered that the positive effect registered on the volume of the bread was the result of removing the embryo hull components out of the whole flour. As previously shown, the bran particles might interfere with gas cells, affecting the porosity and specific volume bread [42].

3.4. Antioxidant Activity of the Breads

The total phenol contents and antioxidant activity of quinoa, sorghum, millet and rice breads are reported in Table 4. The total phenol content of the breads varied from 70.34 to 398.42 mg FAE/100 g d.w. The highest total phenol content was registered in case of bread prepared with quinoa flour, while the lowest content was obtained for bread prepared with rice flour.

Table 4. Total phenol contents and antioxidant activity of quinoa, sorghum, millet and rice breads.

Properties	Breads Prepared With			
	Quinoa Flour	Sorghum Flour	Millet Flour	Rice Flour
Total phenol content, mg ferulic acid equiv/100 g d.w.	398.42 ± 0.15	387.16 ± 0.11	180.09 ± 0.10	70.34 ± 0.10
DPPH-radical scavenging activity, %	32.85 ± 0.11	35.01 ± 0.10	19.24 ± 0.10	10.50 ± 0.11
FRAP, μmoli Fe^{2+}/g d.w.	2.53 ± 0.09	3.67 ± 0.09	2.06 ± 0.07	0.98 ± 0.05
TEAC, μmoli Trolox/g d.w.	19.32 ± 0.12	54.51 ± 0.12	15.07 ± 0.10	6.97 ± 0.10

The antioxidant properties of breads were evaluated by measuring the DPPH-radical scavenging activity, and by TEAC and ABTS and FRAP methods. Re et al. [43] mentioned that the TEAC method quantifies both lipophilic and hydrophilic antioxidants, comprising carotenoids, hydroxycinnamates, and flavonoids, while the DPPH-radical scavenging activity is mainly due to the phenolic compounds [25]. In order to directly measure the antioxidants or reductants in samples the FRAP assay was used.

All methods applied for assessing the antioxidant properties indicated that the antioxidant ability of the bread samples decreased in the following order: sorghum > quinoa > millet > rice (Table 4). Sorghum and quinoa breads had higher value of FRAP, indicating the presence of Fe^{2+} chelating agents in these samples.

The results reported in the literature for the antioxidant activity of different types of flours and breads are very different due to many factors related both to the investigated samples, such as the genotype of the grains and environmental conditions, and to the techniques used for preparing the extracts and quantification of the antioxidant activity [13]. Therefore, is rather difficult to compare our results with those of other authors. Xu et al. [3] studied the antioxidant properties of bread prepared with blends of wheat flour and quinoa flour and found that the phenolic content, ABTS and DPPH-RSA increased with increase of the levels of quinoa flours. They reported a loss of total polyphenol after baking. Similarly, when testing the possibility of using amaranth, quinoa, buckwheat and wheat for obtaining nutritionally enhanced gluten-free formulations, Alvarez-Jubete et al. [44] reported the overall decrease of the antioxidant activity and total phenol content following bread making. Anyway, new substance with antioxidant properties can be synthesized during baking, such as Maillard reaction products, which can be mainly found in the bread crust. Jan et al. [45] observed the increase of the antioxidant activity of cookies during baking process with increasing the sugar level, temperature and time, most probably due to the formation of melanoidins. The antioxidant activity of cookies was stable after baking at 180 °C suggesting the completion of the reactions resulting in molecules with radical scavenging ability. Furthermore, the phenolics breakdown or other degradation reactions might result in products responsible for the increase antioxidant activity. Lindenmeier and Hofmann [46] reported the increase of the antioxidant activity up to five times with the increase of baking temperature and time. They explained this increase by the formation of the antioxidant compound pronyl-L-lysine. Moreover, they found higher antioxidant activity in crust in comparison to the crumb and untreated flour.

Khan et al. [14] reported for the free phenolic acids of red and white sorghum flours values of 81.19 and 150.67 μg GAE/g d.w., respectively. Yousif et al. [17] found that the breads prepared from blends of wheat flour and 40% sorghum flour (red and white, respectively) had the free phenolic contents of 1.09–0.49 mg GAE/g d.w. Additionally, Yousif et al. [17] reported that the addition of red or white sorghum flour to the wheat flour increased the polyphenolic content and the antioxidant capacity of the bread. Khan et al. [14] noted that the phenolic acids prevailing in the extract were ferulic acid, p-hydroxibenzoic acid and salicylic acid. Besides these phenolic acids, the red sorghum flour contained flavonoids such as anthocyanins (luteolinidin and apigeninidin). Thermal processing of cereals usually causes the reduction of the flavonoid contents to different extents, depending on the food matrix and on the intensity of the thermal treatment [44]. In particular, Khan et al. [14] noted that anthocyanins were less stable than phenolic acids during cooking. There are other phytochemicals, such as carotenoids, that contribute to the antioxidant activity of the samples [14]. Lopez-Contreras et al. [47] studied the antioxidant activity of ten sorghum genotypes and reported variation of the FRAP, ABTS and DPPH values from 3.94 to 63.34, from 44.21 to 121.73 and from 1.83 to 65.73 µmol Trolox/g, respectively, and of total phenol content values from 1.408 to 12.348 µg catechin equivalents/g.

Chandrasekara and Shahidi [15] found higher free phenolic contents in finger millet (411–610 mg FAE/100 g), compared to the pearl and proso millets (168 mg FAE/100 g and 140 mg FAE/100 g, respectively). They registered different flavonoid contents of 203–228, 49 and 140 mg/100 g catechin equivalent in the phenolic extracts obtained from finger, pearl and proso millet, respectively. Chethan and Malleshi [48] showed that the main phenolic acids in millets are ferulic acid, gallic acid, p-hydroxibenzoic acid, cumaric acid and proto-catechuic acid. Sreeramulu et al. [49] found higher antioxidant activity (FRAP of 471.71 µmol/g, DPPH-RSA and total phenol content of 1.73 and 373.15 mg Trolox/g, respectively) in finger millet compared to the rice (FRAP of 60.93–67.48 µmol/g, DPPH-RSA and total phenol content of 0.49–1.23 and 47.64–56.61 mg Trolox/g, respectively).

Previous studies mentioned a significant correlation between phenolics and antioxidant activity in quinoa [50]. Large variation of the total phenol content of quinoa seed from 28.49 to 1.59 mg gallic acid/100 g d.w. was reported by Miranda et al. [12]. Later study of Pellegrini et al. [13] revealed the phenolic profile of quinoa flour; the most abundant phenolic acids were 4-hydroxybenzoic, syringic acid, vanillic acid, gallic acid, ferulic acid and p-cumaric acid, while the most dominant flavonoids were neohesperidin, kaempfenol and isoquercitin. In addition to the polyphenols and flavonoids which are know the possess antioxidant properties, Aloisi et al. [50] also mentioned the contribution of the proteins, including the 11S fraction, to the antioxidant activity of the quinoa seeds. Other bioactive compounds like carotene, vitamins, tocopherols etc., might also contribute to the antioxidant properties of quinoa flour [50].

The total phenol content and antioxidant properties varied significantly between breads prepared with quinoa, sorghum, millet and rice flour. Moreover, between the total phenol content and DPPH, FRAP and TEAC were registered strong correlations ($p < 0.05$) of 0.99, 0.88 and 0.71, respectively. These results indicate that the antioxidant properties of breads are mainly due to presence of phenols and flavonoids.

3.5. Starch Properties of Breads

The contents of resistant and digestible starches are shown in Table 5. The quinoa bread had the highest amount of resistant starch of 3.28% d.w., while the rice bread had the highest amount of digestible starch of 81.48% d.w.

De la Hera et al. [51] prepared and characterized the gluten-free bread based on rice flour and reported resistant starch of 0.89–1.96 g/100 g and rapidly digestible starch of 82.07–96.54 g/100 g. They mentioned that the lowest values were obtained when coarse flour was used for samples preparation and dough had low hydration.

Table 5. Rapidly and slowly digestible starch of quinoa, sorghum, millet and rice breads.

Properties	Breads Prepared With			
	Quinoa Flour	Sorghum Flour	Millet Flour	Rice Flour
Resistant starch, % d.w.	3.28 ± 0.07	2.79 ± 0.06	1.83 ± 0.05	1.40 ± 0.05
Rapidly digestible starch, g/100 g starch d.w.	48.31 ± 0.10	60.53 ± 0.10	63.26 ± 0.11	73.23 ± 0.11
Slowly digestible starch, g/100 g starch d.w.	15.50 ± 0.11	11.86 ± 0.10	6.51 ± 0.10	8.25 ± 0.10
Digestible starch, g/100 g starch d.w.	63.81 ± 0.11	72.39 ± 0.11	69.77 ± 0.11	81.48 ± 0.11

When analyzing the red and white sorghum flour, Khan et al. [14] reported resistant starch contents of 2.95 and 2.21% d.w., respectively, higher than in durum wheat semolina (0.42% d.w.). The higher content of resistant starch of the sorghum flours could be the result of the inhibitory effect exerted on the digestive enzyme by the polyphenols or could be due to the interactions occurring between starch and proteins impeding the efficient enzyme recognition of the specific substrate [52]. The cooked pasta prepared with 40% red sorghum flour and with 40% white sorghum flour had 1.44 and 1.16% d.w. resistant starch, respectively.

Yousif et al. [17] found that wheat flour supplementation with red and white sorghum flour resulted in the increase of the resistant starch content of the bread samples from 30.1 g/100 dry starch up to 38.9 and 36.4 g/100 dry starch, respectively. The same authors reported lower rapidly digestible starch contents in the wheat bread samples containing red and white sorghum flour (27.7 and 29.0 g/100 dry starch), than in wheat bread (38.4 g/100 dry starch). Additionally, Yousif et al. [17] noted no correlation between the damaged starch content and starch digestibility of the sorghum containing breads. The activity of the in vitro digestive enzymes might have been limited when incorporating the sorghum flour into the wheat breads, through impeding the efficient starch gelatinization or through inhibition exerted by sorghum polyphenolics [52], therefore reducing the rapidly digestible starch levels. In fact, Taylor and Emmambux [52] mentioned no particular chemical or structural characteristics of sorghum starch, which might result in slow digestibility. It appears that the reduced digestibility arises from the presence of sorghum polyphenols and from the interactions between sorghum starch and proteins which are highly cross-linked through di-sulphide bridges [53].

Xu et al. [3] demonstrated that the wheat flour substitution with quinoa flour results in bread samples containing higher contents of slowly digestible starch and resistant starch and reduced in vitro digestibility with lower estimated glycemic index. The increase of the percentage of quinoa flour from 0 to 15% in wheat flour increased the resistant starch content of bread from 21.83 to 31.94%, and decreased the rapidly digestible starch from 54.41 to 39.23% [3]. The authors explained that the high amounts of dietary fibers and polyphenols found in quinoa flour might have contributed to the decreased starch digestibility in bread with quinoa flour. The polyphenols might inhibit the activity of the amylases, by impeding the contact between enzymes and starch. Quinoa seeds appeared to be rich in insoluble simple or highly polymerized phenols which can be associated with the carbohydrates [50]. In addition, as suggested by Li et al. [54], the enzyme assisted starch hydrolysis might be significantly affected by the presence of high proportion of short chains of amylopectin.

4. Conclusions

The composition and physical properties, resistant starch, antioxidant activity and total phenolic content of quinoa, sorghum and millet revealed the huge potential of these flours to be used for developing functional gluten-free bread. All these flours have higher contents of proteins, fat and fiber compared to the rice flour, usually used in the gluten-free products as main ingredient. Compared to the rice bread, the quinoa bread had higher specific volume and lower crumb firmness, while sorghum and millet bread had lower specific volume and higher crumb firmness. All breads had the total phenol contents and antioxidant properties higher compared to the rice bread. Quinoa bread had the best physical properties, and the highest total phenol content. Additionally, quinoa bread had the highest resistant starch content among all investigated bread samples. These results indicate that quinoa demonstrated the highest potential to be used for obtaining high-quality gluten-free breads.

Author Contributions: Conceptualization, I.A. and I.B.; methodology, I.B.; validation, I.A.; formal analysis, I.B.; investigation, I.B. and I.A.; writing—original draft preparation, I.B.; writing—review and editing, I.A.; supervision, I.A. All authors have read and agreed to the published version of the manuscript.

Funding: This research received no external funding.

Acknowledgments: The authors thank to Vitănescu Maricel for supplying the quinoa seeds.

Conflicts of Interest: The authors declare no conflict of interest.

References

1. Elgeti, D.; Nordlohne, S.D.; Foste, M.; Besl, M.; Linden, M.H.; Heinz, V.; Jekle, M.; Becker, T. Volume and texture improvement of gluten-free bread using quinoa white flour. *J. Cereal Sci.* **2014**, *59*, 41–47. [CrossRef]
2. Taylor, J.R.N.; Schober, T.J.; Bean, S.R. Novel food and non-food uses for sorghum and millets. *J. Cereal Sci.* **2006**, *44*, 252–271. [CrossRef]
3. Xu, X.; Luo, Z.; Yang, Q.; Xiao, Z.; Lu, X. Effect of quinoa flour on baking performance, antioxidant properties and digestibility of wheat bread. *Food Chem.* **2019**, *294*, 87–95. [CrossRef]
4. Bazile, D.; Bertero, H.D.; Nieto, C. *State of the Art Report on Quinoa around the World in 2013*; FAO, CIRAD: Santiago, Chile, 2015; p. 603.
5. Ruiz, K.B.; Biondi, S.; Oses, R.; Acuña-Rodríguez, I.S.; Antognoni, F.; Martinez-Mosqueira, E.A.; Coulibaly, A.; Canahua-Murillo, A.; Pinto, M.; Zurita, A.; et al. Quinoa biodiversity and sustainability for food security under climate change. A review. *Agron. Sustain. Dev.* **2014**, *34*, 349–359. [CrossRef]
6. Collar, C.; Angioloni, A. Pseudocereals and teff in complex breadmaking matrices: Impact on lipid dynamics. *J. Cereal Sci.* **2014**, *59*, 145–154. [CrossRef]
7. Vega-Galvez, A.; Miranda, M.; Vergara, J.; Uribe, E.; Puente, L.; Martinez, E.A. Nutrition facts and functional potential of quinoa (*Chenopodium quinoa* willd), an ancient Andean grain: A review. *J. Sci. Food Agric.* **2010**, *90*, 2541–2547. [CrossRef] [PubMed]
8. Srichuwong, S.; Curti, D.; Austin, S.; King, R.; Lamothe, L.; Hernandez, H.G. Physicochemical properties and starch digestibility of whole grain sorghums, millet, quinoa and amaranth flours, as affected by starch and non-starch constituents. *Food Chem.* **2017**, *233*, 1–10. [CrossRef]
9. Navruz-Varli, S.; Sanlier, N. Nutritional and health benefits of quinoa (*Chenopodium quinoa* Willd). *J. Cereal Sci.* **2016**, *69*, 371–376. [CrossRef]
10. Pereira, E.; Encina-Zelada, C.; Barrosa, L.; Gonzales-Barrona, U.; Cadaveza, V.; Ferreira, I. Chemical and nutritional characterization of *Chenopodium quinoa* Willd (quinoa) grains: A good alternative to nutritious food. *Food Chem.* **2019**, *280*, 110–114. [CrossRef]
11. Vidueiros, S.M.; Curti, R.N.; Dyner, L.M.; Binaghi, M.J.; Peterson, G.; Bertero, H.D.; Pallaro, A.N. Diversity and interrelationships in nutritional traits in cultivated quinoa (*Chenopodium quinoa* Willd.) from Northwest Argentina. *J. Cereal Sci.* **2015**, *62*, 87–93. [CrossRef]
12. Miranda, M.; Vega-Galvez, A.; Lopez, J.; Parada, G.; Sanders, M.; Aranda, M.; Uribe, E.; Di Scala, K. Impact of air-drying temperature on nutritional properties, total phenolic content and antioxidant capacity of quinoa seeds (*Chenopodium quinoa* Willd.). *Ind. Crop. Prod.* **2010**, *32*, 258–263. [CrossRef]
13. Pellegrini, M.; Gonzalesb, R.L.; Riccia, A.; Fontechac, J.; Lopez, J.F.; Alvarez, J.A.P.; Martos, M.V. Chemical, fatty acid, polyphenolic profile, techno-functional and antioxidant properties of flours obtained from quinoa (*Chenopodium quinoa* Willd) seeds. *Ind. Crop. Prod.* **2018**, *111*, 38–46. [CrossRef]
14. Khan, I.; Yousif, A.; Johnson, S.K.; Gamlath, S. Effect of sorghum flour addition on resistant starch content, phenolic profile and antioxidant capacity of durum wheat pasta. *Food Res. Int.* **2013**, *54*, 578–586. [CrossRef]
15. Chandrasekara, A.; Shahidi, F. Content of insoluble bound phenolics in millets and their contribution to antioxidant capacity. *J. Agric. Food Chem.* **2010**, *58*, 6706–6714. [CrossRef]
16. Ragaee, S.; Abdel-Aal, E.-S.M.; Noaman, M. Antioxidant activity and nutrient composition of selected cereals for food use. *Food Chem.* **2006**, *98*, 32–38. [CrossRef]
17. Yousif, A.; Nhepera, D.; Johnson, S. Influence of sorghum flour addition on flat bread in vitro starch digestibility, antioxidant capacity and consumer acceptability. *Food Chem.* **2012**, *134*, 880–887. [CrossRef]

18. Nascimento, A.C.; Mota, C.; Coelho, I.; Gueifao, S.; Santos, M.; Matos, A.S.; Gimenez, A.; Lobo, M.; Samman, N.; Castanheira, I. Characterisation of nutrient profile of quinoa (*Chenopodium quinoa*), amaranth (*Amaranthus caudatus*), and purple corn (*Zea mays* L.) consumed in the North of Argentina: Proximates, minerals and trace elements. *Food Chem.* **2014**, *148*, 420–426. [CrossRef]

19. ASRO. *Romanian Standards Catalog for Cereal and Milling Products Analysis*; SR ISO 712:2005, SR 91:2007 and SR ISO 2171/2002; ASRO: Bucharest, Romania, 2008.

20. AACC International. *Approved Methods of Analysis*, Methods 46–11.02, 76–31.01 and 32–40.01, 11th ed.; American Association of Cereal Chemists International: St. Paul, MN, USA, 2000.

21. Asp, N.G.; Johansson, C.G.; Hallmer, H.; Siljestrom, M. Rapid enzymatic assay of insoluble and soluble dietary fiber. *J. Agric. Food Chem.* **1983**, *31*, 476–482. [CrossRef] [PubMed]

22. Gibson, T.S.; Solah, V.A.; McCleary, B.V. A procedure to measure amylose in cereal starches and flours with concanavalin A. *J. Cereal Sci.* **1997**, *25*, 111–119. [CrossRef]

23. Godon, B.; Wilhm, C. *Primary Cereal Processing a Comprehensive Sourcebook*; VCH: New York, NY, USA, 1994.

24. Banu, I.; Stoenescu, G.; Ionescu, V.; Aprodu, I. Physicochemical and rheological analysis of flour mill streams. *Cereal Chem. J.* **2010**, *87*, 112–117.

25. Aprodu, I.; Banu, I. Antioxidant properties of wheat mill streams. *J. Cereal Sci.* **2012**, *56*, 189–195. [CrossRef]

26. Miao, M.; Jiang, B.; Zhang, T.; Jin, Z.; Mu, W. Impact of mild acid hydrolysis on structure and digestion properties of waxy maize starch. *Food Chem.* **2011**, *126*, 506–513. [CrossRef]

27. Fairbanks, D.J.; Burgener, K.W.; Robison, L.R.; Andersen, W.R.; Ballon, E. Electrophoretic characterization of quinoa seed proteins. *Plant Breed.* **1990**, *104*, 190–195. [CrossRef]

28. Mokrane, H.; Amoura, H.; Belhaneche-Bensemra, N.; Courtin, C.M.; Delcour, J.A.; Nadjemi, B. Assessment of Algerian sorghum protein quality [*Sorghum bicolor* (L.) Moench] using amino acid analysis and in vitro pepsin digestibility. *Food Chem.* **2010**, *121*, 719–723. [CrossRef]

29. Hamaker, B.R.; Bugusu, B.A. In Proceedings of the Overview: Sorghum Proteins and Food Quality, Pretoria, South Africa, 2–4 April 2003.

30. United States Department of Agriculture Agricultural Research Service National Nutrient Database for Standard Reference Legacy Release, Basic Report. Available online: https://ndb.nal.usda.gov/ndb/foods/show/20035?fgcd=&manu=&format=&count=&max=25&offset=&sort=default&order=asc&qlookup=QUINOA+CEREAL%2C+UPC%3A+729955573994&ds=&qt=&qp=&qa=&qn=&q=&ing= (accessed on 13 April 2020).

31. Devi, P.B.; Vijayabharathi, R.; Sathyabama, S.; Malleshi, N.G.; Priyadarisini, V.B. Health benefits of finger millet (*Eleusine coracana* L.) polyphenols and dietary fiber: A review. *J. Food Sci. Tech.* **2014**, *51*, 1021–1040. [CrossRef]

32. Kalinova, J.; Moudry, J. Content and quality of protein in proso millet (*Panicum miliaceum* L.) varieties. *Plant Food Hum. Nutr.* **2006**, *61*, 45–49. [CrossRef]

33. Rooney, L.W. Sorghum and pearl millet lipids. *Cereal Chem.* **1978**, *55*, 584–590.

34. Jan, S.; Ghoroi, C.; Saxena, D.C. Effect of particle size, shape and surface roughness on bulk and shear properties of rice flour. *J. Cereal Sci.* **2017**, *76*, 215–221. [CrossRef]

35. Kurek, M.A.; Karp, S.; Wyrwisz, J.; Niu, Y. Physicochemical properties of dietary fibers extracted from gluten-free sources: Quinoa (*Chenopodium quinoa*), amaranth (*Amaranthus caudatus*) and millet (*Panicum miliaceum*). *Food Hydrocoll.* **2018**, *85*, 321–330. [CrossRef]

36. Lamothe, L.M.; Srichuwong, S.; Reuhs, B.L.; Hamaker, B.R. Quinoa (*Chenopodium quinoa* W.) and amaranth (*Amaranthus caudatus* L.) provide dietary fibres high in pectic substances and xyloglucans. *Food Chem.* **2015**, *167*, 490–496. [CrossRef]

37. Lai, V.; Lu, S.; Hsien He, W.; Chen, H.H. Non-starch polysaccharide compositions of rice grains with respect to rice variety and degree of milling. *Food Chem.* **2006**, *101*, 1205–1210. [CrossRef]

38. Chung, H.J.; Liu, Q.; Lee, L.; Wei, D. Relationship between the structure, physicochemical properties and in vitro digestibility of rice starches with different amylose contents. *Food Hydrocoll.* **2011**, *25*, 968–975. [CrossRef]

39. Kraithong, S.; Lee, S.; Rawdkuen, S. Physicochemical and functional properties of Thai organic rice flour. *J. Cereal Sci.* **2018**, *79*, 259–266. [CrossRef]

40. Marston, K.; Khouryieh, H.; Aramouni, F. Effect of heat treatment of sorghum flour on the functional properties of gluten-free bread and cake. *LWT-Food Sci. Technol.* **2016**, *65*, 637–644. [CrossRef]

41. Rosell, C.M. Enzymatic manipulation of gluten-free breads. In *Gluten-Free Food Science and Technology*; Gallagher, E., Ed.; Wiley-Blackwell: Oxford, UK, 2009; pp. 83–98.

42. Schober, T.J. Manufacture of gluten-free specialty breads and confectionery products. In *Gluten-Free Food Science and Technology*; Gallagher, E., Ed.; Wiley-Blackwell: Oxford, UK, 2009; pp. 130–180.

43. Re, R.; Pellegrini, N.; Proteggente, A.; Pannala, A.; Yang, M.; Evans, C.R. Antioxidant activity applying an improved ABTS radical cation decolorization assay. *Free. Radic. Bio. Med.* **1999**, *26*, 1231–1237. [CrossRef]

44. Alvarez-Jubete, L.; Wijngaard, H.; Arendt, E.K.; Gallagher, E. Polyphenol composition and in vitro antioxidant activity of amaranth, quinoa buckwheat and wheat as affected by sprouting and baking. *Food Chem.* **2010**, *119*, 770–778. [CrossRef]

45. Jan, K.N.; Panesar, P.S.; Singh, S. Optimization of antioxidant activity, textural and sensory characteristics of gluten-free cookies made from whole Indian quinoa flour. *LWT-Food Sci. Technol.* **2015**, *93*, 573–582. [CrossRef]

46. Lindenmeier, M.; Hofmann, T. Influence of baking conditions and precursor supplementation on the amounts of the antioxidant pronyl-L-lysine in bakery products. *J. Agric. Food Chem.* **2004**, *52*, 350–354. [CrossRef] [PubMed]

47. Lopez-Contreras, J.J.; Zavala-Garcia, F.; Urias-Orona, V.; Martinez-Avila, G.C.; Rojas, R.; Nono-Medina, G. Chromatic, phenolic and antioxidant properties of sorghum bicolor genotypes. *Not. Bot. Hort. Agrobot. Cluj-Napoca* **2005**, *43*, 366–370. [CrossRef]

48. Chethan, S.; Malleshi, N.G. Finger millet polyphenols: Optimization of extraction and the effect of pH on their stability. *Food Chem.* **2007**, *105*, 862–870. [CrossRef]

49. Sreeramulu, D.; Reddy, C.V.K.; Manchala, R. Antioxidant activity of commonly consumed cereals, millets, pulses and legumes in India. *Indian J. Biochem. Biophys.* **2009**, *46*, 112–115.

50. Aloisi, I.; Parrotta, L.; Ruiz, K.B.; Landi, C.; Bini, L.; Cai, G.; Biondi, S.; Del Duca, S. New insight into quinoa seed quality under salinity: Changes in proteomic and amino acid profiles, phenolic content, and antioxidant activity of protein extracts. *Front. Plant Sci.* **2016**, *7*, 656. [CrossRef]

51. De la Hera, E.; Rosell, C.M.; Gomez, M. Effect of water content and flour particle size on gluten-free bread quality and digestibility. *Food Chem.* **2014**, *151*, 526–531. [CrossRef] [PubMed]

52. Taylor, J.R.N.; Emmambux, M.N. Developments in our understanding of sorghum polysaccharides and their health benefits. *Cereal Chem. J.* **2010**, *87*, 263–271. [CrossRef]

53. Wong, J.H.; Lau, T.; Cai, N.; Singh, J.; Pedersen, J.F.; Vensel, W.H.; Hurkman, W.J.; Wilson, J.D.; Lemaux, P.G.; Buchanan, B.B. Digestibility of protein and starch from sorghum (Sorghum bicolor) is linked to biochemical and structural features of grain endosperm. *J. Cereal Sci.* **2009**, *49*, 73–82. [CrossRef]

54. Li, G.; Zhu, F. Quinoa starch: Structure, properties, and applications. *Carbohydr. Polym* **2018**, *181*, 851–861. [CrossRef] [PubMed]

Publisher's Note: MDPI stays neutral with regard to jurisdictional claims in published maps and institutional affiliations.

applied sciences

MDPI

Article

Effect of By-Products from Selected Fruits and Vegetables on Gluten-Free Dough Rheology and Bread Properties

Fairouz Djeghim [1], Hayat Bourekoua [1], Renata Różyło [2,*], Agata Bieńczak [3], Wojciech Tanaś [4] and Mohammed Nesreddine Zidoune [1]

[1] Equipe de Transformation et Elaboration de Produits Agro-alimentaires (TEPA), Laboratoire de Nutrition et Technologie Alimentaire (LNTA), Institut de la Nutrition, de l'Alimentation et des Technologies Agro-Alimentaires (INATAA), Université Frères Mentouri-Constantine 1, Route de Ain El-Bey, 25000 Constantine, Algeria; fairouzdje@live.fr (F.D.); bourekoua.hayat@umc.edu.dz (H.B.); zidounem@yahoo.fr (M.N.Z.)

[2] Department of Food Engineering and Machines, University of Life Sciences in Lublin, 28 Głęboka Str., 20-612 Lublin, Poland

[3] Department of Foodstuff Technics and Technology, Łukasiewicz Research Network – Industrial Institute of Agricultural Engineering, 31 Starołęcka Str., 60-963 Poznań, Poland; agata.bienczak@pimr.lukasiewicz.gov.pl

[4] Department of Agricultural Forestry and Transport Machines, University of Life Sciences in Lublin, 28 Głęboka Str., 20-612 Lublin, Poland; wojciech.tanas@up.lublin.pl

* Correspondence: renata.rozylo@up.lublin.pl

Abstract: The aim of the study was to investigate the effect of using various by-products (orange and apple pomace, tomato peel, pepper peel, prickly pear peel, and prickly pear seed peel) on the dough rheology and properties of gluten-free bread. The by-products were incorporated into a gluten-free bread formulation based on corn and chickpea flours (2/1 w/w). Different levels of each by-product (0, 2.5, 5, and 7.5% in the basic replacement) were tested. Wheat bread and gluten-free bread without the addition of by-products were used as controls. The results indicated that the by-products increased the maximum dough height, the total CO_2 production, and CO_2 retention coefficient compared to unenriched gluten-free dough. The highest K-value consistency coefficient was observed for the dough enriched with the prickly pear peel. The addition of by-products significantly improved ($p < 0.0001$) the specific volume of gluten-free bread, with values increasing from 1.48 to 2.50 cm^3/g. The hierarchical cluster analysis and the constellation plot showed four groups: the wheat bread group, the second group containing the gluten-free control bread, the group with bread enriched by pomace, and the group with bread enriched with peels, exhibit the same effect on gluten-free bread and the peels exhibit the same effect on gluten-free bread.

Keywords: gluten-free bread; pomace; peel; rheology; bread properties

Citation: Djeghim, F.; Bourekoua, H.; Różyło, R.; Bieńczak, A.; Tanaś, W.; Zidoune, M.N. Effect of By-Products from Selected Fruits and Vegetables on Gluten-Free Dough Rheology and Bread Properties. *Appl. Sci.* **2021**, *11*, 4605. https://doi.org/10.3390/app11104605

Academic Editor: Silvia Mironeasa

Received: 18 April 2021
Accepted: 12 May 2021
Published: 18 May 2021

1. Introduction

Global market data have shown 10.4% compound annual growth rate in gluten-free sales between 2015 and 2020 because of the incidence of celiac disease and other gluten-associated allergies [1,2]. The gluten-free market today provides a variety of foods that can be eaten safely by patients suffering from celiac disease [3]. Bread is one of the major staple foods [4,5]. In general, breads formulated with gluten-free raw materials have low nutritional properties, poor taste, and are of inferior quality. In the absence of gluten, dough presents poor rheological properties (viscoelasticity) and is unable to develop a protein network, affecting the final quality of gluten-free bread [6,7].

The replacement of gluten is one of the most challenging problems in food technology [8]. The use of alternative ingredients including starch, hydrocolloids, protein, enzymes, emulsifiers, and fibers in the preparation of gluten-free bread can improve the texture, mouthfeel, acceptability, shelf life, and nutritional properties of the products. In

133

addition, the viscoelastic properties of gluten can be imitated [9–12]. Similar effects can be expected when adding fruit and vegetable by-products to bread [13].

Some by-products from fruit and vegetable processing are appropriate sources of nutrients and functional ingredients for gluten-free products, and could be used as low-cost ingredients. Fruit and vegetable by-product processing includes pomace, peel, and seed fractions, which are good sources of functional substances. These products offer beneficial bioactive compounds such as carotenoids, enzymes, polyphenols, oils, vitamins, dietary fibers, amino acids, and proteins for gluten-free products [14–16]. Dried fruit pomace can be used in bakery products to substitute flour, sugar, or fat, increasing the amount of fiber and antioxidants and reducing the energy consumption [8]. Each fruit and vegetable contains approximately 5–50% of peel/skin by-products, which are rich in cellulose, hemicellulose, and lignin as major constituents and may also contain other functional groups of lignin, including aldehydes, ketones, alcohols, carboxyl, hydroxide, and phenols [17]. In addition, antioxidant components, flavonoids, minerals, and vitamins are also available in such by-products [18–20].

Many studies are in progress to evaluate the effects of fruit and vegetable by-products (such as orange peel, pomegranate peel, mango peel, grape peel, and potato peel) on wheat bread to improve the texture and quality of the final products [21–23], but only a few studies have been conducted on the utilization of fruit and vegetable peels for gluten-free bread products.

Thus, in this study we evaluated the rheological and gluten-free properties of bread prepared with corn and chickpea flours and enriched with two different types of pomace (from orange and apple) and four different types of peels (from tomato, pepper, prickly pear, and prickly pear seeds). These breads were compared with the non-enriched gluten-free bread and wheat bread.

2. Materials and Methods

2.1. Breadmaking Materials

Soft white wheat flour was obtained from Kenza (Constantine, Algeria), and contained 12.01% moisture, 0.52% ash, 2.08% fat, 7.9% protein content, 77.04% carbohydrates, and 0.45% fiber. Corn flour (Bio Aglut, Constantine, Algeria) had 9.96% moisture, 0.83% ash, 1.8% fat, 7.9% protein content, 76.31% carbohydrates, and 3.2% fiber. Chickpea flour (9.22% moisture, 2.67% ash, 4.35% fat, 22% protein content, 58.33% carbohydrates and 3.43% fiber) was obtained after grinding dried chickpea grains provided by CCLS (Constantine, Algeria). Salt (ENAsel, Setif, Algeria) and instant dry yeast (Saf-instant, Maisons-Alfort, France) were purchased from a local market.

2.2. By-Product Materials

The raw materials for the research were by-products of selected fruits and vegetables including oranges *(Citrus sinensis)*, apples *(Malus domestica)*, tomatoes *(Solanum lycopersicum)*, peppers *(Capsicum annum)*, and prickly pears *(Opuntia ficus-indica)*. These by-products were sourced from Algerian food industries.

The orange pomace (peel, a small amount of pulp and seeds) was obtained after juice extraction from N'Gaous-Conserves (SPA, Batna, Algeria).

Apple, tomato, and pepper pomace were collected from Maison Latina (Chelghoum Laid- Mila, Algeria). Apple pomace (peel, a small amount of pulp, and seeds) came from the processing of apple jam. Tomato pomace (peels and seeds) was derived from the production of tomato paste. The paprika pomace from green sweet pepper and red-hot pepper (peels and seeds) was obtained after the preparation of *Hmiss* (a local Algerian product).

The dried peel of prickly pear seeds was formed during a process of separating the prickly pear seeds from the pulp (Figure 1), which was supplied by Nopaltec (Souk Ahras, Algeria). Fresh, ripe red-yellow prickly pear fruits were harvested from private farms (Aïn beïda Ahriche, Mila, Algeria).

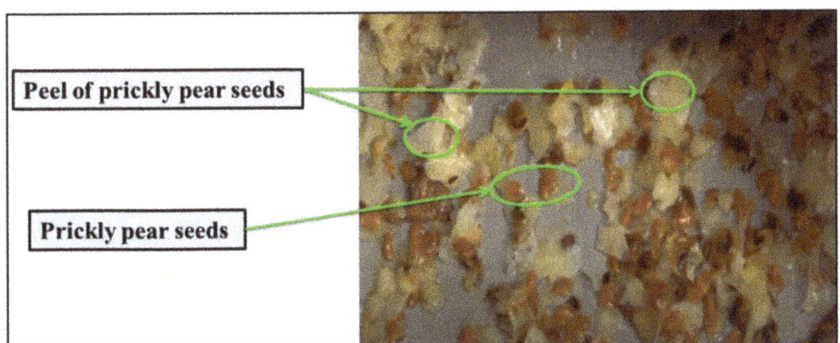

Figure 1. Seeds and seed peels of prickly pear during the processing of separation.

2.2.1. Preparation of By-Product Samples

The orange and apple by-products were dried at 40 °C in a ventilated dryer (MAXEI, S.A. ARRAS MAXEI, Type MC 100, Arras, France). Then, the dried pomace was manually separated from the seeds. The tomato and pepper by-products were dried at 40 °C in a fluid bed dryer (Retsch. TG 200, Haan, Germany). Then, the seeds were separated from the pomace by sieving (1100 μm).

The prickly pear fruits were washed several times using distilled water to remove the thorns, placed on a sieve to drain out surface water, and manually peeled with a knife. The obtained peel (46.60 ± 2.40% of peel) was cut into small pieces, and then the samples were dried in the ventilated dryer at 35 °C. The peel of prickly pear seeds donated by the food industry was already dried. The weight percent of each by-product fraction is shown in Table 1.

Table 1. The proportion of pulp, peel, and seeds in selected fruit and vegetable by-products.

Product	By-Product	Waste Content (%)
Orange *	Peel + pulp	94.50
	Seeds	5.50
Apple *	Peel + pulp	91.22
	Seeds	8.78
Tomato *	Peel	52.13
	Seeds	47.87
Pepper *	Peel	61.5
	Seeds	38.50
Prickly pear **	Peel	46.60
	Seeds + pulp	53.40

* Dried pomace. ** Fresh pomace.

All dried seedless by-product samples (Figure 2) including dried orange pomace (DOP), dried apple pomace (DAP), dried pepper peel (DPP), dried tomato peel (DTP), dried prickly pear peel (DPPP), and dried prickly pear seed peel (DPPSP) were milled to a particle size of 500 μm. The by-product powders were stored at room temperature (25 ± 5 °C) until further use.

2.2.2. Physico-Chemical Analysis of By-Product Samples

Proximate Composition

The moisture, ash, fat, and protein contents of the by-product powders (pomace and peels) were determined by the standard AOAC method [24]. The moisture content was determined by AOAC method 926.12, the ash content by the AOAC method 942.05, and the protein content by the AOAC method 960.52. The fiber content was determined according to the Weende method [25] using a raw fiber extractor. The total carbohydrates were calculated by subtracting: 100 − (%water + %ash + % total fat + % total protein + % total fiber). The chemical composition was determined 3 times.

Figure 2. Dried fruit and vegetable by-product samples. DOP: dried orange pomace; DAP: dried apple pomace; DPP: dried pepper peel; DTP: dried tomato peel; DPPP: dried prickly pear peel; DPPSP: dried prickly pear seed peel.

Extraction of Pectins from By-Product Samples

Extraction was carried out on powders from dried pomace and peels. Each sample (2 g) was mixed with 40 mL of hydrochloric acid (0.1 N), heated at 90 °C for 45 min, and then cooled to room temperature. The insoluble material was then removed by filtration through a nylon strainer. The filtrate was dispersed in 2 volumes of ethanol 95% (V/V); pectin was precipitated overnight at room temperature away from light. The precipitate was collected by filtration through a nylon strainer, washed twice with 70% ethanol, and centrifuged (20 min; 10,000 tr/min, 10 °C). The supernatant was then discarded and the obtained wet pectin was dried at 65 °C until constant weight [26–28]. The percentage yield of pectin samples was calculated as follows [29]:

$$Pectin\ yield\ (\%) = \frac{Obtained\ product}{Initial\ by\ product\ powder} \times 100$$

Water-Holding Capacity

The water-holding capacity was calculated as the water weight which is retained by 1 g of dry by-product sample after soaking and centrifugation [30]. The water-holding capacity (WHC) of samples was determined by mixing 1 g of by-product powder with 15 mL of distilled water in a centrifuge tube. The mixture was vortexed (VELP Scientifica, ZX^3) for 30 s and allowed to hydrate overnight. The suspension was then centrifuged (Sigma 3-30K, Osterode am Harz, Germany) at 15,000× g for 20 min. The supernatant was discarded, and WHC was expressed as gram water retention per gram dry powder [31,32].

2.3. Characteriscics of the Bread Dough

The wheat control dough was made from wheat flour and without by-products. The gluten-free control dough was made with corn flour and chickpea flour, also without by-products. The fortified gluten-free doughs were prepared by adding by-products at different levels (2.5, 5, and 7, 5% w/w). Doughs were prepared according the same process of breadmaking. Then, the properties of the dough were examined with the help of Chopin's rheofermentometer and a rheoviscosimeter.

2.3.1. Rheofermentometric Analysis

A Chopin rheofermentometer (F3 Chopin, Villeneuve La Garenne Cedex, France) was used to measure the gas production and dough development parameters during the fermentation process. A piece of dough (200 g), prepared as described in the breadmaking process, was placed in the rheofermentometer basket, and the piston was placed on dough without the 2 kg of cylindrical weight for gluten-free dough [33–35]. The dough was then fermented for 90 min at 37 °C. The recorded parameters were: Maximum height of dough (H_m), time at a maximum height of dough (T_1), maximum height of gaseous release (H'_m), time of maximum gas formation (T'_1), height of dough at the end of the test (h), weakening coefficient (W) = (H_m-h)/H_m, total CO_2 production (V_t), volume of CO_2 loss (V_l), volume of CO_2 retained (V_r), and CO_2 retention coefficient (R). The analyses were performed in duplicate.

Moreover, to isolate the effect of yeast activity, the adjusted maximum height (H_m^{adj}) was calculated according to the method of Altuna and Ribotta [36].

$$H_m^{adj} = \left(\frac{H_m}{V_t} \right) \times V_{t0}$$

where V_{t0} is the total volume of the gas obtained from the control dough.

2.3.2. Rheological Measurements of Dough (Flow Test)

The flow test of dough was measured with a rheoviscosimeter. All dough ingredients except the yeast were mixed with an overhead mixer (Heidolph RZR 2020, Schwabach, Germany) at 210 rpm. The dough was then left for 20 min before being placed in a Haake VT 550 rheometer (Haake MessTechnik Gmbh Co, Karlsruhe, Germany) to release the residual stress. The dough samples were placed on a parallel plate geometry, the plate diameter was 50 mm (Haake, PK 5, 0.5 grads), and the gaps between the plates were of 4.5 mm. Fine-grained sandpaper was glued to the top and bottom plates to limit possible slippage between the plates and the dough [37–39]. The excess dough was trimmed and the edges coated with paraffin oil to prevent dehydration of the sample. The dough was left for 5 min before starting the test to ease the stress and stabilize the temperature [40]. The temperature was maintained at 25 °C by circulating water and the rheological data were analyzed with HAAKE Rheo Win software version 2.09 over a shear rate range of 1.2–200 s^{-1} for 3 min. The number of measuring points was 100; each test was performed in triplicate. The apparent viscosity was determined using the low power law model (Ostwald de Waele) [35,41]:

$$\eta_{ap} = K.\gamma^{n-1}$$

where η_{ap}: apparent viscosity (Pa.s); γ: shear rate (s^{-1}); and *n*: flow behavior index. K (consistency index, Pa.s^n) represents the stress required to obtain a shear rate of 1 s^{-1}.

2.4. Bread-Making Process

2.4.1. Pre-Hydration of By-Products

By-product powders were soaked in distilled water overnight at room temperature. The amount of soaking in water required for this differed depending on the results of the water-holding capacity for each sample (Table 2). According to Chen and Rubenthaler [42], pre-hydration of the by-product reduces the absorption of water necessary to form the dough and reduces the water competition between the dough ingredients and the dry by-product.

Table 2. Formulations of wheat and gluten-free bread.

	Wheat CTRL1	GF CTRL 2	2.5% B-P	5% B-P	7.5% B-P
Wheat flour (g)	100	0	0	0	0
Corn (g)	0	66.67	65	63.33	61.67
Chickpea flour (g)	0	33.33	32.5	31.67	30.83
DOP/ DAP/ DPP/ DTP/ DPPP/DPPSP (g)	0	0	2.5	5	7.5
Water for dough (mL)	63.25	100	97.5	95	92.5
Water for pre-hydration (mL)	0	0	$2.5 \times X$	$5 \times X$	$7.5 \times X$
Salt(g)	2	2	2	2	2
Yeast (g)	2	2	2	2	2

CTRL: control; DOP: dried orange pomace; DAP: dried apple pomace; DTP: dried tomato peel; DPP: dried pepper peel; DPPP: dried prickly pear peel; DPPSP: dried prickly pear seed peel; B-P: by-product; X: water-holding capacity (WHC).

2.4.2. Preparation of Bread

Bread with 100 g of wheat flour was prepared as control of wheat bread (Wheat CTRL 1). Gluten-free bread made from a mixture of corn flour and chickpea flour at a ratio of 2/1 (*w/w*) and with 0% *w/w* addition of by-products represented the gluten-free control bread (GF CTRL 2). Gluten-free breads were prepared and enriched with pomace and peel powders.

Gluten-free bread recipes were prepared by replacing the base flour with 0, 2. 5, 5, or 7.5% (*w/w*) by-products (DOP, DAP, DPP, DTP, DPPP, or DPPSP). All by-product powders were used after their pre-hydration. The flour, yeast, by-products, and water were mixed for 1 min at medium speed. Then salt, previously dissolved in water, was gradually added to the mixture. The dough was mixed for an additional 6 min; then, the obtained dough was divided, molded, placed in an aluminum foil pan ($9 \times 7 \times 3$ cm), and fermented at 37 °C. The fermentation time varied depending on the T_1—the time at which dough attains the maximum height (H_m) in rheofermentometer measurements. The baking test was carried out in an aluminum foil pan in an electric oven at 230 °C for 30 min. After baking, the bread was cooled at room temperature. The wheat control bread was prepared without any additives, as was the gluten-free control bread. The formulations for each type of bread are presented in Table 2.

2.5. Bread Quality Evaluation

The analysis of wheat and gluten-free bread was performed after 1 h of baking and included: weight loss, specific volume, moisture content, pH of bread, color, and crumb cell analysis.

Weight loss (WL) is the difference between the weight of the dough and the weight of the bread just after baking dividing by the weight of dough [43].

2.5.1. Specific Volume

The loaf volume was calculated by the rapeseed displacement method (10.05) [44], and the specific volume (cm^3/g) was calculated by dividing the volume by the weight.

2.5.2. Moisture Content

Moisture content in bread was determined according to the ICC 110/1 method [45].

2.5.3. Measurement of pH Values

The bread was ground into a fine powder; 10 g of this powder was mixed with 100 mL of distilled water which had an initial pH of 5.68 ± 0.16. The mixture was then placed on an orbital shaker (Heidolph polymax 1040, Schwabach, Germany) and stirred for 30 min at room temperature. The pH was measured in the supernatant solution after a resting time of 10 min [46,47].

2.5.4. Color Analysis

The color of the crust and the crumb of the bread were determined according to the method of He and Pei [48] using the Color Grab color-extracting application (version 3.6.1, 2017, Loomatix Ltd., Munchen, Germany). To ensure that the color capture was not affected by ambient light, a closed polystyrene box (39 × 17 × 28 cm) was used, integrated with a 1.2 W 5 V white LED to obtain an evenly scattered light on top of the sample. The *CIE-L*a*b** color space mode was chosen. This is a mathematical color model based on the sensitivity of the human visual spectrum [49], where L^* represents the lightness, a^* represents green − /red+, and b^* represents blue − /yellow+. Color measurements were taken at 5 different locations on the same crust or crumb.

2.5.5. Crumb Cell Analysis

Image processing was performed using the Image J software (1.43u; Java1.7.0-2132 bit, Wayne Rasband, National Institute of Mental Health (NIH), Bethesda, MD, USA). The bread was cut on a horizontally into 2 slices. The slice images were captured and saved in TIFF format. Each image was cropped to keep only the bread crumb and then converted to 8-bit grayscale to obtain black (crumb pores) and white (crumb) thresholds. The measured parameters were as follows: cell number/mm^2, average pore size, pore area fraction, perimeter, and solidity.

2.6. Statistical Analysis

Statistical analysis was performed using the statistical JMP Trial 15(SAS Institute Inc., Cary, NC, USA). The significance of the difference between the mean values was estimated using one-way analysis of variance (ANOVA), and the means were compared using the Tukey's test ($p \leq 0.05$). Principal component analysis (PCA) and hierarchical cluster analysis (HCA) were carried out to establish the correlation between dough properties and bread quality. The HCA was based on the Ward method using Euclidean distances to show the similarities and differences between the bread samples. The by-product addition percentage parameter was excluded from the HCA. The results of PCA and HCA are presented in graphical form.

3. Results and Discussion

3.1. Physicochemical Properties of By-Products

3.1.1. Proximate Composition

Table 3 shows the proximate composition of the by-products. The moisture content of the samples ranged from 5.23% to 11.41%; protein content ranged from 2.66% to 8.48%, fat from 1.03% to 4.37%, and the fiber and ash contents ranged from 7.41% to 52.90% and from 1.51% to 19.60%, respectively. Carbohydrate content ranged from 16.88% to 64.36%. All the present by-products had a different composition ($p < 0.05$). The highest moisture content was found for DAP and the highest values of ash and protein content were noted for DPPP. These values are higher than those found by Anwar and Sallam [18]. When it comes to fat, DOP had a high value compared to other by-products. DTP, DPPSP, and DPP presented high fiber contents compared to other by-products (greater than 49%). It was found that DAP was rich in carbohydrates; the proximate composition of DAP was confirmed by Skinner and Gigliotti [50]. Differences between the proximate composition of by-products obtained in this study and the results of other studies may be due to geographic location, season, variety, ripeness, or processing.

Table 3. Physico-chemical characteristics of by-products (per 100 g).

	Moisture (%)	Protein (%)	Fat (%)	Fiber (%)	Ash (%)	Carbohydrates (%)	WHC (g/g d.w)	Pectin (%)
DOP	9.90 ± 0.00 [a]	5.40 ± 0.12 [a]	4.37 ± 0.56 [a]	12.41 ± 0.13 [a]	3.76 ± 0.01 [a]	64.16	7.42 ± 0.27 [c]	21.92 ± 1.11 [a]
DAP	11.41 ± 0.28 [b]	3.66 ± 0.23 [b]	3.16 ± 0.80 [c]	15.90 ± 0.02 [b]	1.51 ± 0.08 [b]	64.36	6.84 ± 1.17 [c]	8.65 ± 0.03 [b]
DPP	6.04 ± 0.37 [d]	4.38 ± 0.20 [d]	2.46 ± 0.03 [bc]	49.00 ± 0.1 [d]	2.92 ± 0.00 [c]	35.20	4.61 ± 0.15 [b]	5.76 ± 0.15 [d]
DTP	5.23 ± 0.38 [c]	7.87 ± 0.05 [c]	3.31 ± 0.35 [c]	52.90 ± 0.16 [c]	3.61 ± 0.05 [a]	27.08	5.40 ± 0.34 [b]	3.91 ± 0.24 [c]
DPPP	10.29 ± 0.03 [a]	8.48 ± 0.09 [e]	1.69 ± 0.13 [db]	7.41 ± 0.26 [e]	19.60 ± 0.15 [d]	52.53	4.83 ± 0.33 [b]	7.14 ± 0.07 [e]
DPPSP	7.87 ± 0.04 [e]	7.00 ± 0.23 [f]	1.03 ± 0.07 [d]	50.80 ± 0.30 [f]	16.42 ± 0.34 [e]	16.88	4.59 ± 0.04 [b]	9.85 ± 0.20 [f]
P	0.0000	<0.0001	0.0022	0.0000	0.0000		0.0001	0.0000
F	181,63	93,580.682	15.68	961,272.49	4720.53		19.28	508.60

DOP: dried orange pomace; DAP: dried apple pomace; DTP: dried tomato peel; DPP: dried pepper peel; DPPP: dried prickly pear peel; DPPSP: dried prickly pear seed peel; WHC: water-holding capacity; *P*: *p*-value probability; *F*: F-value Fisher. a–f letters indicate a statistical different of means in the same column. ($p < 0.05$).

3.1.2. Water-Holding Capacity

As shown in Table 3, DOP presented the greatest water-holding capacity (7.42 g/g d.w), followed by DAP and DTP. There was no difference between the WHC of DTP, DPP, DPPP, and DPPSP.

The water-holding capacity of DOP was found to be higher than that reported by Ocen and Xu [51]. The water-holding capacity of DAP was previously demonstrated by Bchir and Rabetafika [52]. Various factors can affect the water-holding capacity of the powder, such as pH, the presence of hydrophilic components such as fiber and free hydroxyl groups, and flour porosity [53–55]. The substances with high WHC can be used as a functional ingredient to promote beneficial health effects, reduce calories, and change the viscosity and texture of processed foods [56,57].

3.1.3. Pectin

As shown in Table 3, DOP had the highest pectin content (21.92%); a lower value was noted for DTP from 3.91%. As reported by Wang and Chen [58] the citrus peel is rich in pectin. Similarly, Sundarraj and Ranganathan [59] showed that citrus peel and apple pomace are the main sources of pectin. The pectin content in DPPP is 9.856%, which is a higher value than that obtained by Anwar and Sallam [18]; this difference may be due to the ripening stage of the fruit and the extraction method. The addition of pectin to bakery products can change their texture, WHC, and consistency behavior. The addition of pectin can also extend the shelf life and be used to replace fat due to their high oil-holding capacity, thus reducing calories without losing taste [60].

3.2. Gas Production and Dough Development Parameters during Fermentation

Table 4 shows the effect of by-product addition on the fermentation characteristics of doughs. Gluten-free dough with DOP, DAP, DPP, DTP, DPP, and DPPSP was characterized by a higher maximum dough height (H_m) compared to the gluten-free control dough, except for the dough with 2.5% DOP and 7.5% DPP, while this parameter was lower for the control wheat dough. Overall, the gluten-free dough with DAP, DTP, and DPPP had higher maximum dough heights (H_m) than the other samples.

The maximum height of gas release (H'_m) of the gluten-free dough increased with the addition of selected dried peels at various levels, except for the 5% and 7.5% DOP addition. The time (T_1) to achieve the maximum dough development ranged from 25 min to 84 min. Overall, all by-products at different levels reduced the time at maximum height (T_1) compared to the wheat dough (at 63 min) and gluten-free dough (at 84 min). The increase in the H'_m value and the decrease in the T_1 value could be related to the accelerated kinetics of CO_2 formation after adding the dried peels.

The time of maximum gas formation (T'_1) was influenced differently by each dried peel at different levels of addition. The same T'_1 value (45 min) was demonstrated for the gluten-free control dough and the DOP dough with 2.5% and 5% addition. The time of porosity (Tx) of all samples with by-products appeared earlier than the Tx of the control wheat dough (55.30 mn) and the gluten-free control dough (31.30 mn). The maximum dough height (H_m) of all samples was reached after the Tx. Verheyen, Jekle [61] indicated that the time when gas started to escape from the dough Tx and its retention capacity was related to the CO_2 volume and the dough rheology. The addition of 7.5% DOP reduced the final dough height (h) to 0.9 mm.

Loss of dough volume W (weakening coefficient) after 90 min was high for all samples compared to the gluten-free and wheat controls, except for the 2.5% DPPSP addition for which the T_1 increased (61:30 mn). The lower weakening coefficient (3.4%) and the higher final height (h) (5.6 mm) indicated that gluten-free dough with 2.5% DPPSP produced a dough with a stronger structure compared to all other samples.

Table 4. Dough development and gas production parameters of wheat bread and gluten-free bread with and without added by-products.

	Addition Level	H_m	H'_m	T_1	T'_1	T_x	h	W	$V_{t\,CO2}$	$V_{l\,CO2}$	V_{rCO2}	R_{CO2}	H^{adj}_m
	%	mm	mm	min	min	min	mm	%	mL	mL	mL	%	mm
Wheat CTRL 1	0	34.6	77.4	63:0	43:30	55:30	32.6	5.8	939	37	902	96.1	34.6
GF CTRL 2	0	4.1	96.3	84:0	45:00	31:30	3.9	4.9	874	259	615	70.4	4.10
CTRL 2+DOP	2.5	4.0	98.1	30:0	45:00	30:00	3.2	20.0	979	80	899	91.9	3.57
	5	5.0	91.9	28:3	45:00	25:30	1.6	68.0	943	89	854	90.6	4.63
	7.5	6.6	85.3	31:3	57:00	28:30	0.9	86.4	953	91	862	90.5	6.05
CTRL 2 +DAP	2.5	6.5	104.7	39:0	49:30	27:00	2.4	63.1	1151	144	1006	87.5	4.94
	5	7.8	105.0	43:30	55:30	31:30	2.4	69.2	1179	132	1047	88.8	5.78
	7.5	7.5	103.9	30:0	54:00	28:30	3.3	56.0	1221	164	1058	86.6	5.37
CTRL 2+DPP	2.5	5.9	106.8	52:30	43:30	24:00	5.0	15.3	1083	107	975	90.1	4.76
	5	6.4	103.0	36:00	36:00	24:00	4.0	37.5	1004	114	890	88.7	5.57
	7.5	3.9	100.3	39:00	39:00	25:30	2.3	41.0	996	100	896	90.0	3.42
CTRL 2+DTP	2.5	8.2	98.8	39:00	40:30	27:00	1.0	87.8	1014	98	917	90.3	7.07
	5	6.5	102.2	31:30	36:00	27:00	1.4	78.5	969	96	874	90.1	5.86
	7.5	7.7	97.6	25:30	33:00	25:30	1.6	79.2	937	90	848	90.4	7.18
CTRL 2+DPPP	2.5	7.6	108.3	48:00	46:30	22:30	4.0	47.4	1118	103	1015	90.8	5.94
	5	7.8	109.1	36:00	42:00	24:00	3.6	53.8	1136	148	988	87.0	6.00
	7.5	5.8	105.7	58:30	46:30	24:00	4.8	17.2	1166	158	1008	86.4	4.35
CTRL 2+DPPSP	2.5	5.8	105.6	61:30	43:30	28:30	5.6	3.4	1070	100	970	90.6	4.74
	5	4.7	102.8	43:30	37:30	24:00	3.3	29.8	993	105	887	89.4	4.14
	7.5	5.7	106.3	31:30	33:00	27:00	4.6	19.3	969	97	872	90.0	5.14

DOP: dried orange pomace; DAP: dried apple pomace; DTP: dried tomato peel; DPP: dried pepper peel; DPPP: dried prickly pear peel; DPPSP: dried prickly pear seed peel; CTRL: control; GF: gluten-free; H_m: maximum height of dough; T_1: time at maximum height of dough; H'_m: maximum height of gaseous release; T'_1: the time of maximum gas formation; h: height of dough at the end of the test; W: weakening coefficient = $(H_m - h)/H_m$; V_t: total CO_2 production volume; V_l: volume of CO_2 loss; V_r: volume of CO_2 retained; R: retention coefficient of CO_2; H_m^{adj}: adjusted maximum height.

The addition of selected dried peels at various levels to gluten-free doughs increased the total volume of CO_2 production (V_t), decreased the volume of CO_2 loss (V_l), and increased the volume of CO_2 retained (V_r) as well as the CO_2 retention coefficient (R), which means that the addition of by-products improves the production and the retention capability of CO_2 compared to the non-enriched gluten-free dough.

Significant gas production and retention have a positive effect on the final quality of the bread, as reported by Martínez and Díaz [34]. The CO_2 retention capacity depends on the ingredients used and internal structure of the dough. The adjusted maximum height (H^{adj}_m) of all samples was found to be lower than H_m after the yeast activity effect was isolated, which explains that the Hm was related to yeast activity and not only to dough rheology, where the selected by-products accelerated yeast activity.

The by-products used contained fermentable sugars such as glucose and fructose [20,62] and minerals such as K^+, Ca^{2+}, and Zn^{2+} [63,64]. The yeast cells consume the fermenting sugars in the dough directly from the by-products or after pre-hydrolyzing the corn starch or chickpea flour, and produce carbon dioxide and ethanol, which are responsible for acidifying the dough during fermentation and the growth phase in the oven. On the other hand, the presence of the Ca^{2+} in fermentation media allows for maximum fermentation efficiency through increased alcohol production [65,66].

3.3. Rheological Properties of Dough

The coefficient of consistency (K), flow behavior index (n), r-value (statistical correlation coefficient), and chi-square ($\chi2$) are presented in Table 5. The power-law model showed low chi-square ($\chi2$) and high values of the correlation coefficient (r) for all samples. The value of consistency coefficient ranged between 103 and 1049 Pa.s^n. The highest values of the K consistency coefficient were observed for DPPP addition and then DPPSP. The lowest values were found in the samples of DPP dough. There was no statistically significant difference between the K consistency coefficient of the control bread and DOP, DAP, and DPP at the 2.5% and 5% additive levels and between the K of DTP at the 5%

additive level and DPPSP at the 2.5% and 5% additive levels. As the amount of DOP, DAP, DPP, or DTP increased in the dough, the K consistency coefficient decreased. This could be related to the pre-hydration of the by-products before adding them to the dough, where the water absorption and the consistency of the dough are reduced. The flow behavior index ranged from −0.436 to 0.33. In accordance with Ronda, Pérez-Quirce [35], the n values of gluten-free dough ranged between 0.25 and 0.35. Then, values of flow behavior index of whole dough were less than 1, indicating pseudo-plasticity (shear thinning) of the samples. In the case of pseudoplastic substances, the viscosity decreases with the increase in the shear rate because the connections between the device components break down as a result of shearing, which, under the influence of shear, decomposes the interactions between the components of the system [67]. Increasing levels of DOP, DPP, and DPPP additive reduced the value of the flow behavior n index of these samples.

Table 5. Rheological parameters of wheat bread and gluten-free bread with and without added by-products.

	Addition Level (%)	K (Pa.s^n)	n	χ^2	r
WheatCTRL 1	0	141.90 ± 25.74 [e]	0.33 ± 0.04 [a]	0.25 ± 0.06	0.97 ± 0.007
GF CTRL 2	0	173.30 ± 20.08 [de]	0.29 ± 0.02 [ab]	0.47 ± 0.26	0.97 ± 0.007
DOP	2.5	216.60 ± 134.49 [de]	0.08 ± 0.01 [bcdefg]	1.77 ± 0.40	0.93 ± 0.01
	5	221.70 ± 48.93 [de]	−0.11 ± 0.09 [abcde]	5.21 ± 3.76	0.88 ± 0.04
	7.5	134.55 ± 9.83 [e]	-0.25 ± 0.04 [efg]	0.94 ± 0.14	0.98 ± 0.004
DAP	2.5	337.30 ± 95.74 [bcde]	0.10 ± 0.06 [abcde]	0.54 ± 0.27	0.98 ± 0.01
	5	291.40 ± 22.84 [de]	-0.12 ± 0.02 [fg]	4.88 ± 2.67	0.89 ± 0.06
	7.5	159.60 ± 21.64 [e]	0.03 ± 0.02 [abcdef]	4.37 ± 0.05	0.87 ± 0.01
DPP	2.5	289.65 ± 130.74 [cde]	0.05 ± 0.05 [abcde]	0.86 ± 0.62	0.97 ± 0.02
	5	182.76 ± 67.76 [e]	0.17 ± 0.04 [abcd]	1.73 ± 1.19	0.87 ± 0.05
	7.5	103.71 ± 39.73 [e]	0.23 ± 0.08 [abc]	0.28 ± 0.17	0.98 ± 0.01
DTP	2.5	955.40 ± 323.29 [ab]	−0.20 ± 0.00 [defg]	2.83 ± 0.46	0.94 ± 0.01
	5	229.05 ± 80.40 [de]	0.14 ± 0.01 [cdefg]	1.15 ± 1.43	0.95 ± 0.06
	7.5	125.75 ± 20.7 [e]	0.162 ± 0.08 [abcde]	0.05 ± 0.03	0.99 ± 0.00
DPPP	2.5	822.70 ± 37.62 [abcd]	−0.211 ± 0.24 [defg]	4.93 ± 5.68	0.92 ± 0.07
	5	1049.00 ± 548.71 [a]	−0.376 ± 0.16 [fg]	8.17 ± 2.11	0.92 ± 0.01
	7.5	895.85 ± 106.84 [abc]	−0.436 ± 0.06 [g]	4.92 ± 1.84	0.93 ± 0.02
DPPSP	2.5	399.0 ± 285.81 [abcde]	0.033 ± 0.12 [abcdef]	1.12 ± 1.03	0.96 ± 0.02
	5	434.35 ± 63.00 [abcde]	-0.078 ± 0.17 [abcdefg]	4.53 ± 3.01	0.90 ± 0.03
	7.5	480.05 ± 24.54 [abcde]	−0.03 ± 0.18 [abcdefg]	2.31 ± 3.10	0.95 ± 0.06
P		< 0.0001	< 0.0001		
F		6.48	7.50		

DOP: dried orange pomace; DAP: dried apple pomace; DTP: dried tomato peel; DPP: dried pepper peel; DPPP: dried prickly pear peel; DPPSP: dried prickly pear seed peel; CTRL: control; GF: gluten-free; K: consistency coefficient; n: flow behavior index; r: statistical correlation coefficient; χ^2: chi-square; P: p-value probability; F: F-value Fisher; a–g letters indicate a statistical different of means in the same column ($p < 0.05$).

According to Cheremisinoff [68] and Ronda, Pérez-Quirce [35], the coefficient of consistency k and the flow behavior index n depend on temperature, pressure, and formulation. While k is more temperature-sensitive than n on the non-Newtonian liquid food [69], the suspension becomes less viscous as temperature increases and more viscous when pressurized [70]. The flow behavior index varies depending on the formulation, in particular the hydration of the dough. Some samples showed a negative value of the flow behavior index n. There are very few studies that explain the negative values for the flow behavior index. These negative values can be attributed to the presence of slippage in the fluid, viscosity dispersion, or molecular degradation of the sample [71–73].

3.4. Characteristics of the Quality of Bread

The weight loss, specific volume, pH-value, and moisture content of various breads are presented in Table 6.

Table 6. Basic properties of wheat bread and gluten-free bread with and without added by-products.

	Level Addition (%)	WL (%)	V_{sp} (cm^3/g)	pH-Value	Moisture Content (%)
Wheat CTRL 1	0	25 ± 0.015 [bcd]	3.39 ± 0.02 [a]	5.65 ± 0.02 [gh]	27.21 ± 0.01 [k]
GF CTRL 2	0	18.09 ± 0.22 [gh]	1.48 ± 0.22 [i]	5.76 ± 0.12 [c]	29.60 ± 0.23 [h]
DOP	2.5	18.33 ± 2.35 [h]	2.00 ± 0.03 [f]	5.59 ± 0.06 [k]	31.13 ± 0.06 [a]
	5	26.11 ± 0.78 [ab]	2.25 ± 0.02 [c]	5.26 ± 0.23 [n]	30.19 ± 0.42 [d]
	7.5	27.77 ± 1.57 [a]	2.46 ± 0.05 [b]	5.21 ± 0.50 [o]	30.02 ± 0.50 [f]
DAP	2.5	23.88 ± 0.78 [cde]	2.02 ± 0.08 [f]	5.68 ± 0.16 [ef]	30.24 ± 0.04 [d]
	5	26.66 ± 3.14 [ab]	2.04 ± 0.01 [f]	5.58 ± 0.03 [l]	30.20 ± 0.16 [d]
	7.5	25.55 ± 0.00 [bc]	2.16 ± 0.10 [ef]	5.51 ± 0.01 [m]	30.42 ± 0.01 [c]
DPP	2.5	20.45 ± 0.37 [f]	1.69 ± 0.01 [h]	5.68 ± 0.36 [fg]	29.57 ± 0.36 [h]
	5	20.45 ± 0.29 [f]	2.04 ± 0.06 [f]	5.57 ± 0.20 [l]	29.63 ± 0.29 [h]
	7.5	20.51 ± 0.33 [f]	2.10 ± 0.10 [de]	5.63 ± 0.40 [i]	30.72 ± 0.12 [b]
DTP	2.5	23.88 ± 0.78 [cde]	2.21 ± 0.02 [cd]	5.57 ± 0.08 [l]	29.01 ± 0.20 [i]
	5	23.33 ± 0.00 [de]	2.10 ± 0.10 [ef]	5.66 ± 0.10 [h]	29.57 ± 0.08 [h]
	7.5	19.42 ± 0.007 [fg]	2.20 ± 0.00 [cd]	5.70 ± 0.01 [e]	29.57 ± 0.40 [h]
DPPP	2.5	19.05 ± 0.01 [fg]	1.83 ± 0.00 [g]	5.73 ± 0.40 [d]	30.00 ± 0.03 [e]
	5	19.28 ± 0.08 [fg]	2.50 ± 0.04 [b]	5.80 ± 0.14 [b]	30.35 ± 0.09 [c]
	7.5	22.76 ± 0.16 [e]	2.15 ± 0.004 [de]	5.68 ± 0.03 [fg]	29.70 ± 0.21 [g]
DPPSP	2.5	20.32 ± 0.10 [f]	1.72 ± 0.08 [h]	5.61 ± 0.08 [i]	28.71 ± 0.12 [j]
	5	20.84 ± 0.48 [f]	1.72 ± 0.04 [h]	5.61 ± 0.04 [i]	30.20 ± 0.05 [d]
	7.5	20.9 ± 0.26 [f]	2.02 ± 0.20 [f]	5.87 ± 0.06 [a]	30.22 ± 0.03 [d]
P		< 0.0001	< 0.0001	< 0.0001	< 0.0001
F		22.34	181.02	1056.020	10211.024

DOP: dried orange pomace; DAP: dried apple pomace; DTP: dried tomato peel; DPP: dried pepper peel; DPPP: dried prickly pear peel; DPPSP: dried prickly pear seed peel; CTRL: control; GF: gluten-free; WL: weight loss; V_{sp}: specific volume; P: p-value probability; F: F-value Fishera–o letters indicate a statistical different of means in the same column ($p < 0.05$).

3.4.1. Weight Loss

Table 6 shows that the weight loss was greater for all formulas compared to the gluten-free control bread ($p < 0.05$). There was no significant difference in the weight loss values of bread containing different amounts of DPP and DPPSP by-products. The weight loss of gluten-free bread increased with increasing levels of DOP, DAP, and DPPP, but decreased with increasing levels of DTP. At high DOP, DAP, or DPPP values, the water cannot be retained in the bread. This may be due to the water-holding capacity of the by-products. Weight loss before and after baking of bread is generally specifically associated with water loss. The addition of pre-hydration by-products increased the water loss during baking, which increased the weight loss of the bread [52]. Compared to the gluten-free control bread, the percent weight loss was greater for the wheat control bread, the DOP bread, and the DAP bread at 5% and 7.5% additions. Furthermore, DOP bread with 2.5% addition resulted in little weight loss (18.33%). Regarding the WL results, the samples with DPPP proved to be more effective in reducing weight loss (19.05–22.76%). In the study of Milde, Ramallo [74], low weight loss was sought to ensure a moisture content that impeded the dehydration of bread and to reduce the hardness.

3.4.2. Specific Volume of Bread

The specific volume of gluten-free bread varies depending on the ingredients and the preparation process [75]. The addition of by-products influences the specific volume of bread to a different extent ($p < 0.0001$). The specific volume of gluten-free bread ranged from 1.48 to 2.50 cm^3/g. All formulas had a much higher specific volume compared to the gluten-free control bread (1.48 cm^3/g), but a lower specific volume compared to the wheat control bread (3.39 cm^3/g)

The specific volume was the highest for DPPP bread at 5% (2.50 cm^3/g), followed by DOP (2.46 cm^3/g) at 7.5%, and the lowest specific volume was recorded for DPP bread (1.69 cm^3/g) with an addition of 2.5%. These results are supported by O'shea et al. [76] with 4% addition of orange pomace. There were no significant differences between the specific volume of DOP bread at 2.5%, DAP bread at 2.5%, and DPP bread at 5% addition.

Overall, the specific volume increased with increasing levels of by-products, with a maximum found at 7.5% (w/w). These results are supported by Arslan et al. [77], who studied the effect of powdered guava pulp on gluten-free bread, Singh et al. [78], who assessed the effect of dietary fiber from black carrot pomace on gluten-free rice muffins, and TÜrKEr et al. [79], who investigated the effects of the green banana peel on gluten-free cakes. According to these authors, adding more than 7.5% by-products reduced the specific volume of gluten-free products. Parra et al. [75] reported that the right balance between the amount of apple pomace and water allowed for a gluten-free bread with an acceptable specific volume to be obtained.

3.4.3. pH Value of Bread

The pH value of bread ranged from 5.21 to 5.87. Except for 5% DPPP bread and 7.5% DPPSP bread, all formulas had a significantly lower pH value than gluten-free control bread ($p < 0.05$). The pH values were similar for DPP bread at 7.5% and DPPSP bread at 2.5% and 5% additive levels. DTP and DPPSP increased the pH value of bread.

The pH of the gluten-free bread was reduced with the addition of DOP and DAP. Similar results were obtained by Majzoobi et al. [47] for carrot pomace powder, where the pH value of gluten-free cakes was reduced due to the presence of organic acid [80], amino acids [81,82], and other acidic components.

3.4.4. Moisture Content of Bread

As shown in Table 6, the bread moisture content ranged from 27.21% to 31.13%. The control wheat bread had the lowest value of moisture. The addition of DOP, DAP, DPPP, and DPPSP increased the moisture content of gluten-free bread. A lower value (28.71%) was noted for DPPSP with 2.5% addition. The moisture content of the gluten-free bread was gradually increased with the gradual addition of DPP and DPPSP and decreased by the addition of DOP. The moisture content measurements showed no significant differences between gluten-free control bread, DPP bread with 2.5% and 5% addition, and DTP bread with 5% and 7.5% addition. Increased bread moisture contributes to an increase in the bread weight.

3.4.5. Crust and Crumb Color of Bread

The results concerning the color of the crust and crumbs are presented in Table 7. A golden brown crust and creamy white bread crumbs are the most important factors indicating the quality of a bakery product to consumers [5]. The addition of by-products had a significant ($p < 0.05$) effect on the color characteristics of crust and crumbs.

The lightness component (L^* for crumb and crust) decreased with the addition of by-products compared to the control wheat and gluten-free bread. For each by-product, the lightness of the crust and crumb color deceased as the level of the additive increased.

Table 7. Crust and crumb colors of wheat bread and gluten-free bread with and without added by-products.

	Addition Level (%)	Crust Color			Crumb Color		
		$L*$	$a*$	$b*$	$L*$	$a*$	$b*$
Wheat CTRL1	0	81.06 ± 2.23 [a]	4.78 ± 1.69 [cdef]	45.58 ± 1.11 [k]	91.62 ± 2.09 [a]	−2.8 ± 0.51 [hi]	24.18 ± 3.54 [n]
GF CTRL 2	0	74.6 ± 3.95 [bc]	4.4 ± 1.54 [def]	53.96 ± 2.06 [bcdef]	80.42 ± 2.62 [bc]	−1.64 ± 0.70 [gh]	45.36 ± 1.30 [de]
DOP	2.5	69.46 ± 9.63 [cd]	2.075 ± 1.69 [f]	47.84 ± 1.96 [ijk]	66.86 ± 0.96 [gh]	−0.64 ± 0.42 [fg]	34.42 ± 1.49 [kl]
	5	58.82 ± 3.02 [fgh]	14.52 ± 1.89 [bc]	52.68 ± 3.91 [defg]	68.4 ± 2.27 [fg]	−0.12 ± 0.94 [ef]	35.24 ± 2.05 [jk]
	7.5	62.68 ± 2.42 [efg]	10.48 ± 2.32 [bcdef]	51.88 ± 2.30 [defgh]	65.84 ± 1.32 [gh]	1.76 ± 1.46 [cd]	40.36 ± 1.28 [fgh]
DAP	2.5	66.58 ± 6.88 [de]	9.58 ± 2.93 [bcdef]	54.25 ± 3.32 [bcde]	64.12 ± 1.55 [hi]	0.00 ± 1.33 [ef]	33.2 ± 1.47 [klm]
	5	62.12 ± 2.05 [efgh]	8.54 ± 3.06 [bcdef]	50.88 ± 3.35 [fghi]	61.2 ± 2.1 [ijk]	0.49 ± 1.08 [def]	31.66 ± 1.06 [lm]
	7.5	63.56 ± 3.72 [ef]	5.82 ± 1.89 [bcdef]	46.36 ± 2.60 [jk]	60.74 ± 1.95 [k]	1.08 ± 0.81 [de]	31.66 ± 1.06 [m]
DPP	2.5	70.88 ± 4.12 [cd]	10.32 ± 2.48 [bcdef]	56.1 ± 1.04 [abc]	72.56 ± 6.52 [de]	3.48 ± 1.49 [b]	44.14 ± 2.17 [e]
	5	71.08 ± 3.32 [cd]	12.88 ± 2.29 [bcde]	54.64 ± 1.38 [abcd]	78.94 ± 1.47 [c]	3.32 ± 1.02 [b]	47.48 ± 1.22 [bc]
	7.5	66.84 ± 5.78 [de]	15.48 ± 2.64 [b]	53.2 ± 1.97 [cdefg]	70.5 ± 1.54 [ef]	9.00 ± 0.44 [a]	48.58 ± 1.04 [b]
DTP	2.5	57.52 ± 1.29 [h]	13.36 ± 0.89 [bcde]	52.14 ± 1.68 [defgh]	63.88 ± 1.71 [hij]	1.82 ± 1.25 [cd]	38.92 ± 2.03 [hi]
	5	56.7 ± 2.28 [gh]	14.04 ± 1.57 [bcd]	54.5 ± 2.17 [bcd]	60.86 ± 1.21 [jk]	3.98 ± 0.89 [b]	42.04 ± 2.12 [f]
	7.5	62.68 ± 2.7 [efg]	6.9 ± 0.81 [bcdef]	54.66 ± 0.86 [abcd]	74.06 ± 1.52 [d]	3.06 ± 0.97 [bc]	51.18 ± 1.58 [a]
DPPP	2.5	70.56 ± 3.13 [cd]	12.7 ± 2.07 [a]	57.66 ± 1.75 [a]	80.66 ± 1.97 [bc]	−1.88 ± 0.60 [gh]	46.06 ± 1.24 [cde]
	5	69.66 ± 3.59 [cd]	7.46 ± 1.42 [bcdef]	56.64 ± 2.46 [ab]	80.48 ± 2.99 [bc]	3.56 ± 1.22 [i]	46.24 ± 1.79 [cd]
	7.5	56.86 ± 9.17 [gh]	14.58 ± 1.74 [bc]	51.22 ± 3.91 [efghi]	60.72 ± 2.15 [k]	3.70 ± 1.81 [b]	47.02 ± 1.13 [bcd]
DPPSP	2.5	80.34 ± 2.73 [ab]	3.36 ± 1.18 [ef]	50.82 ± 4.77 [fghi]	82.2 ± 2.05 [b]	−1.86 ± 0.82 [gh]	41.04 ± 0.79 [fg]
	5	62.18 ± 6.19 [efgh]	12.84 ± 1.5 [bcde]	50.68 ± 1.40 [ghi]	74.46 ± 3.28 [d]	0.25 ± 0.36 [def]	39.62 ± 0.71 [gh]
	7.5	73.28 ± 4.81 [c]	6.28 ± 3.00 [bcdef]	49.32 ± 0.75 [hij]	77.74 ± 2.39 [c]	−0.24 ± 0.87 [f]	37.06 ± 1.35 [ij]
P		0.0000	0.0000	0.0000	0.0000	0.0000	0.0000
F		11.81	2.72	8.26	63.74	39.66	86.24

DOP: dried orange pomace; DAP: dried apple pomace; DTP: dried tomato peel; DPP: dried pepper peel; DPPP: dried prickly pear peel; DPPSP: dried prickly pear seed peel; CTRL: control; GF: gluten-free. P: p-value probability. F: F-value Fisher. a–n letters indicate a statistical different of means in the same column ($p < 0.05$).

Increased levels of DOP, DPP, DTP, and DPPP allowed increasing yellowness (*b**) and redness (*a**) of the gluten-free bread crumb. The gluten-free color of the bread crumb supplemented with by-products was directly correlated with the ingredients used in the production of the dough.

The color of bread depends on the formulation or baking condition. Maillard reactions and caramelization of the crust are responsible for the changes in the color parameters between the crust and the crumb and the development of brown color on the surface at high temperature, but the color given to the ingredients used in bread formulation may mask this color [5,55,83,84].

3.4.6. Crumb Structure of Bread

The number of cells/mm^2, average size, pore area fraction, perimeter, circularity, and solidity of crumb cellular structure obtained from image analysis are shown in Table 8.

The number of cells/mm^2 ranged from 78 to 308 cells/mm^2. A higher number of cells/mm^2 is demonstrated for DPPP and DPP at 2.5% followed by DPPSP at 7.5%, and a lower number was demonstrated for DOP at 5% and DAP at 2.5%. A higher average size was obtained for the wheat control bread (1.672 mm). The average crumb size increased significantly with the addition of DAP, DPP, and DTP and decreased with the addition of DPPSP. Crumbs of the wheat control bread and DOP and DPPP at 5% were characterized by a large perimeter.

Gluten-free bread with a large number of cells is characterized by an aerated structure against the crumbs with a small number of cells. According to Jafari et al. [85] the lower number of cells and the higher average size reflected the aerated structure. Solidity is a measure of the shape of disorder gas cells. The solidity value is lower for irregular shape of gas cells and higher for regular shape [86,87]. The solidity of samples ranged from 0.85 to 0.92. The shape of the gas cells DPP at 5% and DPPSP at 2.5% had a regular structure and uniform shape (solidity = 0.91 and 0.92, respectively) and greater roundness (circularity = 0.91 and 0.93, respectively) compared to control gluten-free bread. The crumb appearance of control and enriched bread is shown in Figure 3.

Table 8. Crumb structure of wheat bread and gluten-free bread with and without added by-products.

		Number of Cells/mm^2	Average Cell Size (mm^2)	Area Fraction (%)	Perimeter	Circularity	Solidity
WheatCTRL1	0	203 ± 8 [bcd]	1.672 ± 0.26 [a]	18.25 ± 1.34 [a]	4.32 ± 0.27 [a]	0.81 ± 0.007 [i]	0.85 ± 0.006 [f]
GFCTRL 2	0	239 ± 13 [abc]	0.46 ± 0.12 [cde]	12.85 ± 3.18 [bc]	2.28 ± 0.29 [defgh]	0.87 ± 0.01 [cdefg]	0.87 ± 0.00 [cdef]
DOP	2.5	122 ± 23 [ef]	0.55 ± 0.19 [cde]	5.85 ± 2.33 [hi]	2.64 ± 0.50 [cdef]	0.85 ± 0.02 [gh]	0.86 ± 0.00 [def]
	5	78 ± 6 [f]	0.91 ± 0.14 [b]	7.25 ± 0.49 [efgh]	3.08 ± 0.06 [bc]	0.86 ± 0.01 [cdefg]	0.881 ± 0.01 [cd]
	7.5	100 ± 0 [f]	0.67 ± 0.04 [bc]	5.9 ± 0.57 [ghi]	2.80 ± 0.15 [bcd]	0.86 ± 0.01 [hi]	0.85 ± 0.01 [def]
DAP	2.5	86 ± 6 [f]	0.49 ± 0.16 [cde]	3.65 ± 0.92 [i]	2.15 ± 0.15 [efgh]	0.89 ± 0.03 [bcde]	0.89 ± 0.03 [bc]
	5	116 ± 20 [f]	0.61 ± 0.09 [cd]	6.57 ± 2.21 [fghi]	2.67 ± 0.25 [cde]	0.86 ± 0.02 [efgh]	0.87 ± 0.01 [def]
	7.5	125 ± 21 [ef]	0.65 ± 0.17 [bc]	7.97 ± 0.61 [efgh]	2.80 ± 0.32 [cd]	0.85 ± 0.02 [gh]	0.86 ± 0.01 [ef]
DPP	2.5	308 ± 25 [a]	0.34 ± 0.05 [e]	11.65 ± 1.91 [bcd]	2.02 ± 0.20 [gh]	0.88 ± 0.02 [cdef]	0.88 ± 0.01 [cd]
	5	200 ± 2 [cde]	0.36 ± 0.02 [de]	6.90 ± 0.99 [fghi]	1.87 ± 0.17 [h]	0.91 ± 0.00 [ab]	0.91 ± 0.00 [ab]
	7.5	239 ± 35 [abc]	0.37 ± 0.02 [de]	9.25 ± 0.78 [defg]	2.08 ± 0.07 [fgh]	0.89 ± 0.01 [bc]	0.89 ± 0.00 [bc]
DTP	2.5	210 ± 9 [bcd]	0.46 ± 0.12 [cde]	9.30 ± 1.25 [def]	2.35 ± 0.25 [defgh]	0.87 ± 0.02 [cdef]	0.88 ± 0.00 [cde]
	5	262 ± 37 [ab]	0.48 ± 0.08 [cde]	11.95 ± 0.59 [bcd]	2.50 ± 0.06 [defg]	0.85 ± 0.02 [gh]	0.86 ± 0.01 [f]
	7.5	242 ± 58 [abc]	0.50 ± 0.03 [cde]	10.30 ± 3.11 [cde]	2.35 ± 0.01 [defgh]	0.90 ± 0.00 [bc]	0.89 ± 0.00 [bc]
DPPP	2.5	308 ± 8 [a]	0.52 ± 0.05 [cde]	14.10 ± 0.42 [bc]	2.46 ± 0.15 [defgh]	0.88 ± 0.01 [bcdef]	0.88 ± 0.00 [bcd]
	5	175 ± 7 [de]	0.89 ± 0.11 [b]	14.55 ± 1.20 [b]	3.37 ± 0.22 [b]	0.86 ± 0.01 [defgh]	0.87 ± 0.01 [def]
	7.5	190 ± 8 [cd]	0.49 ± 0.19 [cde]	11.05 ± 2.33 [cd]	2.39 ± 0.52 [defgh]	0.85 ± 0.01 [fgh]	0.87 ± 0.01 [def]
DPPSP	2.5	115 ± 76 [f]	0.59 ± 0.31 [cde]	5.20 ± 0.85 [hi]	2.37 ± 0.63 [defgh]	0.93 ± 0.02 [a]	0.92 ± 0.00 [a]
	5	223 ± 56 [bcd]	0.44 ± 0.04 [cde]	10.45 ± 1.06 [cde]	2.20 ± 0.01 [efgh]	0.90 ± 0.03 [bcd]	0.89 ± 0.02 [bc]
	7.5	302 ± 29 [a]	0.43 ± 0.09 [cde]	13.05 ± 1.20 [bc]	2.11 ± 0.24 [fgh]	0.87 ± 0.01 [cdefg]	0.88 ± 0.00 [cd]
P		< 0.0001	0.001	< 0.0001	0.004	0.000	0.001
F		12.15	9.16	10.34	8.00	7.414	6.653

DOP: dried orange pomace; DAP: dried apple pomace; DTP: dried tomato peel; DPP: dried pepper peel; DPPP: dried prickly pear peel; DPPSP: dried prickly pear seed peel; CTRL: control; GF: gluten-free. *P*: *p*-value probability. *F*: F-value Fisher. a–i letters indicate a statistical different of means in the same column ($p < 0.05$).

3.5. Cluster Analysis

A hierarchical cluster analysis and constellation plot were performed to identity analogies of the effects of by-products on the properties of the bread (Figure 4). The constellation plot consisted of four clusters. Group A included one cluster (C1) containing only the wheat control bread, and group B included three clusters: C2 (gluten-free control bread), C3, and C4 (gluten-free bread containing different by-products). The C3 cluster contained the highest counts of similar bread. The similarity between DTP, DPP, DPPP, and DPPSP at the additive levels of 5% and 7.5% was observed in the C3 cluster and the gluten-free bread with the addition of DOP, DAP, and DPPPS at the 2.5% additive level in the C4 cluster. The wheat control bread and gluten-free control bread differed highly from those with all samples containing by-products. The wheat control bread was characterized by higher values of Vsp, average size, perimeter, H_m, h, Tx, and R_{CO2}, and the control gluten-free bread was characterized by a long time at the maximum height of dough T_1, a high value of $V_{1\,CO2}$, and a low value of Vs, average size, H_m, and R_{CO2} (Figure 5).

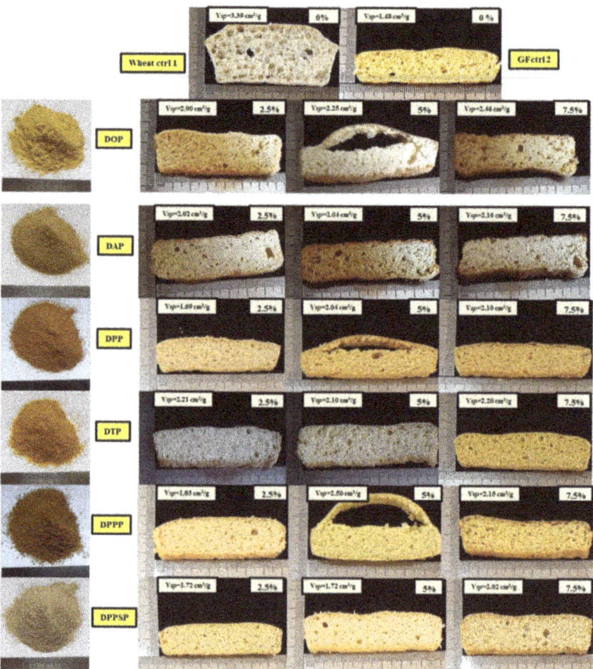

Figure 3. Crumb appearance of wheat bread and gluten-free bread with and without added by-products. DOP: dried orange pomace; DAP: dried apple pomace; DTP: dried tomato peel; DPP: dried pepper peel; DPPP: dried prickly pear peel; DPPSP: dried prickly pear seed peel; CTRL: control; GF: gluten-free; V*sp*: specific volume.

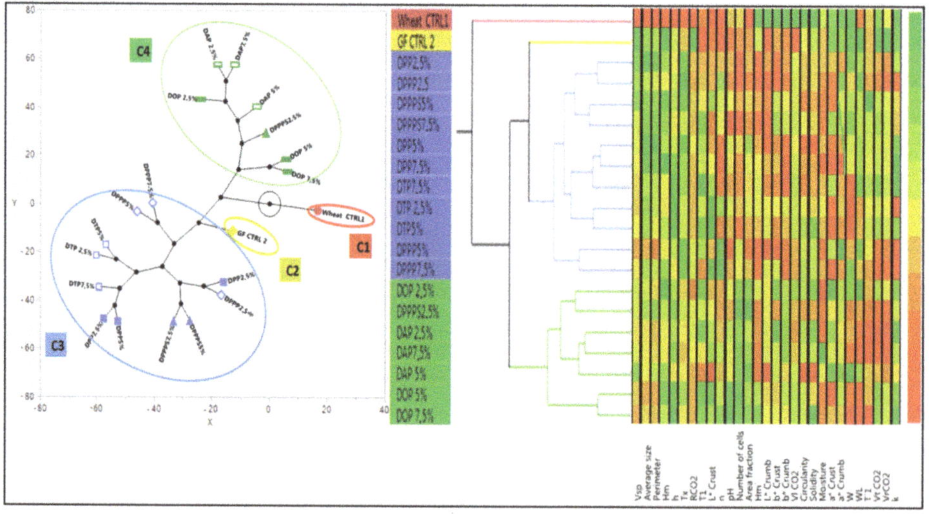

Figure 4. Constellation plot and hierarchical clustering of the investigated wheat and gluten-free bread samples based on the Ward method (level increase: orange, level decrease: green). DOP: dried orange pomace; DAP: dried apple pomace; DTP: dried tomato peel; DPP: dried pepper peel; DPPP: dried prickly pear peel; DPPSP: dried prickly pear seed peel; CTRL: control; GF: gluten-free.

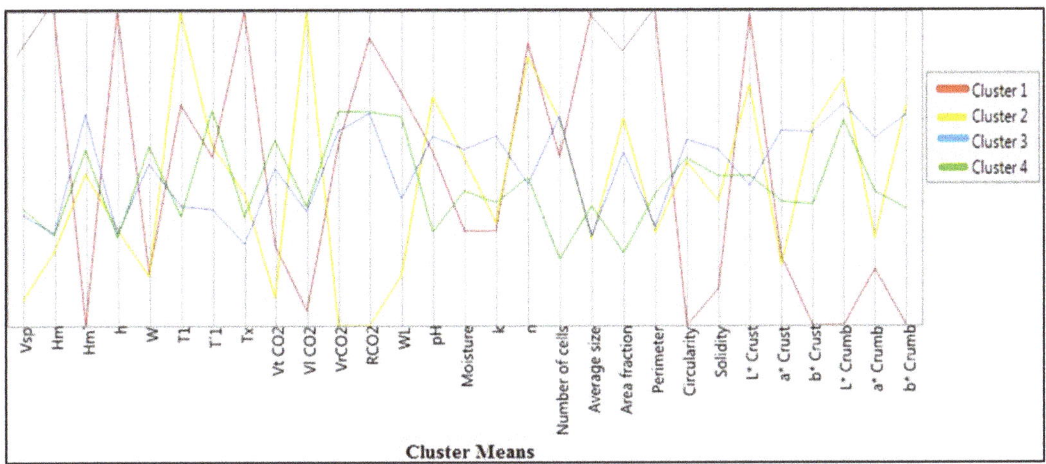

Figure 5. Cluster standard deviations (Cluster 1: wheat control; Cluster 2: gluten-free control; Cluster 3: DOP (2.5%, 5%, 7.5%), DAP (2.5%, 5%, 7.5%), and DPPSP 2.5%; Cluster 4: DPP (2.5%, 5%, 7.5%), DTP (2.5%, 5%, 7.5%), DPPP (2.5%, 5%, 7.5%), and DPPSP (5%, 7.5%)).

3.6. Multivariate Analysis of the Rheofermentometer and Rheological Parameters of Dough and Bread Quality

Figure 6a presents the correlation circle obtained from the principal component analysis on the rheofermentometer and the rheological parameters of dough and bread qualities measured in wheat and gluten-free samples. The first and second principal components justified 49.3% of the variability. Variables with high eigenvectors on principal component 1 were Tx, H_m, average size, perimeter, h, Vsp, H'_m, and L^* crumb (Figure 6a). On PC1, Tx, H_m, average size, perimeter, h, and Vsp were negatively correlated with H'_m and highly significant ($p < 0.001$) and positive correlations were noted between H_m and h ($r = 0.94$), followed by the Tx and H_m pairs ($r = 0.89$), Tx and h ($r = 0.89$) and H_m and Vsp ($r = 0.81$). Variables with high eigenvectors on principal component 2 were W, WL, L^* crust, T_1, pH, and H%. Negative correlations were observed between variables W and WL with L^* crust, T_1, and pH. Highly significant ($p = 0.001$) and negative correlations were found between pH and WL ($r = -0.67$), T_1 and % ($r = -0.62$), and T_1 and W ($r = -0.67$). A scatter plot for bread (wheat and gluten-free bread) is shown in Figure 6b. Principal component analysis confirmed the hierarchical results of the cluster analysis, in which the wheat control bread and the gluten-free control bread were distinguished from gluten-free bread containing by-products. The wheat control bread showed higher Tx, H_m, average size, and h; gluten-free control bread showed higher T_1, pH, and $V_{1\,CO2}$; gluten-free bread prepared with DOP and DAP (all percentage) and DPPPS at 2.5% presented higher W, T'_1, Vr, and CO_2; and gluten-free bread containing DPP, DTP, DPPP, and DPPPS at 5 and 7.5% showed higher H'_m, K, circularity, solidity, a^* crumb, and a^* crust (Figure 5).

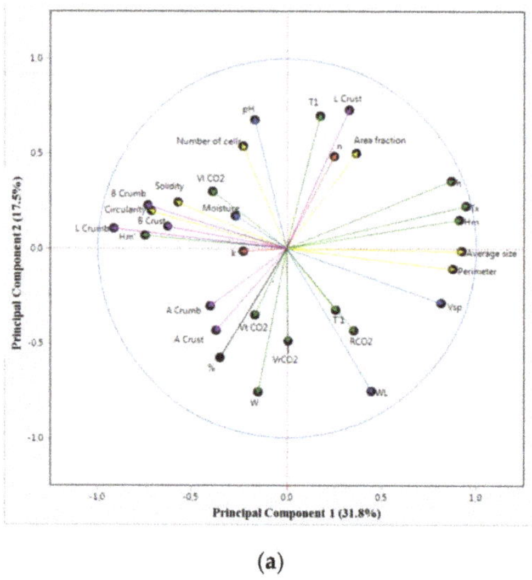
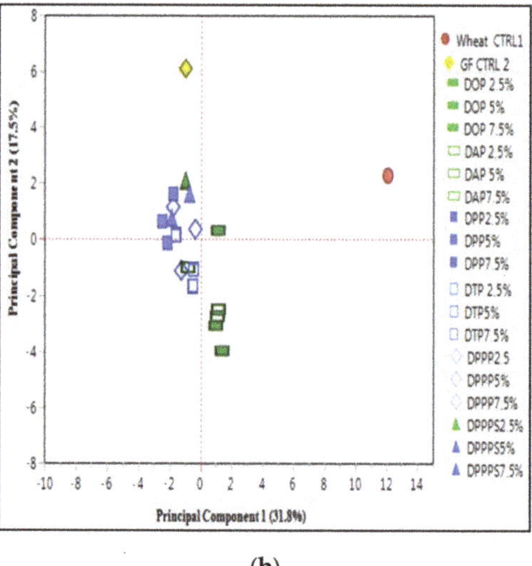

(a)   (b)

Figure 6. Multiple factor analysis correlating the percentage of by-products, dough rheological properties, and bread properties. (**a**) Map of parentage of by-products (in black), rheofermentometer parameters (in green) and rheological parameters of dough (in red), bread qualities (in blue), crumb structure of bread (in yellow), and crust and crumb colors (in purple). (**b**) Map of the distribution of the 20 types of bread.

4. Conclusions

Gluten-free bread based on the corn/chickpea formula was developed by adding different types of by-products (pomace of orange and apple; peels of tomato, pepper, and prickly pear; and prickly pear seeds). The rheology of the dough was measured and the properties of bread were evaluated. It was concluded that orange pomace, apple pomace, and peel of prickly pear seeds at 2.5% w/w induced the same effect on gluten-free bread. Pepper, tomato, prickly pear, and prickly pear seed peels with 5% and 7.5% addition had the same effect on gluten-free bread. Overall, all by-products increased Vsp, H_m, H'_m, V_{tCO2}, R_{CO2}, WL, and circularity, and decreased T_1, T_x, n, area fraction, L^* crust, and L^* crumb as compared to control gluten-free bread. The addition of prickly pear peel at a 5% addition level can result in bread with good characteristics. These by-products could be used as economical and inexpensive improvers for gluten-free bread.

Author Contributions: This study was designed by F.D. and M.N.Z. The analysis was done by F.D. and the manuscript was written by F.D. and H.B. and corrected by R.R., A.B., and W.T. All authors have read and agreed to the published version of the manuscript.

Funding: This research was funded by La Direction Générale de la Recherche Scientifique et du Développement Technologique (DGRSDT).

Institutional Review Board Statement: Not applicable.

Informed Consent Statement: Not applicable.

Data Availability Statement: The data presented in this study are available on request from the corresponding author R.R.

Acknowledgments: F. Djeghim acknowledges the Institut de la Nutrition, de l'Alimentation et des Technologies Agro-Alimentaires (INATAA), Algeria, the University of Life Sciences in Lublin, Poland, and the Centre de Recherche en Biotechnologie, Constantine, Algeria.

Conflicts of Interest: The authors have no conflict of interest to declare.

References

1. El Khoury, D.; Balfour-Ducharme, S.; Joye, I.J. A review on the gluten-free diet: Technological and nutritional challenges. *Nutrients* **2018**, *10*, 1410. [CrossRef] [PubMed]
2. Naqash, F.; Gani, A.; Gani, A.; Masoodi, F. Gluten-free baking: Combating the challenges-A review. *Trends Food Sci. Technol.* **2017**, *66*, 98–107. [CrossRef]
3. Melini, F.; Melini, V.; Luziatelli, F.; Ruzzi, M. Current and forward-looking approaches to technological and nutritional improvements of gluten-free bread with legume flours: A critical review. *Compr. Rev. Food Sci. Food Saf.* **2017**, *16*, 1101–1122. [CrossRef] [PubMed]
4. Hager, A.-S.; Wolter, A.; Czerny, M.; Bez, J.; Zannini, E.; Arendt, E.K.; Czerny, M. Investigation of product quality, sensory profile and ultrastructure of breads made from a range of commercial gluten-free flours compared to their wheat counterparts. *Eur. Food Res. Technol.* **2012**, *235*, 333–344. [CrossRef]
5. Kurek, M.; Wyrwisz, J. The application of dietary fiber in bread products. *J. Food Process. Technol.* **2015**, *6*, 447–450. [CrossRef]
6. Bourekoua, H.; Gawlik-Dziki, U.; Różyło, R.; Zidoune, M.N.; Dziki, D. Acerola fruit as a natural antioxidant ingredient for gluten-free bread: An approach to improve bread quality. *Food Sci. Technol. Int.* **2020**, *27*, 13–21. [CrossRef]
7. Matos, M.E.; Rosell, C.M. Understanding gluten-free dough for reaching breads with physical quality and nutritional balance. *J. Sci. Food Agric.* **2015**, *95*, 653–661. [CrossRef] [PubMed]
8. Salehi, F. Improvement of gluten-free bread and cake properties using natural hydrocolloids: A review. *Food Sci. Nutr.* **2019**, *7*, 3391–3402. [CrossRef] [PubMed]
9. Bourekoua, H.; Różyło, R.; Benatallah, L.; Wójtowicz, A.; Łysiak, G.; Zidoune, M.N.; Sujak, A. Characteristics of gluten-free bread: Quality improvement by the addition of starches/hydrocolloids and their combinations using a definitive screening design. *Eur. Food Res. Technol.* **2017**, *244*, 345–354. [CrossRef]
10. Collar, C. Gluten-free dough-based foods and technologies. In *Sorghum and Millets*; Elsevier: Amsterdam, The Netherlands, 2019; pp. 331–354. [CrossRef]
11. O'Shea, N.; Arendt, E.; Gallagher, E. State of the art in gluten-free research. *J. Food Sci.* **2014**, *79*, R1067–R1076. [CrossRef]
12. Roman, L.; Belorio, M.; Gomez, M. Gluten-Free breads: The gap between research and commercial reality. *Compr. Rev. Food Sci. Food Saf.* **2019**, *18*, 690–702. [CrossRef] [PubMed]
13. Gómez, M.; Martinez, M.M. Fruit and vegetable by-products as novel ingredients to improve the nutritional quality of baked goods. *Crit. Rev. Food Sci. Nutr.* **2018**, *58*, 2119–2135. [CrossRef] [PubMed]
14. Bedrníček, J.; Jirotková, D.; Kadlec, J.; Laknerová, I.; Vrchotová, N.; Tříska, J.; Samková, E.; Smetana, P. Thermal stability and bioavailability of bioactive compounds after baking of bread enriched with different onion by-products. *Food Chem.* **2020**, *319*, 126562. [CrossRef] [PubMed]
15. Kumar, H.; Bhardwaj, K.; Sharma, R.; Nepovimova, E.; Kuča, K.; Dhanjal, D.S.; Verma, R.; Bhardwaj, P.; Sharma, S.; Kumar, D. Fruit and vegetable peels: Utilization of high value horticultural waste in novel industrial applications. *Molecules* **2020**, *25*, 2812. [CrossRef] [PubMed]
16. Patel, S.N.; Sharma, M.; Lata, K.; Singh, U.; Kumar, V.; Sangwan, R.S.; Singh, S.P. Improved operational stability of d-psicose 3-epimerase by a novel protein engineering strategy, and d-psicose production from fruit and vegetable residues. *Bioresour. Technol.* **2016**, *216*, 121–127. [CrossRef]
17. Pathak, P. Medicinal Properties of Fruit and Vegetable Peels. *Adv. Bioeng.* **2020**, 115–128. [CrossRef]
18. Anwar, M.; Sallam, E. Utilization of prickly pear peels to improve quality of pan bread. *Arab J. Nucl. Sci. Appl.* **2016**, *49*, 151–163.
19. Coman, V.; Teleky, B.-E.; Mitrea, L.; Martău, G.A.; Szabo, K.; Călinoiu, L.-F.; Vodnar, D.C. Bioactive potential of fruit and vegetable wastes. *Adv. Food Nutr. Res.* **2020**, *91*, 157–225. [CrossRef]
20. Salim, N.; Abdelwaheb, C.; Rabah, C.; Ahcene, B. Chemical composition of Opuntia ficus-indica (L.) fruit. *Afr. J. Biotechnol.* **2009**, *8*, 1623–1624.
21. Babiker, W.A.; Sulieman, A.M.E.; Elhardallou, S.B.; Khalifa, E. Physicochemical properties of wheat bread supplemented with orange peel by-products. *Int. J. Food Sci. Nutr.* **2013**, *2*, 1–4. [CrossRef]
22. Mironeasa, S.; Mironeasa, C. Dough bread from refined wheat flour partially replaced by grape peels: Optimizing the rheological properties. *J. Food Process. Eng.* **2019**, *42*, e13207. [CrossRef]
23. Pathak, D.; Majumdar, J.; Raychaudhuri, U.; Chakraborty, R. Characterization of physicochemical properties in whole wheat bread after incorporation of ripe mango peel. *J. Food Meas. Charact.* **2016**, *10*, 554–561. [CrossRef]
24. AOAC. *Official Methods of Analysis*; Association of Official Analytical Chemists: Rockville, MD, USA, 2005.
25. Henneberg, W.; Stohmann, F. *Beiträge zur Begründung einer Rationellen Fütterung der Wiederkäuer*; Braunschweig: Schwetschke, Germany, 1864; Volume 2.
26. Abid, M.; Cheikhrouhou, S.; Renard, C.M.; Bureau, S.; Cuvelier, G.; Attia, H.; Ayadi, M. Characterization of pectins extracted from pomegranate peel and their gelling properties. *Food Chem.* **2017**, *215*, 318–325. [CrossRef] [PubMed]
27. Doukani, K.; Tabak, S. Profil Physicochimique du fruit" Lendj"(*Arbutus unedo* L.). *Nat. Technol.* **2015**, *51*, 53–66.
28. Liew, S.Q.; Chin, N.L.; Yusof, Y.A. Extraction and characterization of pectin from passion fruit peels. *Agric. Agric. Sci. Procedia* **2014**, *2*, 231–236. [CrossRef]
29. Torralbo, D.; Batista, K.; Di-Medeiros, M.; Fernandes, K. Extraction and partial characterization of Solanum lycocarpum pectin. *Food Hydrocoll.* **2012**, *27*, 378–383. [CrossRef]

30. McConnell, A.; Eastwood, M.; Mitchell, W. Physical characteristics of vegetable foodstuffs that could influence bowel function. *J. Sci. Food Agric.* **1974**, *25*, 1457–1464. [CrossRef]

31. Ang, J.F. Water retention capacity and viscosity effect of powdered cellulose. *J. Food Sci.* **1991**, *56*, 1682–1684. [CrossRef]

32. Rabetafika, H.N.; Bchir, B.; Aguedo, M.; Paquot, M.; Blecker, C. Effects of processing on the compositions and physicochemical properties of fibre concentrate from cooked fruit pomaces. *Food Bioproc. Technol.* **2014**, *7*, 749–760. [CrossRef]

33. Gómez, M.; Talegón, M.; De La Hera, E. Influence of mixing on quality of gluten-free bread. *J. Food Qual.* **2013**, *36*, 139–145. [CrossRef]

34. Martínez, M.M.; Díaz, Á.; Gómez, M. Effect of different microstructural features of soluble and insoluble fibres on gluten-free dough rheology and bread-making. *J. Food Eng.* **2014**, *142*, 49–56. [CrossRef]

35. Ronda, F.; Pérez-Quirce, S.; Villanueva, M. Rheological properties of gluten-free bread doughs: Relationship with bread quality. In *Advances in Food Rheology and Its Applications*; Elsevier: Amsterdam, The Netherlands, 2017; pp. 297–334. [CrossRef]

36. Altuna, L.; Ribotta, P.D.; Tadini, C.C. Effect of a combination of enzymes on dough rheology and physical and sensory properties of bread enriched with resistant starch. *LWT Food Sci. Technol.* **2015**, *64*, 867–873. [CrossRef]

37. Dai, S.; Qi, F.; Tanner, R. Interpreting shear creep data for bread dough using a damage function model. *Appl. Rheol.* **2011**, *21*. [CrossRef]

38. Hicks, C.I.; See, H. The rheological characterisation of bread dough using capillary rheometry. *Rheol. Acta* **2010**, *49*, 719–732. [CrossRef]

39. Tanner, R.I.; Qi, F.; Dai, S.-C. Bread dough rheology and recoil: I. Rheology. *J. Non-Newton. Fluid Mech.* **2008**, *148*, 33–40. [CrossRef]

40. Witczak, M.; Korus, J.; Ziobro, R.; Juszczak, L. The effects of maltodextrins on gluten-free dough and quality of bread. *J. Food Eng.* **2010**, *96*, 258–265. [CrossRef]

41. Sabanis, D.; Lebesi, D.; Tzia, C. Effect of dietary fibre enrichment on selected properties of gluten-free bread. *LWT Food Sci. Technol.* **2009**, *42*, 1380–1389. [CrossRef]

42. Chen, H.; Rubenthaler, G.; Leung, H.; Baranowski, J. Chemical, physical, and baking properties of apple fiber compared with wheat and oat bran. *Cereal Chem.* **1988**, *65*, 244–247.

43. Keskin, S.O.; Sumnu, G.; Sahin, S. Bread baking in halogen lamp–microwave combination oven. *Food Res. Int.* **2004**, *37*, 489–495. [CrossRef]

44. AACC. *American Association of Cereal Chemists. Approved Methods of the AACC*; The Association: St Paul, MN, USA, 2000.

45. ICC. *Standard Methods of the International Association for Cereal Science and Technology*; International Association for Cereal Chemistry: Vienna, Austria, 1996.

46. Majzoobi, M.; Poor, Z.V.; Jamalian, J.; Farahnaky, A. Improvement of the quality of gluten-free sponge cake using different levels and particle sizes of carrot pomace powder. *Int. J. Food Sci. Technol.* **2016**, *51*, 1369–1377. [CrossRef]

47. Majzoobi, M.; Vosooghi Poor, Z.; Mesbahi, G.; Jamalian, J.; Farahnaky, A. Effects of carrot pomace powder and a mixture of pectin and xanthan on the quality of gluten-free batter and cakes. *J. Texture Stud.* **2017**, *48*, 616–623. [CrossRef] [PubMed]

48. He, X.; Pei, Q.; Xu, T.; Zhang, X. Smartphone-based tape sensors for multiplexed rapid urinalysis. *Sens. Actuators B Chem.* **2020**, *304*, 127415. [CrossRef]

49. Chen, C.; Ren, M. The significance of license plate location based on Lab color space. In Proceedings of the 2nd International Conference on Information, Electronics and Computer, Wuhan, China, 7–9 March 2014; Atlantis Press: Paris, France, 2014.

50. Skinner, R.C.; Gigliotti, J.C.; Ku, K.-M.; Tou, J.C. A comprehensive analysis of the composition, health benefits, and safety of apple pomace. *Nutr. Rev.* **2018**, *76*, 893–909. [CrossRef]

51. Ocen, D.; Xu, X. Effect of citrus orange (Citrus sinensis) by-product dietary fiber preparations on the quality characteristics of frozen dough bread. *Am. J. Food Technol.* **2013**, *8*, 43–53. [CrossRef]

52. Bchir, B.; Rabetafika, H.N.; Paquot, M.; Blecker, C. Effect of pear, apple and date fibres from cooked fruit by-products on dough performance and bread quality. *Food Bioprocess Technol.* **2014**, *7*, 1114–1127. [CrossRef]

53. Yılmaz, E.; Emir, D.D. Compositional and functional characterisation of poppy seed (Papaver somniferum L.) press cake meals. *Qual. Assur. Saf. Crop Foods* **2017**, *9*, 141–151. [CrossRef]

54. Hosseini, S.S.; Khodaiyan, F.; Kazemi, M.; Najari, Z. Optimization and characterization of pectin extracted from sour orange peel by ultrasound assisted method. *Int. J. Biol. Macromol.* **2019**, *125*, 621–629. [CrossRef]

55. Boubaker, M.; Omri, A.E.; Blecker, C.; Bouzouita, N. Fibre concentrate from artichoke (Cynara scolymus L.) stem by-products: Characterization and application as a bakery product ingredient. *Food Sci. Technol. Int.* **2016**, *22*, 759–768. [CrossRef]

56. O'shea, N.; Rößle, C.; Arendt, E.; Gallagher, E. Modelling the effects of orange pomace using response surface design for gluten-free bread baking. *Food Chem.* **2015**, *166*, 223–230. [CrossRef] [PubMed]

57. Grigelmo-Miguel, N.; Martín-Belloso, O. Characterization of dietary fiber from orange juice extraction. *Food Res. Int.* **1998**, *31*, 355–361. [CrossRef]

58. Wang, X.; Chen, Q.; Lü, X. Pectin extracted from apple pomace and citrus peel by subcritical water. *Food Hydrocoll.* **2014**, *38*, 129–137. [CrossRef]

59. Sundarraj, A.A.; Ranganathan, T.V. A review-Pectin from Agro and industrial waste. *Int. J. Appl. Environ. Sci.* **2017**, *12*, 1777–1801.

60. Dalal, N.; Neeraj, V.B.; Dhakar, U. Potential of fruit and vegetable waste as a source of pectin. *IJCS* **2020**, *8*, 3085–3090. [CrossRef]

61. Verheyen, C.; Jekle, M.; Becker, T. Effects of Saccharomyces cerevisiae on the structural kinetics of wheat dough during fermentation. *LWT Food Sci. Technol.* **2014**, *58*, 194–202. [CrossRef]

62. Sudha, M. Apple pomace (by-product of fruit juice industry) as a flour fortification strategy. In *Flour and Breads and Their Fortification in Health and Disease Prevention*; Elsevier: Amsterdam, The Netherlands, 2011; pp. 395–405. [CrossRef]
63. Gowe, C. Review on potential use of fruit and vegetables by-products as a valuable source of natural food additives. *Food Sci. Qual. Manag.* **2015**, *45*, 47–61.
64. Chiocchetti, G.D.M.E.; Fernandes, E.A.D.N.; Bacchi, M.A.; Pazim, R.A.; Sarriés, S.R.V.; Tome, T.M. Mineral composition of fruit by-products evaluated by neutron activation analysis. *J. Radioanal. Nucl. Chem.* **2013**, *297*, 399–404. [CrossRef]
65. Walker, G.M.; Stewart, G.G. Saccharomyces cerevisiae in the production of fermented beverages. *Beverages* **2016**, *2*, 30. [CrossRef]
66. Struyf, N.; Van der Maelen, E.; Hemdane, S.; Verspreet, J.; Verstrepen, K.J.; Courtin, C.M. Bread dough and baker's yeast: An uplifting synergy. *Compr. Rev. Food Sci. Food Saf.* **2017**, *16*, 850–867. [CrossRef]
67. Zhu, F.; Sakulnak, R.; Wang, S. Effect of black tea on antioxidant, textural, and sensory properties of Chinese steamed bread. *Food Chem.* **2016**, *194*, 1217–1223. [CrossRef]
68. Cheremisinoff, N.P. *Practical Fluid Mechanics for Engineers & Scientists*; CRC Press: Boca Raton, FL, USA, 1990.
69. Krokida, M.; Maroulis, Z.; Saravacos, G. Rheological properties of fluid fruit and vegetable puree products: Compilation of literature data. *Int. J. Food Prop.* **2001**, *4*, 179–200. [CrossRef]
70. Ramsey, M. Rheology, Viscosity, and Fluid Types. *Pract. Wellbore Hydraul. Hole Clean.* **2019**, 217–237.
71. Padmanabhan, M.; Bhattacharya, M. Flow behavior and exit pressures of corn meal under high-shear–high-temperature extrusion conditions using a slit diea. *J. Rheol.* **1991**, *35*, 315–343. [CrossRef]
72. Fraiha, M.; Biagi, J.D.; Ferraz, A.C.d.O. Rheological behavior of corn and soy mix as feed ingredients. *Food Sci. Technol.* **2011**, *31*, 129–134. [CrossRef]
73. Drozdek, K.D.; Faller, J.F. Use of a dual orifice die for on-line extruder measurement of flow behavior index in starchy foods. *J. Food Eng.* **2002**, *55*, 79–88. [CrossRef]
74. Milde, L.B.; Ramallo, L.A.; Puppo, M.C. Gluten-free bread based on tapioca starch: Texture and sensory studies. *Food Bioproc. Technol.* **2012**, *5*, 888–896. [CrossRef]
75. Parra, A.F.R.; Ribotta, P.D.; Ferrero, C. Apple pomace in gluten-free formulations: Effect on rheology and product quality. *Int. J. Food Sci. Technol.* **2015**, *50*, 682–690. [CrossRef]
76. O'shea, N.; Ktenioudaki, A.; Smyth, T.; McLoughlin, P.; Doran, L.; Auty, M.; Arendt, E.; Gallagher, E. Physicochemical assessment of two fruit by-products as functional ingredients: Apple and orange pomace. *J. Food Eng.* **2015**, *153*, 89–95. [CrossRef]
77. Arslan, M.; Rakha, A.; Khan, M.R.; Zou, X. Complementing the dietary fiber and antioxidant potential of gluten free bread with guava pulp powder. *J. Food Meas. Charact.* **2017**, *11*, 1959–1968. [CrossRef]
78. Singh, J.P.; Kaur, A.; Singh, N. Development of eggless gluten-free rice muffins utilizing black carrot dietary fibre concentrate and xanthan gum. *J. Food Sci. Technol.* **2016**, *53*, 1269–1278. [CrossRef]
79. Türker, B.; Savlak, N.; Kaşıkcı, M.B. Effect of green banana peel flour substitution on physical characteristics of gluten-free cakes. *Curr. Res. Nutr. Food Sci. J.* **2016**, *4*, 197–204. [CrossRef]
80. Khosravi, F.; Iranmanesh, B.; Olia, S.S.S.J. Determination of Organic Acids in Fruit juices by UPLC. *Int. J. Life Sci.* **2015**, *9*, 41–44. [CrossRef]
81. Tezcan, F.; Uzaşçı, S.; Uyar, G.; Öztekin, N.; Erim, F.B. Determination of amino acids in pomegranate juices and fingerprint for adulteration with apple juices. *Food Chem.* **2013**, *141*, 1187–1191. [CrossRef] [PubMed]
82. Gómez-Ariza, J.; Villegas-Portero, M.; Bernal-Daza, V. Characterization and analysis of amino acids in orange juice by HPLC–MS/MS for authenticity assessment. *Anal. Chim. Acta* **2005**, *540*, 221–230. [CrossRef]
83. Martins, Z.; Pinho, O.; Ferreira, I. Fortification of Wheat Bread with Agroindustry By-Products: Statistical Methods for Sensory Preference Evaluation and Correlation with Color and Crumb Structure. *J. Food Sci.* **2017**, *82*, 2183–2191. [CrossRef] [PubMed]
84. Capriles, V.D.; Arêas, J.A.G. Novel approaches in gluten-free breadmaking: Interface between food science, nutrition, and health. *Compr. Rev. Food Sci. Food Saf.* **2014**, *13*, 871–890. [CrossRef]
85. Jafari, M.; Koocheki, A.; Milani, E. Functional effects of xanthan gum on quality attributes and microstructure of extruded sorghum-wheat composite dough and bread. *LWT Food. Sci. Technol.* **2018**, *89*, 551–558. [CrossRef]
86. Goriewa-Duba, K.; Duba, A.; Wachowska, U.; Wiwart, M. An evaluation of the variation in the morphometric parameters of grain of six Triticum species with the use of digital image analysis. *Agronomy* **2018**, *8*, 296. [CrossRef]
87. Alba, K.; Rizou, T.; Paraskevopoulou, A.; Campbell, G.M.; Kontogiorgos, V. Effects of blackcurrant fibre on dough physical properties and bread quality characteristics. *Food Biophys.* **2020**, 313–322. [CrossRef]

applied sciences

Article

Nutritional, Sensory, Texture Properties and Volatile Compounds Profile of Biscuits with Roasted Flaxseed Flour Partially Substituting for Wheat Flour

Simona Maria Man [1], Laura Stan [2,*], Adriana Păucean [1,*], Maria Simona Chiş [1], Vlad Mureşan [1], Sonia Ancuţa Socaci [2], Anamaria Pop [1] and Sevastiţa Muste [1]

[1] Department of Food Engineering, Faculty of Food Sciences and Technology, University of Agricultural Sciences and Veterinary Medicine of Cluj-Napoca, 3-5 Mănăştur Street, 400372 Cluj-Napoca, Romania; simona.man@usamvcluj.ro (S.M.M.); simona.chis@usamvcluj.ro (M.S.C.); vlad.muresan@usamvcluj.ro (V.M.); anamaria.pop@usamvcluj.ro (A.P.); sevastita.muste@usamvcluj.ro (S.M.)

[2] Department of Food Sciences, Faculty of Food Sciences and Technology, University of Agricultural Sciences and Veterinary Medicine of Cluj-Napoca, 3-5 Mănăştur Street, 400372 Cluj-Napoca, Romania; sonia.socaci@usamvcluj.ro

* Correspondence: laurastan@usamvcluj.ro (L.S.); adriana.paucean@usamvcluj.ro (A.P.)

Abstract: The study aimed at assessing effects of partial replacement (0–40%) of wheat flour with roasted flaxseed flour (RFSF) on the quality attributes of biscuits. Nutritional, antioxidative, volatile and sensory properties, as well as texture analysis and the contents of macroelements and microelement were studied. Increasing RFSF content in biscuits resulted in a significant increase ($p < 0.05$) in protein (from 8.35% to 10.77%), fat (from 15.19% to 28.34%) and ash (from 1.23% to 2.60%) while the hardness and spread factor of the biscuits decreased with the increased level of roasted flaxseed flour. Moreover, the addition of 40% RFSF registered a positive influence on the fibre content of the final baked biscuits, increasing its value about 6.7-fold than in the control sample. Total phenolic content, antioxidant activity and biscuits' aroma volatile profile increased their amounts with RFSF addition. The nutritional, textural and sensorial results of the present study demonstrated that 25% RFSF could be added in the biscuits manufacturing without affecting the biscuits aftertaste, offering promising healthy and nutritious alternative to consumers.

Keywords: roasted flaxseed flour; GC/MS; aroma; antioxidant activity; sensory evaluation; fiber; macro and microelements; biscuits

Citation: Man, S.M.; Stan, L.; Păucean, A.; Chiş, M.S.; Mureşan, V.; Socaci, S.A.; Pop, A.; Muste, S. Nutritional, Sensory, Texture Properties and Volatile Compounds Profile of Biscuits with Roasted Flaxseed Flour Partially Substituting for Wheat Flour. *Appl. Sci.* **2021**, *11*, 4791. https://doi.org/10.3390/app11114791

Academic Editor: Silvia Mironeasa

Received: 27 April 2021
Accepted: 21 May 2021
Published: 23 May 2021

Publisher's Note: MDPI stays neutral with regard to jurisdictional claims in published maps and institutional affiliations.

1. Introduction

Bakery and pastry products are consumed worldwide in large quantities on a daily basis and play an important role in human nutrition. Among bakery products, biscuits are the most popular amongst consumers as a rich source of carbohydrates and fats, but low in proteins and dietary fibers along with quite a good shelf life [1]. Generally, biscuits are made with refined wheat flour which is deficient in some essential amino acids (lysine, tryptophan, threonine, methionine and histidine) and other nutrients (fiber, minerals and vitamins) [2–4]. Nowadays there is a high demand for high nutritional value foods. In the current Western society's diet, the intake of ω-6 fatty acids is generally higher compared to the intake of ω-3 fatty acids and this leads to unfavorable nutritional consequences. Therefore, there is a need on the market for innovative products that meet the nutritional requirements of consumers. Some authors suggest that diet supplementation with flaxseed (*Linum usitatissimum*) offers potential health benefits in cases like cardiovascular risk, severe hyperlipidemia, certain types of cancers and other metabolic disorders [5–9]. Flaxseed contain protein (20 g/100 g), dietary fiber (28 g/100 g), fat (41 g/100 g), moisture (6.5 g/100 g), minerals (2.4 g/100 g) and carbohydrates (28.9%), being recognized as an important oilseed and fiber crop [10–12]. It ought to be emphasized that the linseed chemical composition may vary depending on the genetic characteristics, growing conditions and crop management practices. Furthermore, flaxseed has a unique fatty acid profile. It is high in polyunsaturated fatty acids (73 g/100 g of total fatty acids), moderate in monounsaturated fatty acids (18 g/100 g), and low in saturated fatty acids (9 g/100 g). The content of linoleic acid is about 16 g/100 g of total fatty acids while α-linolenic acid (ALA) reaches about 57 g/100 g [13]. Tocopherols (20–70 mg/100 g) and carotenoids (~5.7 mg/100 g) are also found in flaxseed oil, therefore flaxseed is considered an important ancient medicine and functional food ingredient [14].

Flaxseed is one of the most important oilseed crops for industrial as well as for food and feed purposes, being part of the human diet for thousands of years, and more recently it has been used as a source of nutraceuticals [10,11,15]. The bioactive components in flaxseed that provide health benefits include α-linolenic acid, lignans and dietary fiber [16,17], proteins and soluble mucilage [8,11,16,18–21]. The flaxseed mucilage may be considered a food hydrocolloid due to its composition, which consists of a mixture of neutral arabinoxylans and acidic rhamnose-containing polysaccharides [22]. Evenmore, antioxidant activity of the lignans and the presence of different types of phenolic compounds such as phenolic acids, flavonoids, phenylpropanoids and tannins [23] may reduce the risk of cardiovascular diseases and contribute to the anticancer activity [12,24–26], particularly hormone-dependent cancers such as prostate and breast [27].

The behavior of proteins in a food system is affected by their techno-functional properties which are mainly dependent on the following factors: the structure of the protein, their hydration mechanisms for solubility and water or oil retention capacity, rheological characteristics for viscosity and gelation, and their interfacial properties for emulsions and foams [28]. Flaxseed proteins have real potential to be used as techno-functional food ingredient in several food products particularly in breads, meat emulsions, and sauces [26]. Thus, flaxseed can be incorporated into diet as ground flaxseed or flaxseed oil. Flaxseed was previously incorporated in various bakery products such as cookies and biscuits [26,29–33] bread [13,22,34,35] and cakes [15,36]. The results of these studies showed that a substitution levels up to 20% led to good product acceptability [29].

Consumers' trend to seek nutritional foods with remarkable value in the diet is highly directed toward the use of flaxseed as a potential ingredient in foods. The nutrient profile of flaxseed biscuits is a valued option, especially for consumers interested in healthy snack diversification.

The aim of this study was to examine the effects of partial replacement of wheat flour with roasted flaxseed flour on the quality attributes of biscuits, with special attention on their nutritional proprieties and volatile profile. Wheat flour substitution with flaxseed flour in biscuits formulation proposed in this study can be an example of successful collaboration between research and food industry and are in agreement with the confectionery products trends.

2. Materials and Methods

2.1. Materials

All the ingredients used in the biscuit formulations were bought from the local market in Cluj-Napoca, Românâ. The wheat flour (WF) sample was produced by a local mill (Boromir, Deva, Romania) and sold as type 000 according to its ash content following the Romanian classification [37] (0.48% ash content and 29.57% wet gluten). Flaxseeds originating from Ukraine were used to prepare the flaxseed flour as described further. Initially, flaxseeds were roasted at 180 °C for 15 min, then cooled and ground on a GM200 laboratory mill (Grindomix, Retsch GmbH, Haan, Germany) at 10,000 rotations/min for 50 s. To ensure the uniformity of particles, the roasted flaxseed flour (RFSF) was sieved through a 0.8 mm sieve. Four blends of various ratios of both flours were used to prepare the biscuits, as presented in Section 2.2.2.

2.2. Methods

2.2.1. Proximate Composition Analysis

The chemical characteristics were determined using the AACC (2000) methods [38]. Wheat flour (WF), roasted flaxseed flour (RFSF), control biscuits and RFSF incorporated biscuits were analysed for moisture (AACC 44-15.02, 2000), ash (AACC 08-01.01, 2000), fat (AACC 30-25.01, 2000), crude fiber (AACC 32-07.01, 2000). The proteins were measured using the Kjeldahl method (AACC 46-11.02, 2000) using the nitrogen to protein conversion factor of 5.7. Total carbohydrate (%) content was calculated according to Equation (1) according to methods reported by Man et al. [39]:

$$Total\ carbohydrates(\%) = 100 - [moisture(\%) + ash(\%) + proteins(\%) + lipids(\%) + crude\ fiber(\%)] \tag{1}$$

All determinations were made in triplicate.

2.2.2. Preparation of Biscuits

First step was to formulate the four flour blends (100:0, 90:10, 75:25 and 60:40 WF:RFSF, w:w). The flour blends were stored in air-tight containers until further use. The second step was premixing margarine (60% fat) with powdered sugar into a cream base and then mixing that with the flours, instant whole milk powder (26% fat and 26% protein), sodium bicarbonate, vanilla essence and water. The doughs were mixed for seven min in a mixer (KitchenAid® Precise Heat Mixing Bowl, Greenville, OH, USA), at medium speed. Mixing, resting and baking technological parameters are listed in Table 1. The control sample was prepared using the same procedure, but omitting the flaxseed. The dough was rolled out into circular shapes of 0.4 cm thickness and then the biscuits were shaped by stamping with cylindrical shapes into 5.5 cm diameter. After baking (Zanolli oven, Verona, Italy) all biscuits samples were removed from the trays for cooling, then packed in polypropylene bags and sealed until further analysis. Biscuit samples were coded as B0, B10, B25 and B40, respectively, taking into consideration the amount of RFSF (0%, 10%, 25%, 40%) as shown in Table 1. Each batch of biscuits was made the day before the sensory, texture and volatile analysis.

Table 1. Ingredients and technological parameters used in the preparation of biscuits.

Ingredients (g)	Biscuits Samples *			
	B0	B10	B25	B40
Wheat flour (WF)	100	90	75	60
Roasted flaxseed flour (RFSF)	-	10	25	40
Powder milk **	20	20	20	20
Margarine **	40	40	40	40
Sugar powdered **	40	40	40	40
Sodium bicarbonate **	2.5	2.5	2.5	2.5
Vanilla essence **	0.5	0.5	0.5	0.5
Water **	25	25	25	25
Technological parameters				
Mixing time (min)	7	7	7	7
Dough temperature (°C)	22	22.4	23	23.5
Resting time (min)	60	60	60	60
Temperature (°C)	3–4	3–4	3–4	3–4
Baking time (min)	10	10	10	10
Temperature (°C)	200	200	200	200

Note: * B0 = 0% biscuits with 100% wheat flour (control sample); B10 = biscuits with 10% RFSF; B25 = 25% biscuits with 25% RFSF; B40 = biscuits with 40% RFSF. ** The auxiliar materials are reported to 100% of flour blends.

2.2.3. Macro- and Microelements

Macro- and microelements were determined by atomic absorption spectrophotometry (AAS), according to the methods described by Păucean et al. [40] and Chiş et al. [41]. Briefly, 3 g of biscuits were burned at 550 °C in the furnace for 10 h (Nabertherm B150, Lilienthal, Germany). The ash was dissolved in HCl 20% and made up to a final volume of 20 mL in a volumetric flask. The macroelements (K, Ca, Mg) and microelements (Fe, Cu, Zn and Mn) were determined by AAS (Varian 220 FAA instrument, Mulgrave, Victoria, Australia). Mixed standard solutions with P, K, Cu, Zn, Fe, Mn, Ca, Mg (ICP Multielement Standard solution IV CertiPUR) were purchased from Merck (Merck KGaA, Darmstadt, Germany). All chemicals and solvents used in this study were of analytical grade. The results were expressed related to the weight of the fresh biscuits.

2.2.4. Physical Evaluation of Biscuits

The diameter of six biscuits placed edge to edge was measured with a Vernier caliper (0.01 mm accuracy). The average diameter in centimetres was calculated by dividing [42] the value by six [4]. Similarly, the mean thickness of the biscuits was measured by stacking 6 biscuits on top of each other and dividing the value by six. For the spread factor the diameter was divided by thickness. The weight of the biscuits was measured on an analytical balance and calculated as the mean value of the weight of four individual biscuits.

2.2.5. Total Phenolics Content

Total polyphenols content and antioxidant activity were determined as described by Bunea et al. [43] with slight modifications as proposed by Chiş et al. [44]. Briefly, one gram of biscuit sample was extracted three times with 100 mL acidified methanol (85:15 v/v, MeOH:HCl) by maceration under continuous stirring (magnetic stirrer, Velp Scientifica, Usmate Velate, Italy) for 24 h. Total phenolics content of the extracts was determined spectrophotometrically (Folin–Ciocalteu method). Thus, in a 100 mL volumetric flask, 10 mL of methanolic extract was mixed with 5 mL of Folin-Ciocalteu's reagent; 5 mL of 7.5% sodium carbonate solution was added. Distilled water was used to fill up the flask until the graduation marking. Samples were kept in the dark for 90 min, and then, absorbance was read at 760 nm on a model 1700UV/VIS spectrophotometer (Shimadzu Scientific Instruments, Kyoto, Japan).

2.2.6. DPPH Radical Scavenging Activity Assay

The radical scavenging activity was determined spectrophotometrically by the 2,2-diphenyl-1-picrylhydrazyl (DPPH) method as described by Păucean et al. [40]. The phenolic extracts (0.1 mL) were mixed with DPPH solution (3.9 mL), kept in the dark at ambient temperature for 30 min. The absorbance of the mixtures was recorded at 515 nm (1700 UV/VIS spectrophotometer Shimadzu) against a methanol blank. Negative control was prepared using 0.1 mL methanol and 3.90 mL of DPPH. The radical scavenging activity was calculated according to the following Equation (2):

$$RSA[\%] = \frac{Abs_{DPPH} - Abs_{sample}}{Abs_{DPPH}} \times 100 \tag{2}$$

where Abs_{DPPH} = absorbance of DPPH solution and Abs_{Sample} = absorbance of the sample.

2.2.7. Volatile Compounds

Qualitative analyses of the volatile compounds were achieved using "in-tube extraction" coupled with gas-chromatography mass spectrometry (ITEX/GC-MS, Shimadzu GC-MS model QP-2010) [45]. Three grams of each sample were weighed into a 20 mL sealed-cap headspace vial and incubated for 30 min at 70 °C with continuous agitation. The volatile compounds from the gaseous headspace of the sample were repeatedly adsorbed onto a microtrap (ITEX-2TrapTXTA, TA 80/100 mesh, Tenax, Zwingen, Switzerland) using a CombiPAL AOC-5000 autosampler (CTC Analytics, Zwingen, Switzerland). The volatile compounds were further thermally desorbed into the injection port of the GC-MS QP2010 system and separated on a ZB—5 ms column of 30 m × 0.25 mm i.d. and 0.25 μm film thickness (Phenomenex, Torrance, CA, USA). The column oven temperature program was set initially at 40 °C (kept at this temperature for 6 min) and then increased to 50 °C at 2 °C/min, followed by an increase to 240 °C at 7 °C/min (kept at this temperature for 5 min). The temperatures of the ion source, injector, and interface were set at 250 °C. The carrier gas was helium at a flow rate of 1 mL/min. The split ratio was 1:2. The MS detector was operated in the EI mode, using a scan acquisition mode in the range of 40–400 *m/z*. Mass spectra was identified by comparison with a series of standard alkanes—alkane standard solution.

The identification of volatile compounds was performed by comparing their mass spectra with those in the NIST27 and NIST147 libraries and by retention indices drawn from [46,47] (for columns with a similar stationary phase to ZB-5 ms). Afterwards, the mass spectra values were compared and verified with the retention indices drawn from Pherobase [46] and Flavornet [47] databases. The results were expressed as relative percentage of the total peaks area.

2.2.8. Instrumental Analysis of Texture

The textural properties of the biscuits were measured with a CT 3 Texture Analyzer (Brookfield Engineering Labs, Middleboro, MA, USA), equipped with a 10 kg load cell and the 6 mm cylindrical probe (TA41). The textural properties were determined as described by Pop et al. [48] with slight modifications. A compression test was applied to all samples (5 mm target distance, 3 mm/s test and post-test speed, trigger load 5 g). The Brookfield Engineering Labs Texture Pro CT V1.6 software was used to calculate the specific texture parameters.

2.2.9. Sensory Analysis

The sensory evaluation of the biscuits took place in the Laboratory for Sensory Analysis from the authors' faculty, under normal daylight conditions. All samples were evaluated for acceptability using a nine-point hedonic scale. A total number of 52 evaluators (40 females and 12 males) between 19–52 years old participated at the study. The evaluators were students and staff members from the Faculty of Food Science and Technology, selected on basis of their interest, availability and regular consumption of biscuits (at least once per week). Samples were coded anonymously with a three digit code and presented on plastic odour-free plates, in random order. Two replications of each sample were evaluated by each participant. Plain mineral water was provided to clean the palate between samples. Five sensory characteristics were rated on the nine-point hedonic scale: "appearance" (1 = extremely irregular surface, 9 = extremely regular surface), "hardness" and "crispiness" (1 = extremely weak, 9 = extremely strong), "chewiness" (1 = extremely difficult to chew, 9 = extremely easy to chew) and finally, "taste and aroma" (1 = no flaxseed taste and aroma, 9 = extremely strong flaxseed taste and aroma). The attributes of hardness were explained to the evaluators and their definitions were also inserted in the evaluation forms. The evaluators were trained with the scale and the sensory attributes used for the evaluation of biscuit samples (2 sessions × 2 h). As already published by Park et al. [49] the hardness was considered the force needed to bite through the sample by using the front teeth. Crispiness was evaluated as the force and the noise with which the sample breaks down between the molar teeth. Finally, the chewiness was evaluated as the energy necessary to masticate solid foods so that they can be easily swallowed.

2.2.10. Statistical Analysis

The results of three independent ($n = 3$) replicates were expressed as means ± standard deviations (SD). Data of proximate composition, sensory data and instrumental textural analysis were analyzed using Duncan multiple comparison test ($p < 0.05$) using the SPSS version 19 software (IBM Corp., Armonk, NY, USA). Correlation among means of the sensory data was determined using a two-tailed Pearson Correlation test ($p < 0.05$) using Microsoft Office—Excel.

3. Results and Discussions

3.1. Proximate Composition of Flours

The proximate composition of WF and RFSF is reported in Table 2. The data show that the two flours have complementary nutritional profiles. The moisture content of any flour is an important quality criterion for its preservation, packaging and transport as well. WF had eight times more humidity than RFSF (14.55% compared to 1.64%, respectively). The low moisture content for RFSF can be explained by the prior roasting process which removed most of the sample's water content. Comparison of the carbohydrate content of the two samples presented a similar pattern: WF presented high levels of carbohydrates (74.45%), whereas RFSF contained 21.76% of total carbohydrate. However, RFSF presented higher levels of protein (21.40%), fiber (8.75%) and fat content (42.50%) compared to WF.

Table 2. Proximate composition of raw materials.

Parameters	Wheat Flour Type 000 (WF) *	Roasted Flaxseed Flour (RFSF) *
Moisture (%, fw)	14.55 ± 1.20 [b]	1.64 ± 0.04 [a]
Protein (%, fw)	9.12 ± 1.03 [a]	21.40 ± 1.50 [b]
Ash (%, fw)	0.48 ± 0.01 [a]	3.95 ± 0.04 [b]
Fat (%, fw)	1.03 ± 0.20 [a]	42.50 ± 1.50 [b]
Crude fiber (%, fw)	0.37 ± 0.23 [a]	8.75 ± 1.45 [b]
Total carbohydrate (%, fw)	74.45 ± 0.9 [b]	21.76 ± 0.76 [a]

* Values represent mean of three independent determinations ± SD; Mean values followed by the same superscript alphabet in the row are not significantly different at $p < 0.05$ according to Duncan comparison test; fw-fresh weight.

Therefore, a composite flour can be successfully used to obtain biscuits with improved nutritional value. These results are in agreement with those obtained by Kaur et al. [26] and Kelapure [50] who found the fat content of flaxseed flour was 38.2 and 40.5%. The protein, crude fiber, and ash content of RFSF, in our study, was observed to be 21.40%, 8.75%, and 3.95%, respectively. These values correlated well with the values reported earlier by Hussain et al. [19], Kaur et al. [26], Masoodi [29] and Kelapure [50].

The valuable chemical composition of RFSF along with the high carbohydrate content of WF, especially starch, led to a composite flour with good characteristics for biscuit manufacturing. Mamat and Hill [51] showed that flour with high carbohydrate content and low in gluten is highly recommended for biscuit manufacturing. Therefore, we considered that the composite flours with 25% RFSF could be successfully used for further processing. WF contributes to the final characteristics of biscuits with the pasting properties of the starch and with the gluten content. Moreover, the high protein content of RFSF could contribute also to the pasting property on the finished product and to the dough viscosity, as it was reported previously in the development of protein-enriched composite flour for biscuits production [52].

3.2. Volatile Compounds of Raw and Roasted Flaxseed

The purpose of the analysis was to compare the volatile compounds profile of both raw and roasted flaxseed, to explain, based on the identified volatile compounds, consumers' acceptability of roasted seeds. As a result, a total number of 18 aroma compounds were identified by using ITEX/GS-MS and are presented in Table 3 group in six classes of compounds: alcohols, aldehydes, ketones, terpenes, and terpenoids, acids and others.

Table 3. Volatile compounds of raw flaxseeds and roasted flaxseed.

Volatile Compounds	RT (Retention Time, min)	Raw Flaxseeds (%) *	Roasted Flaxseed (%) *	Aroma Characteristics **
		Alcohols		
3-Methylbutan-1-ol	3.904	2.75 ± 0.05 [a]	4.48 ± 0.03 [b]	Whiskey, malt, burnt
2-Methylbutan-1-ol	3.983	3.81 ± 0.21 [a]	3.64 ± 0.032 [a]	Malt
Pentan-1-ol	4.843	3.97 ± 0.11 [b]	1.38 ± 0.21 [a]	Fruit, balsamic
Hexan-1-ol	9.943	23.06 ± 0.03 [b]	3.74 ± 0.11 [a]	Resin, flower, green
Phenol	15.682	3.63 ± 0.03 [a]	2.99 ± 0.61 [a]	Whiskey, malt, burnt
		Aldehydes		
Hexanal	6.059	N.D.	24.24 ± 0.21	Grass, tallow, fat, fruity, sweaty
Benzaldehyde	14.758	5.76 ± 0.11 [a]	10.42 ± 0.32 [b]	Almond, burnt sugar
Nonanal	19.753	1.55 ± 0.08 [a]	2.85 ± 0.11 [b]	Fat, citrus, green
Decanal	22.41	1.82 ± 0.03 [a]	1.25 ± 0.04 [a]	Orange peel, tallow
		Ketones		
Acetophenone	18.532	13.22 ± 0.21 [a]	17.28 ± 0.21 [b]	Must, flower, almond
Benzophenone	31.22	3.18 ± 0.03 [b]	1.85 ± 0.11 [a]	Balsamic
		Terpenes and terpenoids		
α-Pinene	13.358	12.70 ± 0.21 [b]	6.81 ± 0.30 [a]	Pine
β-Pinene	15.343	4.80 ± 0.21 [b]	1.99 ± 0.03 [a]	Pine, resin
D-limonene	17.32	10.72 ± 0.10 [b]	9.85 ± 0.03 [a]	Citrus, mint
Camphene	14.117	2.48 ± 0.21	N.D.	Camphor
		Acids		
Benzoic acid	21.15	1.19 ± 0.21	N.D.	Balsamic
		Others		
2-Methyloctane	9.503	4.51 ± 0.02 [b]	2.80 ± 0.02 [a]	Alkane
2-Pentylfuran	15.97	N.D.	2.95	Fruity

* Each value was the mean of triplicate measurements; N.D.—not detected. ** drawn from [46,47]; Note: [a,b] different superscripts in a row indicate significant differences within samples ($p < 0.05$) according to Duncan comparison test.

It was found that the main volatile compounds of raw flaxseed were hexan-1-ol (23.06%), acetophenone (13.22%), α-pinene (12.70%), D-limonene (10.72%) and on the other hand, the major volatile compounds in the roasted flaxseed were hexanal (24.24%), acetophenone (17.28%), benzahdehyde (10.42%), D-limonene (9.85%), α-pinene (6.81%), 3-methylbutan-1-ol (4.48%) and hexan-1-ol (3.74%), respectively (Table 3).

Compounds with fresh and balsamic aroma characteristics, for example hexan-1-ol, benzophenone, β-pinene, D-limonene, camphene and benzoic acid, were present in significantly higher ($p < 0.05$) amounts in raw flaxseed compared to roasted flaxseed. With respect to hexan-1-ol, a lipid oxidation volatile compound with a significant value in raw flaxseed, its amount could be justified by the activity of dehydrogenase enzyme during storage [53].

In the roasted flaxseed compounds with aromas like malt, burned sugar and almond were found in significantly higher amounts ($p < 0.05$) than in raw flaxseed. Hexanal is the primary oxidation product of linoleic acid [52]. The increase of hexanal in roasted flaxseed reaching a final value of 24.24% may be due to lipid oxidation during thermal treatment [54]. Roasted flaxseeds are a rich source of lipids and in the present study, a total amount of 42.50% total fat was previously mentioned. Furthermore, Wei et al. [53] reported a higher content of α-linolenic acid, the precursor of docosahexaenoic and eicosapentaenoic acids, in flaxseeds, thus explaining the formation of hexanal during the roasting process. Furthermore, acetotophenone (17.28%) was the major volatile compound from the ketone group, while the main alcohol identified was 3-methylbutan-1-ol (4.48%). The presence of aldehydes, ketones and alcohols in the roasted flaxseeds could be justified by thermal reactions during roasting such as non-enzymatic Maillard reactions and sugar caramelization. Likewise, the aforementioned compounds could also be formed as a result of lipid oxidation [54].

3.3. Proximate Composition for Biscuits with RFSF

The proximate composition of the biscuits with RFSF (B0, B10, B25, B40) is summarized in Table 4 Incorporation of the RFSF in biscuits has a significant influence on the major components ($p < 0.05$) and a decrease of the moisture content was observed while fat, ash, protein and fiber content of composite flour mixes increased. RFSF had a significant ($p < 0.05$) effect on the moisture content of the biscuits compared to the control with a decrease in moisture content from 7.34% to 4.92%. Similar patterns were previously noticed by Khouryieh and Aramouni [55] and Kaur et al. [26] who found that as the flaxseed flour concentration increased in the blend, the moisture content of cookies decreased. These results may be due to low moisture content (1.64%) of RFSF. Moreover, a high amount of fat (42.50%) was recorded in RFSF, which also contributed to the hydration of the dough, thus reducing the moisture of the biscuits. Furthermore, the low moisture content of the biscuits allows a longer preservation period, due to minimal microbial or chemical activity.

Table 4. Proximate characteristics of the biscuits samples.

Parameters	Biscuit Samples *			
	B0	B10	B25	B40
Proximate Composition (% fw)				
Moisture (%)	7.34 ± 0.10 [d]	6.50 ± 0.20 [c]	5.7 ± 0.15 [b]	4.92 ± 0.14 [a]
Protein (%)	8.35 ± 0.50 [a]	9.12 ± 0.20 [b]	9.89 ± 0.40 [c]	10.77 ± 0.90 [d]
Ash (%)	1.23 ± 0.02 [a]	1.61 ± 0.05 [b]	2.13 ± 0.07 [c]	2.60 ± 0.02 [d]
Fat (%)	15.19 ± 0.90 [a]	19.06 ± 1.01 [b]	23.88 ± 1.20 [c]	28.34 ± 1.50 [d]
Crude fiber (%)	0.75 ± 0.92 [a]	1.95 ± 0.04 [b]	3.20 ± 0.14 [c]	5.05 ± 0.45 [d]
Total carbohydrate (%)	67.14 ± 0.70 [d]	61.76 ± 0.06 [c]	55.19 ± 0.21 [b]	48.32 ± 0.32 [a]
TPC (mg GAE/100 g, fw)	63.06 ± 0.07 [a]	69.95 ± 0.03 [b]	74.23 ± 0.12 [c]	78.82 ± 0.26 [d]
DPPH (%RSA)	13.57 ± 0.14 [a]	18.89 ± 0.09 [b]	26.41 ± 0.29 [c]	32.03 ± 0.71 [d]

* Values represent mean of three independent determinations ± SD; fw-fresh weight; Mean values followed by the same superscript alphabet in the row are not significantly different at $p < 0.05$ according to Duncan comparison test.

Ash content increased from 1.23% (control biscuits) to 2.60% (sample B40). This might be due to the higher mineral content of flaxseed flour [56,57]. Also, the protein content increased significantly ($p < 0.05$) from 8.35 g/100 g in B0 to 10.77 g/100 g in B40 and fat from 15.19 g/100 g in B0 to 28.34 g/100 g in B40. Moreover, the addition of 40% RFSF registered a positive influence on the fibre content of the final baked biscuits, increasing its value about 6.7-fold compared to the control sample. Improving the nutritional value of pastry products by substituting wheat flour with other fiber-rich sources was also recommended by others [58,59]. This increase could be explained by the fact that roasted flaxseed flour is far higher in fat, protein and crude fiber content compared with wheat flour as mentioned in the proximate analysis section of this article.

The total carbohydrate content was significantly decreased ($p < 0.05$) in biscuits substituted with flaxseed from 61.76% to 48.32% compared to control biscuits (67.14%). These results are similar with the results obtained by Ahmed et al. [60] El-Demery et al. [61] mentioned that as the level of substitution with flaxseed flour increased, all compounds increased except total carbohydrate. Other authors also revealed that the substitution of wheat flour with flaxseed flour resulted in a considerable improvement in protein, crude fiber and ash of biscuit samples [12,16,24,26,50,59,62].

TPC and DPPH of biscuit samples increased with the substitution level of RFSF to WF in the blends. The highest increase in DPPH activity and TPC was exhibited by sample B40 (32.03% RSA and 78.82 mg GAE/100 g) and the lowest (13.67% RSA and 63.06 mg GAE/100 g) was shown by the control sample. Similar results have been obtained by Kaur, et al. [26,33]. This increase can be explained by the fact that flaxseed possesses a very powerful antioxidant system, being particularly rich in lignans, e.g., secoisolariciresinol diglucoside (SDG), which are also present in flaxseed oil. [63]. Besides lignans, flaxseed contains high amounts of phenolic compounds, such as ferulic acid, syringic acid, cinnamic acid, vanillic acid, *p*-coumaric acid and gallic acid [64]. However, although flaxseed enriches the material basis for exerting its antioxidant activity, previous studies have observed differences in the quality characteristics of flaxseed varieties from different regions of the world, indicating some geographical and varietal specificity [64]. Deng et al. [64] reported that the total phenolic contents in studied flaxseed varieties ranged from 109.93 mg GAE/100 g to 246.88 mg GAE/100 g, and the DPPH values ranged from 32.56 mg TE/100 g to 46.22 mg TE/100 g. Even if the roasting process of flaxseeds could produce a slight decrease in the antioxidant activity, the baking process resulted in a significant increase in antioxidative activity due to the Maillard pigments recognized as having a high antioxidative capacity [59].

3.4. Mineral Content of Biscuits

The results of the evaluation of the mineral content of biscuits are displayed in Table 5. The addition of RFSF increased the P, K, Zn, Fe, Ca and Mg contents, meanwhile Cu and Mn could not be identified in the biscuit samples.

Table 5. The concentration of macro and microelements in biscuit samples.

Mineral Content (mg/100 g, fw)	Biscuit Samples *			
	B0	B10	B25	B40
P	27.83 ± 0.03 [a]	35.64 ± 0.39 [b]	43.20 ± 067 [c]	49.78 ± 0.03 [d]
K	240.46 ± 0.89 [a]	275.38 ± 0.76 [b]	313.08 ± 0.55 [c]	356.99 ± 0.99 [d]
Cu	N.D.	N.D.	N.D.	N.D.
Zn	0.71 ± 0.09 [a]	1.12 ± 0.03 [b]	1.54 ± 0.05 [c]	1.97 ± 0.02 [d]
Fe	1.40 ± 0.03 [a]	1.87 ± 0.05 [b]	2.27 ± 0.07 [c]	2.76 ± 0.11 [d]
Mn	N.D.	N.D.	N.D.	N.D.
Ca	164.46 ± 0.83 [a]	195.43 ± 0.79 [b]	223.05 ± 0.85 [c]	256.09 ± 0.533 [d]
Mg	41.36 ± 0.55 [a]	90.16 ± 0.59 [b]	141.47 ± 0.88 [c]	191.33 ± 0.39 [d]

* Values represent mean of three independent determinations ± SD; fw-fresh weight; Mean values followed by the same superscript alphabet in the row are not significantly different at $p < 0.05$ according to Duncan comparison test. N.D.—not determined.

The significant increase of minerals ($p < 0.05$) in biscuits could be justified by the rich minerals content of RFSF. A large body of literature highlighted that flaxseeds are a valuable source of minerals. For instance, Bernacchia et al. [65] showed that P, K, Ca, Mg flaxseed content reached values of 622, 831, 236 and 431 mg/100 g, respectively. Furthermore, Kaur et al. [57] mentioned that flaxseed is rich in Mg which is the second most abundant element in human body, as well as in K, which is the most common macro-mineral with positive effect in reduction of stroke incidence and blood platelets aggregation.

It is important to note that mineral content could vary between flaxseed cultivars and could be influenced by external factors such as soil conditions, fertilizers, water availability, climatic conditions and genetic factors [57].

3.5. Analysis of Volatile Compounds of Biscuit Samples

A total number of 21 volatile compounds were identified in the RFSF enriched biscuits, by means of ITEX/GC-MS technique as shown in Table 6. In all biscuit samples, the main volatile compound from the aldehydes group was hexanal, ranging from 14.76% to 18.39%, meanwhile β-myrcene and D-limonene were the main volatile compounds from terpenes and terpenoids group, ranging from 8.99% to 15.65% and 2.47% to 8.48%, respectively. Acetophenone was the major volatile compound from the ketones group with values between 2.50% to 10.98%, meanwhile, 4-methyloctane reached a final value of 6.92% in the B40 sample. The presence of 4-methyloctane could be justified by the addition of vanilla essence during the biscuit manufacturing [66]. The amount of D-limonene in RFSF increased linearly with increasing levels of RFSF in the final baked products. D-Limonene is responsible for odour perceptions like citrus and mint. Furthermore, flaxseed represent a high source of carotenoids [67] which could be correlated with higher amounts of D-limonene, as reported by Chiş et al. [68]. From the terpenes and terpenoids group apart from D-limonene, β-myrcene enhanced the final odour perception through its balsamic, musty and spice perceptions.

Table 6. Volatile compounds content of the biscuits with roasted flaxseed flour.

Volatile Compounds	RT (Retention Time, min)	Biscuit Samples *				Aroma Characteristics **
		B0	B10	B25	B40	
Alcohols						
Phenol	15.65	1.83 ± 0.03 [b]	1.8 ± 0.02 [b]	1.13 ± 0.03 [a]	3.63 ± 0.04 [c]	Phenol
Aldehydes						
Hexanal	6.07	18.39 ± 0.05 [c]	14.76 ± 0.12 [a]	15.35 ± 0.22 [a]	16.36 ± 0.03 [b]	Grass, tallow, fat
Heptanal	11.926	0.63 ± 0.02 [b]	0.19 ± 0.02 [a]	0.98 ± 0.04 [c]	1.12 ± 0.07 [c]	Fat, citrus
Octanal	16.506	0.97 ± 0.03 [a]	0.87 ± 0.06 [a]	1.20 ± 0.05 [b]	1.46 ± 0.04 [c]	Lemon, green
Nonanal	19.74	5.13 ± 0.22 [c]	3.12 ± 0.11 [b]	1.69 ± 0.05 [a]	2.99 ± 0.03 [b]	Fat, citrus, green
Benzaldehyde	14.744	2.40 ± 0.03 [a]	5.08 ± 0.21 [b]	7.18 ± 0.33 [c]	8.82 ± 0.03 [c]	Almond, burnt sugar
Decanal	22.412	0.59 ± 0.04 [b]	0.45 ± 0.02 [a]	0.62 ± 0.03 [b]	0.90 ± 0.02 [c]	Orange peel, tallow
Ketones						
Heptan-2-one	11.167	0.89 ± 0.04 [a]	1.13 ± 0.02 [a]	3.82 ± 0.06 [b]	4.22 ± 0.08 [b]	Cheese, fruity, ketonic, green banana, with a creamy nuance
Acetophenone	18.524	2.50 ± 0.03 [a]	5.02 ± 0.02 [b]	8.69 ± 0.06 [c]	10.98 ± 0.02 [d]	Must, flower, almond
Nonan-2-one	19.341	1.59 ± 0.02 [b]	1.44 ± 0.03 [b]	0.53 ± 0.05 [a]	1.70 ± 0.04 [b]	Fruity, sweet, waxy, green herbaceous
Benzophenone	31.223	1.26 ± 0.04 [c]	0.72 ± 0.06 [b]	0.91 ± 0.03 [b]	0.44 ± 0.05 [a]	Balsamic
Terpenes and terpenoids						
β-Myrcene	15.951	8.99 ± 0.17 [a]	10.77 ± 0.23 [a]	12.79 ± 0.31 [b]	15.65 ± 0.22 [c]	Balsamic, must, spice
D-Limonene	17.323	2.47 ± 0.03 [a]	4.28 ± 0.04 [b]	6.14 ± 0.07 [c]	8.48 ± 0.03 [d]	Citrus, mint
Acids						
Benzoic acid	21.232	9.62 ± 0.02 [c]	5.91 ± 0.03 [b]	7.15 ± 0.07 [b]	3.02 ± 0.03 [a]	Balsamic
Dodecanoic acid	29.711	2.47 ± 0.03 [a]	4.28 ± 0.04 [b]	6.14 ± 0.07 [c]	8.48 ± 0.03 [d]	Mild fatty, coconut bay oil
Others						
Dimethyl disulfide	4.067	7.29 ± 0.05 [b]	6.67 ± 0.03 [b]	4.06 ± 0.12 [a]	4.27 ± 0.34 [a]	Sulfurous
2,4-Dimethylheptane	7.008	8.18 ± 0.22 [c]	6.86 ± 0.31 [b]	6.81 ± 0.02 [b]	3.16 ± 0.12 [a]	Alkane
4-Methyloctane	9.388	6.55 ± 0.12 [b]	6.05 ± 0.05 [a]	6.75 ± 0.03 [bc]	6.92 ± 0.02 [c]	Alkane
3,7-Dimethyldecane	18.213	8.79 ± 0.15 [b]	14.51 ± 0.21 [c]	5.27 ± 0.23 [a]	2.95 ± 0.08 [a]	Alkane
3,4-Dimethylundecane	18.388	2.60 ± 0.07 [b]	2.52 ± 0.09 [b]	1.20 ± 0.02 [a]	0.68 ± 0.03 [a]	Alkane
Ethyl 2,4-dioxohexanoate	16.369	0.94 ± 0.04 [a]	1.20 ± 0.03 [ab]	1.66 ± 0.05 [b]	N.D.	Apple peel, fruit

* Each value was the mean of triplicate measurements; N.D.—not detected. ** drawn from [46,47]. Note: [a–d] different superscripts in a row indicate significant difference within samples ($p < 0.05$) according to Duncan comparison test.

Acetophenone, from the ketone group, was previously pinpointed as a volatile compound with implication in the overall flavor of flour products which could be formed during Maillard reactions [69]. Its odour perception is pleasant, having musty, flowery and almond characteristics.

The increased amount of heptan-2-one from the ketone group with increasing yield of RFSF could be explained by the chemical composition of flaxseed, rich in tocopherol and ascorbic acid [65]. In this line, Starowicz et al. [70] showed that tocopherol and ascorbic acid significantly increased the peak area of heptan-2-one.

Furthermore, a recent study of Hidalgo and Zamora [71] highlighted that benzaldehyde could be formed through chemical reactions such as lipid oxidation. The reaction involved in the first step is a carbonyl-amine reaction followed by a free radical amino-acid degradation, mainly phenylalanine. It was previously shown that flaxseed is a rich source of 6 essential amino-acids, from which phenylalanine amount could vary in the range from 1.44 mg/100 g up to 66.6 mg/100 g, depending on the flaxseed cultivars [57]. Therefore, the significant differences ($p < 0.05$) between benzaldehyde sample amounts, could be justified by the chemical composition of flaxseed and by lipid oxidation.

It is worth nothing that non-enzymatic Maillard reactions such as the Strecker degradation process could lead to the formation of aldehydes as a result of the reaction of aminoacids with dehydroreductones [54].

Overall, the increased RFSF percentages during biscuits manufacturing, led to significant differences ($p < 0.05$) between aroma volatile compounds, mainly due to the rich chemical composition in lipids, protein, aminoacids, phenols and chemical reactions such as Maillard and lipid oxidation.

3.6. Physical Characteristics of the Biscuits

The evaluation results of the physical characteristics of the biscuits indicate that there was a significant difference ($p < 0.05$) between the control sample and B40 in thickness, diameter and spread factor (Table 7). However, there was no significant difference in the weight of any type of biscuit although it was lower than the control biscuits.

Table 7. Physical characteristics of biscuits.

Biscuits Samples	Physical Parameters *			
	Weight (g)	Thickness (cm)	Diameter (cm)	Spread Factor
B0	12.9 ± 1.09 [a]	0.75 ± 0.30 [a]	5.93 ± 1.20 [a]	7.90 ± 1.90 [c]
B10	12.8 ± 1.10 [a]	0.79 ± 1.00 [ab]	6.08 ± 1.04 [b]	7.70 ± 1.50 [bc]
B25	12.7 ± 1.40 [a]	0.83 ± 0.50 [b]	6.17 ± 1.09 [bc]	7.43 ± 2.10 [b]
B40	12.6 ± 1.25 [a]	0.89 ± 0.90 [c]	6.26 ± 1.30 [c]	7.03 ± 2.90 [a]

* Values represent mean of three independent determinations ± SD. Mean values followed by the same superscript alphabet in the column are not significantly different at $p < 0.05$ according to Duncan comparison test.

The diameter and thickness of biscuit samples (B10, B25 and B40) increased slightly with increasing substitution percentage of RFSF compared with control biscuit (B0). Sample B40 presented the maximum diameter and thickness (6.26 and 0.89 cm). The spread factor is the ratio that depends on the values of the thickness and diameter of the cookies and it is used to determine the quality of flour for producing cookies [26]. The results of the spread ratio of biscuits revealed a significant reduction ($p < 0.05$) in spread ratio from 7.90 to 7.03 cm for B40. With the increase in the concentration of RFSF, the spread factor of biscuits gradually decreased. These results are in the line with the findings of other authors [16,32] who stated that proteins and dietary fibers have more water-binding power. Ganorkar and Jain [16], argue that the presence of more water in the dough, leads to higher dissolution of sugar during mixing and this lowers the initial dough viscosity and the cookie is able to spread at a faster rate during cooking. Moreover, in the opinion of the same authors, an inverse correlation is obtained when the flour components absorb large quantities of water and, as a consequence, reduce the amount of water that is available to dissolve the sugars in the formula. Therefore, the initial viscosity is higher and the biscuits spread less during baking.

3.7. Instrumental Analysis of Texture

The results of the instrumental analysis of texture are presented in Table 8. Instrumental analysis of hardness showed statistically significant differences ($p < 0.05$) between control (B0) and B40. Hardness decreased as the amount of RFSF increased, mean values of 10,655 g (control sample) and 5714 g (B40), respectively. Similar patterns were noticed for the load required to reach the hardness work value. The values of hardness work (mJ) decreased with increasing RFSF from 262.73 mJ to 122.30 mJ. These results are similar to those obtained by Ganorkar and Jain [16] and Omran et al. [32] who reported that the textural parameters were found to decrease with increasing flaxseed flour incorporation. This might be due to an increase in dietary fiber and protein, high water absorbing capacity components as well as due to the high level of fat (42.50%) found in RFSF. These factors contributed to a sticky dough which reduced the extensibility of dough. Previous studies claim that with an increase in fat content, the gluten network gets interrupted thus the physical properties of biscuits are changed and become less hard. At very high fat content the lubricating function is high and a soft texture is obtained. Hence the hardness gradually decreases forming softer biscuits with an increased level of flaxseed flour [16,32,72].

Table 8. Instrumental analysis of texture.

Sample	B0	B10	B25	B40
Hardness cycle 1 (g)	$10{,}655 \pm 21$ [d]	8801 ± 80 [c]	7136 ± 69 [b]	5714 ± 34 [a]
Hardness Work cycle 1 (mJ)	262.73 ± 34.22 [d]	200.05 ± 11.33 [c]	150.8 ± 35.10 [b]	122.30 ± 25.82 [a]

Values represent mean of three independent determinations $\pm$ SD. Mean values followed by the same superscript alphabet in the row are not significantly different at $p < 0.05$ according to Duncan comparison test.

3.8. Sensory Analysis

Before a product becomes commercially available, many tests are conducted with consumers to evaluate the product acceptability. Nowadays, consumers are more health conscious then before and functional foods have an ascending trend on the market. This context brings new challenges upon sensory analysis, transforming it into a more proactive role in producing unique, innovative and functional products. Austria et al. [73] hypothesised that foods that contain significant quantities of flaxseed will be well tolerated by the public.

The mean of hedonic scores and their standard deviation are presented in Table 9. The control sample scored the highest for appearance, hardness, chewiness and aftertaste.

Table 9. Sensory evaluation of the biscuits.

Sample	Appearance	Hardness	Crispiness	Chewiness	Taste and Aroma	Aftertaste	Overall Appreciation
B0	5.83 ± 1.34 [c]	3.58 ± 1.07 [b]	5.38 ± 1.48 [a]	6.15 ± 1.32 [b]	5.77 ± 0.83 [a]	5.02 ± 1.79 [c]	6.88 ± 1.15 [ab]
B10	5.62 ± 1.39 [c]	2.37 ± 0.84 [a]	6.79 ± 1.75 [b]	5.33 ± 1.46 [a]	6.10 ± 0.72 [b]	4.42 ± 1.35 [b]	6.92 ± 1.20 [ab]
B25	4.94 ± 1.38 [b]	2.88 ± 0.92 [ab]	6.56 ± 1.64 [b]	5.50 ± 1.42 [a]	6.42 ± 1.05 [c]	4.23 ± 1.65 [b]	6.75 ± 1.37 [a]
B40	4.35 ± 1.69 [a]	2.65 ± 1.25 [a]	7.12 ± 1.96 [c]	5.40 ± 1.65 [a]	6.87 ± 1.47 [d]	3.90 ± 2.04 [a]	6.94 ± 1.49 [b]

Values represent hedonic scores calculated as mean $\pm$ SD ($n = 52$). Mean values followed by the same superscript alphabet in the column are not significantly different at $p < 0.05$ according to Duncan comparison test.

For the attribute of appearance no significant difference was encountered between the control sample and the B10. However, all tested samples which contained flaxseed flour (B10, B25 and B40) scored progressively lower than the control (B0). This can be explained by the fact that evaluators were able to visually discriminate the samples solely based on their color and degree of irregularities which could be seen on the surface of the biscuit. The decrease of the hedonic appearance values may be due to the progressive decrease amount of moisture and total carbohydrates in the biscuit samples (B0 > B10 > B25 > B40 as can be seen in Table 4) which caused the depreciation of biscuits' surface and darker color during baking. There was a positive correlation between appearance and moisture (0.97, $p < 0.05$), and between appearance and total carbohydrates (0.99, $p < 0.05$). Moreover, the appearance negatively correlated with ash (-0.99, $p < 0.05$), fat (-0.99, $p < 0.05$) and crude fiber (-0.98, $p < 0.05$). The decrease of appearance perception with the increase of the flaxseed content was also recorded previously in studies performed on biscuits [29,32,74] as well as on other bakery products, like cookies [19,75] muffins [76,77] and bread [13].

The texture plays a key role in assessing consumers' acceptance of a food product. The complex changes of the texture during the process of eating (mouth behaviour) can influence the acceptance or rejection of the product. Although most of the times it is unconscious, the consumer's preference toward a product is often due to the texture. The biscuits were evaluated by three different textural attributes: hardness, crispiness and chewiness. Each attribute was clearly explained to the participants and a brief description of them was inserted in the evaluation sheet next to the hedonic scale as stated in the method section. The participants were instructed to evaluate the crispiness as the force and noise created when the sample breaks during chewing the sample between the molar teeth. It was noted that the biscuits with higher amount of RFSF presented lower amount of moisture (Table 4). Negative correlation ($p < 0.05$) was found between the consumer's perception of crispiness and the amount of moisture and total carbohydrates of the biscuits (-0.85 and -0.82, respectively). Similar findings were published by Hussain et al. [19] and Marpalle et al. [13]. This result can be explained by flaxseed's capacity to bind water, which influences the textural properties of the final product [76].

Taste and aroma improved with the increasing amount of the flaxseed in the biscuits. The flaxseed has a unique nutty flavour (Tables 3 and 7) which was found pleasant by the consumers, therefore the taste and aroma characteristics of samples with RFSF scored higher than the control (B0 > B10 > B25 > B40). Moreover, the taste and aroma of the samples was positively correlated with the biscuits' content in protein (0.99, $p < 0.005$), ash (0.99, $p < 0.005$), fat (0.99, $p < 0.005$) and crude fiber (0.99, $p < 0.001$) and negatively correlated with moisture (-0.99, $p < 0.005$) and total carbohydrates (-0.99, $p < 0.01$). Some authors mention a bitter taste in samples prepared with high amounts of flaxseed flour [78]. However, in this study, the nutty flavour was appreciated by the consumers which are accustomed to the taste and flavor of different types of traditional biscuits and cookies prepared with various amounts of nuts.

The aftertaste correlated negatively with the biscuit's content in protein (-0.97, $p < 0.05$), ash (-0.95, $p < 0.05$), fat (-0.96, $p < 0.05$) and crude fiber (-0.95, $p < 0.05$) and correlated positively with moisture (0.97, $p < 0.05$) and total carbohydrates (0.96, $p < 0.05$).

Although the sensory attributes of the samples with RFSF were decreased compared to the control, the aroma and crispiness increased and the significant differences between samples in appearance, hardness, chewiness and aftertaste did not reflect on the overall appreciation of the biscuits.

4. Conclusions

The results of this study reveal the effect of roasted flaxseed flour addition on the physico-chemical parameters, volatile profile, and sensory acceptability of biscuits. The high incorporation of RFSF advantageously influenced the nutritional properties of biscuits as evidenced by significant increases in the fibre contents (about 6.7-fold higher than in control biscuits), proteins and minerals. Furthermore, the addition of RFSF increased the total phenolic content and radical scavenging activity amount, reaching final values for biscuits manufactured with 40% RFSF such as 78.82 mg GAE/100 g and 32.03%, respectively. With respect to the volatile profile of biscuits, the addition of RFSF leads to the formation of aroma compounds such as hexanal, β-myrcene, D-limonene, acetophenone and 4-methyloctane, having odor perceptions such as grass, balsamic, must, citrus mint and alkane. Sensory evaluation indicated that the overall biscuits appreciation was not affected in a significant way ($p < 0.05$) by RFSF addition, meanwhile the aftertaste intensity decreased. Considering the nutritional, textural and sensorial analysis of the final baked goods, we can conclude that 25% RFSF could be successfully used in biscuit manufacturing.

Author Contributions: Conceptualization, S.M.M. and M.S.C.; methodology, V.M., S.A.S., L.S. and A.P. (Anamaria Pop); software, M.S.C., L.S.; validation, S.A.S. and A.P. (Adriana Păucean); formal analysis, M.S.C., S.A.S., S.M.M. and L.S.; writing—original draft preparation, S.M.M., M.S.C. and L.S.; writing—review and editing, A.P. (Adriana Păucean) and L.S.; supervision, L.S., A.P. (Adriana Păucean) and S.M.M.; project administration, S.M.M., S.M. All authors have read and agreed to the published version of the manuscript.

Funding: This research received no external funding.

Acknowledgments: This work was supported by a grant of Romanian Ministry of Research and Innovation, CNCS—UEFISCDI, project number PN-III-P4-ID-PCE2020-1847, within PNCDI III.

Conflicts of Interest: The authors declare no conflict of interest.

References

1. Bala, A.; Gul, K.; Riar, C.S. Functional and sensory properties of cookies prepared from wheat flour supplemented with cassava and water chestnut flours. *Cogent Food Agric.* **2015**, *1*, 1019815. [CrossRef]
2. Khan, A.; Saini, C.S. Effect of roasting on physicochemical and functional properties of flaxseed flour. *Cogent Eng.* **2016**, *3*, 1145566. [CrossRef]
3. Chauhan, A.; Saxena, D.C.; Singh, S. Physical, textural, and sensory characteristics of wheat and amaranth flour blend cookies. *Cogent Food Agric.* **2016**, *2*, 1125773. [CrossRef]
4. Man, S.; Păucean, A.; Muste, S.; Chiș, M.S.; Pop, A.; Calian-Ianoș, I.D. Assessment of amaranth flour utilization in cookies production and quality. *J. Agroaliment. Process. Technol.* **2017**, *23*, 97–103.
5. Katare, C.; Saxena, S.; Agrawal, S.; Prasad, G. Flax Seed: A potential medicinal food. *J. Nutr. Food Sci.* **2012**, *2*, 1000120. [CrossRef]
6. Hussain, S.; Anjum, F.M.; Alamri, M.S.; Mohamed, A.A.; Nadeem, M. Functional flaxseed in baking. *Qual. Assur. Saf. Crop. Foods* **2013**, *5*, 375–385. [CrossRef]
7. Chishty, S.; Bissu, M. Health benefits and nutritional value of flaxseed—A Review. *Indian J. Appl. Res.* **2016**, *6*, 243–245.
8. Soni, R.P.; Katoch, M.; Kumar, A.; Verma, P. Flaxseed—Composition and its health benefits. *Res. Environ. Life Sci.* **2016**, *9*, 310–316.
9. Kanikowska, D.; Korybalska, K.; Mickiewicz, A.; Rutkowski, R.; Kuchta, A.; Sato, M.; Kreft, E.; Fijałkowski, M.; Gruchała, M.; Jankowski, M.; et al. Flaxseed (*Linum usitatissimum* L.) supplementation in patients undergoing lipoprotein apheresis for severe hyperlipidemia—A pilot study. *Nutrients* **2020**, *12*, 1137. [CrossRef]
10. Singh, K.K.; Mridula, D.; Rehal, J.; Barnwal, P. Flaxseed: A potential source of food, feed and fiber. *Crit. Rev. Food Sci. Nutr.* **2011**, *51*, 210–222. [CrossRef] [PubMed]
11. Kajla, P.; Sharma, A.; Sood, D.R. Flaxseed—A potential functional food source. *J. Food Sci. Technol.* **2015**, *52*, 1857–1871. [CrossRef] [PubMed]
12. Shekhara, N.R.; Anurag, A.P.; Prakruthi, M.; Mahesh, M.S. Flax Seeds (*Linum usitatissimmum*): Nutritional composition and health benefits. *IP J. Nutr. Metab. Health Sci.* **2020**, *3*, 35–40. [CrossRef]
13. Marpalle, P.; Sonawane, S.K.; Arya, S.S. Effect of flaxseed flour addition on physicochemical and sensory properties of functional bread. *LWT Food Sci. Technol.* **2014**, *58*, 614–619. [CrossRef]
14. Mohanan, A.; Nickerson, M.T.; Ghosh, S. Oxidative stability of flaxseed oil: Effect of hydrophilic, hydrophobic and intermediate polarity antioxidants. *Food Chem.* **2018**, *266*, 524–533. [CrossRef] [PubMed]
15. Moraes, É.A.; de Souza Dantas, M.I.; de Castro Morais, D.; Oliveira da Silva, C.; Ferreira de Castro, F.A.; Martino, H.S.D.; Ribeiro, S.M.R. Sensory evaluation and nutritional value of cakes prepared with whole flaxseed flour. *Ciência Tecnol. Aliment.* **2010**, *30*, 974–979. [CrossRef]
16. Ganorkar, P.M.; Jain, R.K. Effect of flaxseed incorporation on physical, sensorial, textural and chemical attributes of cookies. *Int. Food Res. J.* **2014**, *21*, 1515–1521.
17. Parikh, M.; Maddaford, T.G.; Austria, J.A.; Aliani, M.; Netticadan, T.; Pierce, G.N. Dietary Flaxseed as a strategy for improving human health. *Nutrients* **2019**, *11*, 1171. [CrossRef] [PubMed]
18. Conforti, F.D.; Davis, S.F. The effect of soya flour and flaxseed as a partial replacement for bread flour in yeast bread. *Int. J. Food Sci. Technol.* **2006**, *41*, 95–101. [CrossRef]
19. Hussain, S.; Anjum, F.M.; Butt, M.S.; Khan, M.I.; Asghar, A. Physical and sensoric attributes of flaxseed flour supplemented cookies. *Turk. J. Biol.* **2006**, *30*, 87–92.
20. Oomah, B.D.; Der, T.J.; Godfrey, D.V. Thermal characteristics of flaxseed (*Linum usitatissimum* L.) proteins. *Food Chem.* **2006**, *98*, 733–741. [CrossRef]
21. Tarpila, S.; Aro, A.; Salminen, I.; Tarpila, A.; Kleemola, P.; Akkila, J.; Aldercreutz, H. The effect of flaxseed supplementation in processed foods on serum fatty acids and enterolactone. *Eur. J. Clin. Nutr.* **2002**, *56*, 157–165. [CrossRef] [PubMed]
22. Codină, G.G.; Istrate, A.M.; Gontariu, I.; Mironeasa, S. Rheological properties of wheat-flaxseed composite flours assessed by mixolab and their relation to quality features. *Foods* **2019**, *8*, 333. [CrossRef]
23. Kasote, D.M. Flaxseed phenolics as natural antioxidants. *Int. Food Res. J.* **2013**, *20*, 27–34.
24. Park, J.B.; Velasquez, M.T. Potential effects of lignan-enriched flaxseed powder on bodyweight, visceral fat, lipid profile, and blood pressure in rats. *Fitoterapia* **2012**, *83*, 941–946. [CrossRef] [PubMed]
25. Goyal, A.; Sharma, V.; Upadhyay, N.; Gill, S.; Sihag, M. Flax and flaxseed oil: An ancient medicine & modern functional food. *J. Food Sci. Technol.* **2014**, *51*, 1633–1653. [CrossRef] [PubMed]
26. Kaur, M.; Singh, V.; Kaur, R. Effect of partial replacement of wheat flour with varying levels of flaxseed flour on physicochemical, antioxidant and sensory characteristics of cookies. *Bioact. Carbohydr. Diet. Fibre* **2017**, *9*, 14–20. [CrossRef]
27. Lowcock, E.C.; Cotterchio, M.; Boucher, B.A. Consumption of flaxseed, a rich source of lignans, is associated with reduced breast cancer risk. *Cancer Causes Control* **2013**, *24*, 813–816. [CrossRef]
28. Moure, A.; Sineiro, J.; Domínguez, H.; Parajó, J.C. Functionality of oilseed protein products: A review. *Food Res. Int.* **2006**, *39*, 945–963. [CrossRef]

29. Masoodi, L.; Bashir, V.A.K. Fortification of Biscuit with Flaxseed: Biscuit production and quality evaluation. *IOSR J. Environ. Sci. Toxicol. Food Technol.* **2012**, *1*, 06–09. [CrossRef]

30. De Moura, C.C.; Peter, N.; Schumacker, B.D.O.; Borges, L.R.; Helbig, E. Biscoitos enriquecidos com farelo de linhaça marrom (*Linum Usitatissiumun* L.): Valor nutritivo e aceitabilidade. *Demetra Aliment. Nutr. Saúde* **2014**, *9*, 71–82. [CrossRef]

31. Bartkiene, E.; Jakobsone, I.; Pugajeva, I.; Bartkevics, V.; Zadeike, D.; Juodeikiene, G. Reducing of acrylamide formation in wheat biscuits supplemented with flaxseed and lupine. *LWT Food Sci. Technol.* **2016**, *65*, 275–282. [CrossRef]

32. Omran, A.A.; Ibrahim, O.S.; Mohamend, Z.E.-O.M. Quality characteristics of biscuit prepared from wheat and flaxseed Flour. *Adv. Food Sci.* **2016**, *38*, 129–138.

33. Kaur, P.; Sharma, P.; Kumar, V.; Panghal, A.; Kaur, J.; Gat, J. Effect of addition of flaxseed flour on phytochemical, physicochemical, nutritional, and textural properties of cookies. *J. Saudi Soc. Agric. Sci.* **2019**, *18*, 372–377. [CrossRef]

34. Xu, Y.; Hall, C.A.; Manthey, F.A. Chemical composition, antioxidant and antihyperglycemic activities of the wild lactarius deliciosus from China. *Molecules* **2019**, *24*, 1357. [CrossRef] [PubMed]

35. Codina, G.G.; Mironeasa, S.; Todosi-Sanduleac, E. Studies regarding the influence of brown flaxseed flour addition in wheat flour of a very good quality for bread making on bread quality. *Bull. Univ. Agric. Sci. Vet. Med. Cluj-Napoca Food Sci. Technol.* **2016**, *73*, 70–76. [CrossRef]

36. Silva, L.M.; Gama de Mendonça, L.; Zambelli, R.A.; Magalhães, A.L.M.; Silva da Costa, C.; Venânces de Souza Leão, L.M.; de Albuquerque Ribeiro de Sá Costa, R. Impact of green pulp banana and flaxseed flour on pound cake quality. *Int. J. Nutr. Food Sci.* **2017**, *6*, 243–249. [CrossRef]

37. Asociatia de Standardizare din Romania. Wheat Flour (original title in Rom.: Făină de grau). Romanian Standard 877:1996, 2 November 2011.

38. AACC. Method 38-12. In *Approved Methods of the American Association of Cereal Chemists*, 11th ed.; American Association of Cereal Chemists: St. Paul, MN, USA, 2000.

39. Man, S.M.; Paucean, A.; Calian, I.D.; Muresan, V.; Chiş, M.S.; Pop, A.; Muresan, A.; Bota, M.; Muste, S. Influence of fenugreek flour (*Trigonella foenum-graecum* L.) addition on the technofunctional properties of dark wheat flour. *J. Food Qual.* **2019**, *2019*, 8635806. [CrossRef]

40. Păucean, A.; Man, S.M.; Chiş, M.S.; Mureşan, V.; Pop, C.R.; Socaci, S.A.; Mureşan, C.C.; Muste, S. Use of pseudocereals preferment made with aromatic yeast strains for enhancing wheat bread quality. *Foods* **2019**, *8*, 443. [CrossRef]

41. Chiş, M.S.; Păucean, A.; Man, S.M.; Vodnar, D.C.; Teleky, B.E.; Pop, C.R.; Stan, L.; Borsai, O.; Kadar, C.B.; Urcan, A.C.; et al. Quinoa sourdough fermented with *Lactobacillus plantarum* ATCC 8014 designed for gluten-free muffins—A powerful tool to enhance bioactive compounds. *Appl. Sci.* **2020**, *10*, 7140. [CrossRef]

42. Seevaratnam, V.; Banumathi, P.; Premalatha, M.R.; Sundaram, S.P.; Arumugam, T. Studies on the preparation of biscuits incorporated with potato flour. *World J. Dairy Food Sci.* **2012**, *7*, 16. [CrossRef]

43. Bunea, A.; Rugină, D.O.; Pintea, A.M.; Sconţa, Z.; Bunea, C.I.; Socaciu, C. Comparative polyphenolic content and antioxidant activities of some wild and cultivated blueberries from romania. *Not. Bot. Horti Agrobot. Cluj Napoca* **2011**, *39*, 70–76. [CrossRef]

44. Chiş, M.S.; Păucean, A.; Stan, L.; Mureşan, V.; Vlaic, R.A.; Man, S.; Muste, S. *Lactobacillus plantarum* ATCC 8014 in quinoa sourdough adaptability and antioxidant potential. *Rom. Biotechnol. Lett.* **2018**, *23*, 13581–13591.

45. Socaci, S.A.; Socaciu, C.; Muresan, C.; Farcas, A.; Tofana, M.; Vicas, S.; Pintea, A. Chemometric discrimination of different tomato cultivars based on their volatile fingerprint in relation to lycopene and total phenolics content. *Phytochem. Anal.* **2014**, *25*, 161–169. [CrossRef] [PubMed]

46. El-Sayed, A.M. The Pherobase: Database of Pheromones and Semiochemicals. 2011. Available online: https://www.pherobase.com/ (accessed on 7 December 2020).

47. Acree, T.; Arn, H. Flavornet and Human Odor Space. Available online: http://www.flavornet.org (accessed on 7 December 2020).

48. Pop, A.; Paucean, A.; Socaci, S.A.; Alexa, E.; Man, S.M.; Muresan, V.; Chis, M.S.; Salanta, L.; Popescu, I.; Berbecea, A.; et al. Quality characteristics and volatile profile of macarons modified withwalnut oilcake by-product. *Molecules* **2020**, *25*, 2214. [CrossRef] [PubMed]

49. Park, J.; Choi, I.; Kim, Y. Cookies formulated from fresh okara using starch, soy flour and hydroxypropyl methylcellulose have high quality and nutritional value. *LWT Food Sci. Technol.* **2015**, *63*, 660–666. [CrossRef]

50. Kelapure, N.N.; Rameshwar, H.; Jaju, R.H.; Amarjeet, N.; Satwase, A.N.; Avdhut, V.; Gutthe, A.V.; Tidke, S.S. Study on quality attributes of flaxseed flour supplemented cookies. *Int. J. Pure Appl. Biosci.* **2018**, *6*, 1439–1445. [CrossRef]

51. Mamat, H.; Hill, S.E. Structural and functional properties of major ingredients of biscuit. Mini Review. *Int. Food Res. J.* **2018**, *25*, 462–471.

52. Devi, K.; Haripriya, S. Pasting behaviors of starch and protein in soy flour-enriched composite flours on quality of biscuits. *J. Food Process. Preserv.* **2014**, *38*, 116–124. [CrossRef]

53. Wei, C.; Zhou, Q.; Han, B.; Chen, Z.; Liu, W. Changes occurring in the volatile constituents of flaxseed oils (FSOs) prepared with diverse roasting conditions. *Eur. J. Lipid Sci. Technol.* **2019**, *121*, 1800068. [CrossRef]

54. Pico, J.; Bernal, J.; Gómez, M. Wheat bread aroma compounds in crumb and crust: A review. *Food Res. Int.* **2015**, *75*, 200–215. [CrossRef]

55. Khouryieh, H.; Aramouni, F. Physical and sensory characteristics of cookies prepared with flaxseed flour. *J. Sci. Food Agric.* **2012**, *92*, 2366–2372. [CrossRef]

56. Cambus, H.; Mikulec, A.; Gambus, F.; Pisulewski, P. Perspectives of linseed utilisation in baking. *Polish J. Food Nutr. Sci.* **2004**, *13*, 21–27.
57. Kaur, M.; Kaur, R.; Gill, B.S. Mineral and amino acid contents of different flaxseed cultivars in relation to its selected functional properties. *J. Food Meas. Charact.* **2017**, *11*, 500–511. [CrossRef]
58. Zouari, R.; Souhail Besbes, S.; Ellouze-Chaabouni, S.; Ghribi-Aydi, D. Cookies from composite wheat–sesame peels flours: Dough quality and effect of *Bacillus subtilis* SPB1 biosurfactant addition. *Food Chem.* **2016**, *194*, 758–769. [CrossRef]
59. Antoniewska, A.; Jarosława Rutkowska, J.; Pineda, M.M.; Agata Adamska, A. Antioxidative, nutritional and sensory properties of muffins with buckwheat flakes and amaranth flour blend partially substituting for wheat flour. *LWT Food Sci. Technol.* **2018**, *89*, 217–223. [CrossRef]
60. Ahmed, M.; Assem, N.H.A.; Nadia, M.; Fahim, J.S. Utilization of flaxseeds in improving bread quality. *Egypt. J. Agric. Res.* **2011**, *89*, 241–250.
61. El-Demery, M.; Mahmoud, K.F.; Bareh, G.F.; Albadawy, W. Effect of fortification by full fat and defatted flaxseed flour sensory properties of wheat bread and lipid profile laste. *Int. J. Curr. Microbiol. App. Sci.* **2015**, *4*, 581–598.
62. Hassan, A.A.; Rasmy, N.M.; Foda, M.I.; Bahgaat, W.K. Production of functional biscuits for lowering blood lipids. *World J. Dairy Food Sci.* **2012**, *7*, 1–20. [CrossRef]
63. Akl, E.M.; Mohamed, S.S.; Hashem, A.I.; Taha, F.S. Biological activities of phenolic compounds extracted from flaxseed meal. *Bull. Natl. Res. Cent.* **2020**, *44*, 27. [CrossRef]
64. Deng, Q.; Yu, X.; Ma, F.; Xu, J.; Huang, F.; Huang, Q.; Sheng, F. Comparative analysis of the in-vitro antioxidant activity and bioactive compounds of flaxseed in China according to variety and geographical origin. *Int. J. Food Prop.* **2018**, *20*, S2708–S2722. [CrossRef]
65. Bernacchia, R.; Preti, R.; Vinci, G. Chemical composition and health benefits of flaxseed. *Austin J. Nutr. Food Sci.* **2014**, *2*, 1045.
66. Petersen, K.; Gmbh, G.; Kg, C. Flavour and fragrance analysis: Wondrous Vanilla. *Column* **2014**, *10*, 11–15.
67. Suri, K.; Singh, B.; Kaur, A.; Yadav, M.P.; Singh, N. Influence of microwave roasting on chemical composition, oxidative stability and fatty acid composition of flaxseed (*Linum usitatissimum* L.) oil. *Food Chem.* **2020**, *326*, 126974. [CrossRef]
68. Chiş, M.S.; Pop, A.; Paucean, A.; Socaci, S.A.; Alexa, E.; Man, S.M.; Bota, M.; Muste, S. Fatty acids, volatile and sensory profile of multigrain biscuits enriched with spent malt rootles. *Molecules* **2020**, *25*, 442. [CrossRef]
69. Chai, D.; Li, C.; Zhang, X.; Yang, J.; Liu, L.; Xu, X.; Du, M.; Wang, Y.; Chen, Y.; Dong, L. Analysis of volatile compounds from wheat flour in the heating process. *Int. J. Food Eng.* **2019**, *15*, 1–13. [CrossRef]
70. Starowicz, M.; Koutsidis, G.; Zieliński, H. Determination of antioxidant capacity, phenolics and volatile Maillard reaction products in rye-buckwheat biscuits supplemented with 3β-D-rutinoside. *Molecules* **2019**, *24*, 982. [CrossRef] [PubMed]
71. Hidalgo, F.J.; Zamora, R. Formation of phenylacetic acid and benzaldehyde by degradation of phenylalanine in the presence of lipid hydroperoxides: New routes in the amino acid degradation pathways initiated by lipid oxidation products. *Food Chem. X* **2019**, *2*, 100037. [CrossRef] [PubMed]
72. Rajiv, J.; Indrani, D.; Prabhasankar, P.; Rao, G.V. Rheology, fatty acid profile and storage characteristics of cookies as influenced by flax seed (*Linum usitatissimum*). *J. Food Sci. Technol.* **2012**, *49*, 587–593. [CrossRef] [PubMed]
73. Austria, J.A.; Aliani, M.; Malcolmson, L.J.; Dibrov, E.; Blackwood, D.P.; Maddaford, T.G.; Guzman, R.; Pierce, G.N. Daily choices of functional foods supplemented with milled flaxseed by a patient population over one year. *J. Funct. Foods* **2016**, *26*, 772–780. [CrossRef]
74. Čukelj, N.; Novotni, D.; Sarajlija, H.; Drakula, S.; Voučko, B.; Ćurić, D. Flaxseed and multigrain mixtures in the development of functional biscuits. *LWT Food Sci. Technol.* **2017**, *86*, 85–92. [CrossRef]
75. Rangrej, V.; Shah, V.; Patel, J.; Ganorkar, P.M. Effect of shortening replacement with flaxseed oil on physical, sensory, fatty acid and storage characteristics of cookies. *J. Food Sci. Technol.* **2015**, *52*, 3694–3700. [CrossRef] [PubMed]
76. Aliani, M.; Ryland, D.; Pierce, G.N. Effect of flax addition on the flavor profile of muffins and snack bars. *Food Res. Int.* **2011**, *44*, 2489–2496. [CrossRef]
77. Ramcharitar, A.; Badrie, N.; Mattfeldt-Beman, M.; Matsuo, H.; Ridley, C. Consumer acceptability of muffins with flaxseed (*Linum usitatissimum*). *J. Food Sci.* **2005**, *70*, s504–s507. [CrossRef]
78. Rathi, P.; Mogra, R. Sensory evaluation of biscuits prepared with flaxseed flour. *Int. J. Food Nutr. Sci.* **2013**, *2*, 1–4.

Article

Optimization of Ingredients for Biscuits Enriched with Rapeseed Press Cake—Changes in Their Antioxidant and Sensory Properties

Aleksandra Szydłowska-Czerniak [1,*], Szymon Poliński [1,2] and Monika Momot [1,3]

[1] Faculty of Chemistry, Nicolaus Copernicus University in Toruń, Gagarina 7, 87-100 Toruń, Poland; szymon.polinski@kopernik.com.pl (S.P.); Monika.Momot@bunge.com (M.M.)
[2] Fabryka Cukiernicza "Kopernik" S.A, Stanisława Żółkiewskiego 34, 87-100 Toruń, Poland
[3] Bunge Europe Research and Development Center, Niepodległości 42, 88-150 Kruszwica, Poland
* Correspondence: olasz@umk.pl

Abstract: The optimum formulation for wheat flour (WF)-based biscuits containing the rapeseed press cake (RPC)—the primary by-product of rapeseed oil production rich in phenolic compounds and different types of fats (rapeseed oil, margarine and coconut oil)—was estimated using the central composite design (CCD) with two factors and response surface methodology (RSM). Effects of partial substitution of WF for RPC (0–40 g) in a total flour blend (100 g) and fats with various amounts of saturated fatty acids (SAFA = 2.3–24.9 g) on antioxidant capacity (AC) and sensory characteristics (color, odor, texture, flavor, overall acceptability, and purchase intent scores) of the novel biscuits were investigated. Conventional solid (liquid)–liquid extraction and ultrasound-assisted extraction (UAE) were applied to extract total antioxidants from main ingredients used for the preparation of doughs as well as the baked biscuits. The AC of biscuits and their components were determined by 2,2-diphenyl-1-picrylhydrazyl (DPPH) assay. The DPPH results were the highest for the RPC flour (DPPH = 15,358–15,630 μmol Trolox (TE)/100 g) and biscuits containing rapeseed oil and 40 g of RPC flour (DPPH = 7395–10,088 μmol TE/100 g). However, these biscuits had lower sensory scores for each attribute and the lowest purchase intent scores. The quadratic response surfaces were drawn from the mathematical models in order to ensure the good quality of the proposed biscuits with RPC. The DPPH results obtained and the mean sensory scores correlate with the predicted values (R^2 = 0.7751–0.9969). The addition of RPC with high antioxidant potential to biscuits and the replacement of margarine or coconut oil by rapeseed oil interfered with their acceptability.

Keywords: rapeseed press cake; fats; biscuits; antioxidant capacity; sensory analysis; consumer acceptance; response surface methodology

Citation: Szydłowska-Czerniak, A.; Poliński, S.; Momot, M. Optimization of Ingredients for Biscuits Enriched with Rapeseed Press Cake—Changes in Their Antioxidant and Sensory Properties. *Appl. Sci.* **2021**, *11*, 1558. https://doi.org/10.3390/app11041558

Academic Editor: Silvia Mironeasa
Received: 31 December 2020
Accepted: 5 February 2021
Published: 9 February 2021

Publisher's Note: MDPI stays neutral with regard to jurisdictional claims in published maps and institutional affiliations.

1. Introduction

For some time, it has been possible to observe the interest of both producers and consumers in the confectionery market in products with the addition of functional ingredients. One of the categories of functional additions are components with antioxidative properties, which can reduce the level of oxidative stress in cells. Food by-products are mainly considered rich and cheap sources of valuable compounds for supplementation of confectionery and bakery products [1–15]. Recently, the effect of different amounts (5–95%) of agro-industrial by-products such as olive stone powder, okara powder, prickly pear peel, fruit pomaces (rosehip, rowanberry, blackcurrant, elderberry, grape, and blueberry), fruit by-products (pineapple central axis, apple endocarp, melon peels, waste left after goji berry concentrate extraction), cocoa shell, soybean meal and fermented soybean meal, defatted chia flour, and defatted sunflower seed flour on the antioxidant capacity (AC) of biscuits, cookies, and muffins has been analyzed by 2,2-diphenyl-1-picrylhydrazyl (DPPH), 2,2′-azinobis-(3-ethylbenzo-thiazoline-6-sulfonic acid) (ABTS), ferric reducing antioxidant potential (FRAP), β-carotene/linoleic acid, phosphomolybdenum complex, oxygen radical absorbance capacity (ORAC), cupric reducing antioxidant capacity (CUPRAC), and Folin-Ciocalteu (F-C) assays [1,3,5–15]. However, there is little information on the optimization of the dough formulations to maximize functional properties, mainly antioxidant potential of confectionery products enriched with agro-industrial by-products. This knowledge is needed for the improvement of production processes and commercial applications. Response surface methodology (RSM) was only applied to optimize antioxidant capacity determined by ABTS method and total phenolic content (TPC) in the formulation of cookies supplemented with blueberry pomace [7]. Moreover, RSM-based models were proposed to study the effect of replacing WF with organic grape flour (13.06–16.74%), interesterified fat concentration (23.96–34.04%), and sucrose content (11.96–22.04%) on the sensory properties (appearance aroma, flavor, texture, overall impression) of the fortified cookies, and to evaluate the predictive ability of these mathematical models to describe general acceptability of final product [4]. The combined impact of three independent variables, carrot pomace powder (10–20 g), finger millet flour (2.5–7.5 g), and baking time (21–25 min) on general acceptability and physiochemical properties (spread ratio, change in color, amounts of moisture, ash, fat, and fiber, hardness) of the biscuits enriched with waste of carrot juice industry was also optimized by the RSM [2].

Additionally, the AC and TPC of cookies with cocoa shell, soy flour, and green banana flour developed by the simplex centroid design were analyzed by ABTS, FRAP, β-carotene/linoleic acid, phosphomolybdenum complex, and Fast Blue methods. This experimental design was effective for optimization of the acceptance of cookies made with cocoa shell, a by-product of the chocolate industry [11].

However, to the best of our knowledge, there are no reports on the evaluation of the effect of rapeseed press cake (RPC) flour and type of fat on antioxidant properties and sensory quality of baked biscuits. Only the impact of cold pressed RPC, RPC fiber isolate, and RPC alkaline extract on the generation of acrylamide and 5-hydroxymethylfurfural in cookies was investigated. The cookies with RPC had a higher concentration of 5-hydroxymethylfurfural, while the alkaline extract from RPC caused a decrease of acrylamide content in the supplemented cookies [16].

Rapeseed is the third most abundant oil plant worldwide (after palm and soya) and the primary oil seed crop in Poland. During rapeseed processing, several wastes are generated, and RPC is one of the major types of residual biomass from the rapeseed oil industry [16,17]. The RPC can provide a viable and economical source of bioactive compounds, because the varieties grown in Poland have an improved nutritional profile (low amounts of erucic acid and glucosinolates). Nevertheless, glucosinolates and products of their degradation present in RPC can create unique and characteristic flavor, thus, they can be applied as food additives for improvement of the sensory characteristics [18]. Recently, with the increasing interest in circular economy and zero waste, there is intense effort to revalorize food by-products. For this reason, the RPC as natural source of proteins, carbohydrates, crude fiber, lipids, minerals, as well as polyphenols, glucosinolates, and isothiocyanates having antioxidant, antimicrobial and anticarcinogenic properties can be developed for confectionery applications [17,18]. On the other hand, an increase in rapeseed production implies larger domestic supplies of RPC, which would affect the stability of the prices of confectionery enriched with RPC.

Therefore, the aim of this study was to create biscuits with new and attractive for the consumer features that would improve the antioxidant potential and would not cause a deterioration in their sensory properties. The present work is focused for the first time on the optimization of the production of functional and acceptable confectionery products fortified with RPC rich in antioxidants and high quality fats. A central composite experimental design (CCD) and the RSM were used for evaluation of the effects of two independent variables (RPC content—a novel ingredient—and saturated fatty acids (SAFA) content) and their interactions on the response variables: AC determined by the modified DPPH assay and sensory characteristics (color, odor, texture, flavor, overall acceptability, and purchase intent scores) of wheat flour (WF)-based biscuits. Moreover, in the present study, DPPH results of extracts obtained from ingredients (RPC, WF, rapeseed oil, margarine, and coconut oil) and the baked biscuits by the conventional solid/liquid–liquid extraction and the ultrasound-assisted extraction (UAE) were compared and discussed.

The hedonic method was conducted to evaluate the acceptance, desirability, and preferences of biscuits without and with RPC containing various types of fats.

2. Materials and Methods

2.1. Reagents and Samples

All reagents were of analytical or high performance liquid chromatography (HPLC) grade. 2,2-Diphenyl-1-picrylhydrazyl radical (DPPH, 95%), 6-hydroxy-2,5,7,8- tetramethylchromane-2-carboxylic acid (Trolox (TE), 97%) and methanol (99.8%) were purchased from Sigma Aldrich (Poznań, Poland). Redistilled water was used for the preparation of solutions.

The RPC, the primary by-product of the rapeseed oil industry, and the refined rapeseed oil, the final product of a conventional technological process, were provided by a local vegetable oil factory and were stored in an airtight closed poly(ethylene terephthalate) (PET) bag and bottle, respectively, at room temperature. All the other baking ingredients in customary quality, such as WF, refined coconut oil, margarine, sugar, salt, baking powder, and non-carbonated spring water were purchased from the local market. All samples in the original packaging were stored at ambient temperature until treatment and further analysis.

2.2. Biscuits Preparation

Biscuits fortified with RPC were prepared on a laboratory scale according to the following recipe. RPC was milled in the mill (Model FW100, Chemland, Stargard, Poland) and sieved through a mesh sieve of 0.355 mm (Retsch Test Sieve, Haan, Germany) to obtain a uniform size of flour.

The composite flour blends were prepared using various combinations of WF and RPC flour in the ratio of 100:0, 80:20, 60:40, respectively. Three commercial fats such as refined rapeseed oil, margarine, and coconut oil, with a declared SAFA content of 7.5 g/100 g, 26 g/100 g and 83 g/100 g, respectively, were used for preparation of dough formulations. Initially, each flour blend (100 g), salt (1.0 g), and baking powder (2.0 g) were thoroughly mixed. Sugar (45 g) was dissolved in hot spring water (22 mL) and cooled. Finally, 30 g of refined rapeseed oil, margarine and coconut oil containing 2.3 g, 7.8 g and 24.9 g of SAFA, respectively was added separately into mixtures of various powdering ingredients and sugar solution. All ingredients were mixed for 6 min in a hand mixer (5-speed, 500 W, Moulinex, Powermix HM610130, Ecully Cedex, France) at medium speed for 6 min with scraping every 2 min to obtain a homogenous mixtures. Afterwards, solid dough formulations were cooled in a refrigerator at 4 °C for 30 min. Each dough type was kneaded and sheeted to a uniform thickness of 5 mm and cut into circular shapes of 6.5 cm diameter.

The dough pieces were placed on a baking tray with baking paper and baked at 180 °C for 18 min in an electrically heated oven (Electrolux, Warszawa, Poland). Biscuit samples were cooled and stored in airtight plastic containers at ambient temperature before AC and sensory analyses were conducted.

2.3. Determination of Antioxidant Capacity

2.3.1. Samples Preparation by Conventional Solid (Liquid)–Liquid Extraction

Portions of RPC (2.0 g), WF (2.0 g), fats (2.0 g), biscuits (5.0 g of five ground biscuits baked from the same dough formulation), and 70% methanol (20, 15, 5, and 20 mL, respectively), were transferred into Erlenmeyer flasks and shaken using a shaker SK-L 330-Pro (Chemland, Stargard, Poland) at room temperature for 30 min.

Each sample was extracted in triplicate, and the residual samples were separated by centrifugation (centrifuge MPW-54, Chemland, Stargard, Poland, $3120\times g$, 15 min). The pooled extracts were filtered using polytetrafluorethylene syringe filters (PTFE, pore size 0.20 μm/diameter 13 mm, Sigma Aldrich, Poznań, Poland) and stored in a refrigerator at 4 °C prior to analysis.

2.3.2. Samples Preparation by Ultrasound-Assisted Extraction

Portions of RPC (2.0 g), WF (2.0 g), fats (2.0 g), biscuits (5.0 g of five ground biscuits baked from the same dough formulation), and 70% methanol (20, 15, 5, and 20 mL, respectively), were transferred into Erlenmeyer flasks and placed in an ultrasonic cleaner bath (5200DTD, Chemland, Poland) with a frequency of 40 kHz, ultrasound input power of 180 W and heating power of 800 W, equipped with a digital timer and temperature controller. The ultrasonic bath's water level was adjusted so that it was slightly higher than the level of samples and solvent into Erlenmeyer flasks. The UAE was performed for 5 min, and the temperature was kept constant at 25 ± 0.3 °C.

The same sample was sonicated in triplicate and centrifuged at $1880 \times g$ for 15 min (centrifuge MPW-54, Chemland, Stargard, Poland). The supernatants were filtered through a PTFE syringe filter (0.20 μm/13 mm) and collected for subsequent determination of the AC.

2.3.3. DPPH Method

The AC of major ingredients and prepared biscuits were analyzed spectrophotometrically according to a modified DPPH procedure described previously [19]. Briefly, 0.01–0.70 mL of methanolic extracts obtained by conventional solid (liquid)/liquid extraction and UAE were added to 1.99–1.30 mL of methanol and 0.5 mL of DPPH methanolic solution (304.0 μmol/L). The changes in color from deep violet to light yellow were measured at 517 nm against a reagent blank (2 mL of methanol + 0.5 mL of DPPH methanolic solution) after 15 min of reaction using a Hitachi U-2900 spectrophotometer (Tokyo, Japan). The DPPH values were expressed as micromoles of Trolox equivalents (TE) per 100 g of the studied samples.

2.4. Sensory Acceptance Test

Sensory analysis of the baked biscuits was performed using an effective acceptance test with 72 untrained panelists (35 male and 37 female) in the age range of 18–69 recruited among students, staff, and professors of the Faculty of Chemistry, Nicolaus Copernicus University in Toruń, Poland. The sensory test was conducted two days after baking trials using a 9-point hedonic scale (1—"disliked extremely", 2—"disliked very much", 3—"disliked moderately", 4—"disliked slightly", 5—"neither liked nor disliked", 6—"liked slightly", 7—"liked moderately", 8—"liked very much", 9—"liked extremely"), according to Stone and Sidel [20]. The participants were asked to assess the following attributes: liking of color, liking of odor, liking of texture, liking of flavor, and overall acceptability. Additionally, the purchase intent was evaluated using a 5-point scale (1—"certainly would not buy" and 5—"certainly would buy"). For this reason, there was a question: "How likely is it that you will buy this product if it will be available at stores?" at the end of the questionnaire card.

Each untrained panelist evaluated a total of eleven biscuits in an odor-free plastic container with a lid labeled with a 3-digit code in a randomized order to avoid an order effect [21]. Warm dark tea was used by the panelists to rinse the mouth between samples testing.

2.5. Determination of Biscuit Physical Properties

The physical properties, such as diameter, thickness, spread ratio, and weight of biscuits were measured according to the procedures described by Mildner-Szkudlarz et al. [8]. Six biscuits were laid edge to edge and measured for diameter (mm). The biscuits were rotated through 90° and the diameter were remeasured. Than the average value was taken. Thickness was measured by stacking six biscuits on top of each other and taking average thickness (mm). Spread ratio was calculated by dividing the value of diameter by value of biscuits' thickness. Biscuits' weight was determined using an analytical electronic balance (precision—0.0001 g, model AS 110.R2, Radom, Poland).

2.6. Statistical Analysis and Experimental Design for Optimization

The AC values of the baked biscuits were analyzed by five-fold determination of each methanolic extract obtained by conventional extraction and UAE within the same day using the modified DPPH method. The results obtained were presented as the mean ± standard deviation (SD). All data were statistically evaluated by the analysis of variance (ANOVA) test. A post hoc Duncan's test was applied for the calculation of the significant differences among mean values of characteristic oil parameters at the probability level $p < 0.05$.

The RSM was applied to study the simultaneous effects of the RPC content and SAFA content in fats used for biscuits preparation on their AC determined by DPPH method and sensory characteristics. The levels of RPC and SAFA for a CCD and RSM were determined on the basis of preliminary experiments carried out and were varied from 0 to 40 g and 2.3–24.9 g, respectively. The experimental design used for the analysis was a CCD with two factors and three levels. In this experimental design, the factor levels were coded using "−1" for the lowest level (0 g and 2.3 g for RPC content and SAFA content, respectively), "1" for the highest level (40 g and 24.9 g for RPC content and SAFA content, respectively), and "0" for neutral (middle) level (20 g and 7.8 g for RPC content and SAFA content, respectively). The experiments consisted of 11 runs with two factors and two replicates of the central point for the estimation of pure error. The effect of the two independent variables (RPC content and SAFA content) on the responses (Y_n, Y_1—$DPPH_{CE}$, Y_2—$DPPH_{UAE}$, Y_3—color, Y_4—odor, Y_5—texture, Y_6—flavor, Y_7—overall acceptability, and Y_8—purchase intent) was modeled using a polynomial response surface. The second-order response function for the experiments was predicted by the following equation:

$$Y_n = \beta_0 + \beta_1 RPC + \beta_2 SAFA + \beta_{11} RPC^2 + \beta_{22} SAFA^2 + \beta_{12} RPC \times SAFA \quad (1)$$

where Y_n is one of the eight responses; RPC and SAFA represent the independent variables; β_0 is the constant; β_1, β_2 are the linear-term coefficients; β_{11}, β_{22} are the quadratic-term coefficients; and β_{12} is the cross-term coefficient.

The fitness of the model was evaluated by the determination coefficient R^2, the fraction of the variation explained by the model, and analysis of variance (ANOVA). The F-test was applied to confirm whether the variance explained by the regression model was significantly larger than the variance of the residual and to evaluate the model lack-of-fit (model error).

The effects of two factors (RPC content and SAFA content) and their interactions on DPPH in methanolic extracts of biscuits obtained by conventional extraction and UAE, as well as sensory characteristics, were displayed in surface and contour plots. The chemometric analyses were constructed using the Statistica 8.0 software (StatSoft, Tulsa, OK, USA). However, Fizz software (Biosystemes, Courtenon, France) was applied for the collection of all sensory data.

3. Results and Discussion

3.1. Antioxidant Properties of Biscuits' Ingredients

A low-cost and simple DPPH assay based on spectrophotometric mixed-mode (having both electron transfer (ET) and hydrogen atom transfer (HAT) mechanisms) was proposed as the most suitable method to determine the AC of ingredients used in the preparation of functional biscuits. This assay was chosen due to the fact that DPPH radical is known to work well with lipophilic (rather than hydrophilic) antioxidants in alcohol solvents [22]. The AC of RPC, WF, rapeseed oil, margarine, and coconut oil was determined by the modified DPPH method after conventional extraction and UAE, and the obtained results are listed in Table 1.

Table 1. Antioxidant capacity of biscuits' ingredients.

Ingredient	Antioxidant Capacity of Methanolic Extracts	
	Conventional Extraction DPPH$_{CE}$ * $\pm$ SD [μmol TE/100 g]	Ultrasound-Assisted Extraction DPPH$_{UAE}$ * $\pm$ SD [μmol TE/100 g]
Rapeseed press cake	15,358 $\pm$ 412 [d,x]	15,630 $\pm$ 612 [d,x]
Wheat flour	126 $\pm$ 7 [b,x]	149 $\pm$ 5 [a,y]
Rapeseed oil	424 $\pm$ 10 [c,x]	456 $\pm$ 12 [c,x]
Margarine	171 $\pm$ 4 [b,x]	217 $\pm$ 3 [b,y]
Coconut oil	87 $\pm$ 2 [a,x]	118 $\pm$ 2 [a,y]

* $n = 5$; SD—standard deviation; Different letters within the same column (a–d) indicate significant differences between DPPH results of biscuits' ingredients. Different letters (x,y) within the same row indicate significant differences between DPPH of each ingredient extract prepared by the conventional extraction (CE) and the ultrasound-assisted extraction (UAE) (one-way ANOVA and Duncan test, $p < 0.05$).

It is noteworthy that the DPPH values of biscuits' ingredients differ significantly from each other (Duncan test). Moreover, DPPH results of WF, margarine, and coconut oil extracts prepared using the classical extraction method over 30 min were significantly lower than those obtained by the UAE for 5 min (Table 1). However, the Duncan test indicated that methanolic extracts from RPC and rapeseed oil after classical extraction, and sonication did not differ significantly in DPPH results.

The UAE permits higher extraction yields in a shorter time, thereby reducing the electrical energy input. An increase in the AC can be explained by the action of the cavitation bubbles generated during the propagation of the acoustic waves. The cavitation bubbles can disrupt the material cell wall, causing solvent diffusion and increasing the release of antioxidant compounds. On the other hand, ultra-sonication contributes to reducing the particle size of the raw material, which increases the surface area. Moreover, the ultrasound waves generate shear force during ultra-sonication thus, the mass transfer of the original material into an extract solution is enhanced [23].

It is evident that the RPC incorporated into the biscuits was the richest source of antioxidants. The methanolic extracts of RPC revealed the highest DPPH values (15,358 and 15,630 μmol TE/100 g for conventional extraction and UAE, respectively). In our previous report the DPPH ranged between 8770 and 33,980 μmol/100 g for ethanolic, methanolic, and aqueous RPC extracts [24]. These differences between the DPPH results for RPC samples can be explained by the influences of genetic, agronomic, environmental, and technological factors, as well as conditions of extract preparation, mainly polarity of the used solvent, which affect the total level of antioxidants.

However, among the evaluated fats, rapeseed oil had the highest DPPH results (424 and 456 μmol TE/100 g after conventional extraction and UAE, respectively), whereas the antioxidant potential of coconut oil determined by the same method was the lowest (87 and 118 μmol TE/100 g after conventional extraction and UAE, respectively) (Table 1).

Also, Casoni et al. [25] observed significantly higher radical scavenging capacity of rapeseed oil (RSC = 35.12%) to scavenge DPPH radical than coconut oils (RSC = 0.46–2.26%). This suggests that rapeseed oil is a rich source of antioxidants, mainly tocopherols (78.51 mg/100 g) and phenolic compounds (5.77 mg/100 g), while coconut oil contains a trace amount of tocopherols (2.90 mg/100 g) and phenolics (less than 0.01 mg/100 g) [26].

Nevertheless, margarine revealed approximately twofold lower DPPH results (171–217 μmol TE/100 g) in comparison with the refined rapeseed oil (DPPH = 424–456 μmol TE/100 g). Although, somewhat lower AC results (126–149 μmol TE/100 g) were determined for WF.

For comparison, WFs used for the preparation of functional shortbread cookies with fruit pomace, and biscuits incorporated with prickly pear peel, revealed significantly higher DPPH values (377–860,000 μmol TE/100 g) [6,14].

3.2. Antioxidant Capacity of Biscuits with Rapeseed Press Cake

The different fats such as rapeseed oil, margarine, and coconut oil with the declared amounts of SAFA at 2.3, 7.8, and 24.9 g, respectively, and increasing RPC powder levels from 0 to 40 g, were used for the preparation of functional biscuits. It can be noted that as the concentration of RPC in the baked products increased, the biscuits became darker in color (Figure 1).

Figure 1. Photograph of biscuits prepared from dough: (**a–c**) without, (**d–f**) with 20 g and (**g–i**) with 40 g of rapeseed press cake using rapeseed oil, margarine, and coconut oil, respectively.

The biscuits fortified with RPC were found to be darker than the control sample without RPC flour. This suggests that RPC flour is a rich source of chlorophyll. Moreover, the darkening of biscuits enriched with RPC can be due to the browning of RPC carbohydrates during baking.

For comparison, with increasing concentrations of by-products, such as olive stone powder (5–15%), white grape pomace (10–30%), goji berry by-product (10–40%), defatted chia flour (5–20%), okara powder (20–40%), prickly pear peel (10–30%), defatted sunflower seed flour (18–36%) in doughs, the color of the enriched biscuits and cookies changed gradually from light brown to brown and finally to dark brown [1,8,10,12–15].

Furthermore, the DPPH values of the baked biscuits increased rapidly with the increasing content of the added RPC (Table 2).

Therefore, the highest DPPH results (8375–10,088 µmol TE/100 g and 4761–7395 µmol TE/100 g) were revealed for methanolic extracts of biscuits with the highest RPC level (40 g) obtained after classical extraction and UAE, respectively. However, the DPPH (535–843 µmol TE/100 g and 598–890 µmol TE/100 g) were the lowest for the studied samples without RPC.

The increase in DPPH values from 20.12 to 30.44%, 3.39–7.55 mmol TE/g, 198–236 mmol TE/g, and 1.5–2.9 mg TE/g was also observed when the biscuit doughs were fortified with increasing amounts of olive stone powder (5–15%), white grape pomace (10–30%), prickly pear peel (10–30%), and defatted sunflower seed flour (18–36%), respectively [1,8,14,15].

Table 2. Antioxidant capacity of the studied biscuits.

Exp.	Independent Variables (Coded Level)		Dependent Variables—Antioxidant Capacity of Methanolic Extracts			
			Conventional Extraction		Ultrasound-Assisted Extraction	
	RPC Content [g]	SAFA Content [g]	DPPH$_{CE}$ [μmol TE/100 g]		DPPH$_{UAE}$ [μmol TE/100 g]	
			Exp.* ± SD	Pred.	Exp.* ± SD	Pred.
1	0 (−1)	2.3 (−1)	843 ± 8 [b,x]	1413	890 ± 13 [b,x]	1220
2	0 (−1)	7.8 (0)	535 ± 5 [a,x]	8	598 ± 27 [a,x]	58
3	0 (−1)	24.9 (1)	575 ± 11 [a,x]	532	610 ± 5 [a,x]	820
4	20 (0)	2.3 (−1)	5588 ± 321 [d,x]	5055	3344 ± 49 [d,y]	3622
5	20 (0)	7.8 (0)	3600 ± 36 [c,x]	3653	2336 ± 16 [c,y]	2378
6	20 (0)	24.9 (1)	3904 ± 168 [c,x]	4185	3004 ± 53 [d,y]	2886
7	40 (1)	2.3 (−1)	10,088 ± 269 [g,x]	10,051	7395 ± 356 [g,y]	6787
8	40 (1)	7.8 (0)	8375 ± 334 [e,x]	8650	4761 ± 217 [e,y]	5461
9	40 (1)	24.9 (1)	9428 ± 421 [f,x]	9190	5807 ± 208 [f,y]	5715
10	20 (0)	7.8 (0)	3705 ± 54 [c,x]	3653	2630 ± 18 [c,y]	2378
11	20 (0)	7.8 (0)	3402 ± 80 [c,x]	3653	2328 ± 25 [c,y]	2378

* n = 5; SD—standard deviation; Exp.—experimental data; Pred.—predicted values; different letters within the same column (a–g) indicate significant differences between the DPPH of the studied biscuits; different letters (x,y) within the same row indicate significant differences between DPPH of biscuits extracts prepared by the conventional extraction (CE) and the ultrasound-assisted extraction (UAE) (one-way ANOVA and Duncan test, $p < 0.05$).

Interestingly, biscuits prepared by replacing coconut oil (lower DPPH = 87–118 μmol TE/100 g) with margarine (higher DPPH = 171–217 μmol TE/100 g) and the same content of RPC had a lower level of antioxidants capable of scavenging the free DPPH radical (Tables 1 and 2). Thus, replacement of fat with a richer in SAFA insignificantly increased the DPPH activity of the baked biscuits (Table 2, Duncan test). This can be explained by the fact that SAFA are strongly resistant to deterioration reactions during thermal oxidation, while antioxidants present in the dough are not consumed for the termination of oxidative processes by trapping free radicals.

Unexpectedly, the DPPH results of fortified biscuits' extracts decreased significantly after ultrasonic treatment (Table 2, Duncan test). A decrease in DPPH values for extracts obtained by UAE indicates that natural antioxidants present in RPC were degraded by free radicals generated upon the ultrasound irradiation [27]. On the other hand, the thickness of the boundary layer (the fortified biscuits–solvent interface) did not sufficiently decrease during mixing by means of ultrasound due to the ultrasonic power dissipation (180 W) used for too short treatment time (5 min × 3 times), and the transfer of antioxidants was limited.

3.3. Hedonic Scale Sensory Evaluation of Biscuits with Rapeseed Press Cake

It can be noted that there were significant differences in consumer acceptability between 11 biscuit samples containing 0, 20 and 40 g of RPC flour, respectively (Table 3, Duncan test).

Therefore, the samples studied can be divided into three groups according to the mean sensory scores for overall acceptability. The baked biscuits with the highest amount of RPC flour (40 g) had the lowest sensory score values for overall acceptance. The most preferred samples by Polish consumers were biscuits prepared without RPC (Table 3).

Table 3. Mean sensory scores for the color, odor, texture, flavor, overall acceptability, and purchase intent of the studied biscuits.

| Exp. | Independent Variables (Coded Level) | | Dependent Variables—Sensory Characteristics | | | | | | | | | | | |
| | RPC Content [g] | SAFA Content [g] | Color [#] | | Odor [#] | | Texture [#] | | Flavor [#] | | Overall Acceptability [#] | | Purchase Intent [##] | |
			Exp.* ± SD	Pred.	Exp.* ± SD	Pred.	Exp.* ± SD	Pred.	Exp.* ± SD	Pred.	Exp.* ± SD	Pred.	Exp.* ± SD	Pred.
1	0 (−1)	2.3 (−1)	6.03 ± 0.56 [c]	6.12	5.77 ± 0.82 [c]	5.82	6.88 ± 0.42 [c]	6.51	6.89 ± 0.78 [c]	6.56	6.79 ± 0.87 [c]	6.60	3.32 ± 0.47 [c]	3.29
2	0 (−1)	7.8 (0)	6.10 ± 0.75 [c]	5.93	5.73 ± 0.53 [c]	5.64	5.52 ± 0.61 [a,b]	5.84	5.85 ± 0.82 [c]	6.14	6.07 ± 0.62 [c]	6.17	3.19 ± 0.79 [c]	3.21
3	0 (−1)	24.9 (1)	5.81 ± 1.14 [b,c]	5.89	6.10 ± 1.11 [c]	6.14	6.10 ± 0.97 [c]	6.15	6.42 ± 1.20 [c]	6.46	6.26 ± 0.45 [c]	6.34	3.33 ± 0.66 [c]	3.34
4	20 (0)	2.3 (−1)	5.56 ± 0.72 [b]	5.48	4.42 ± 0.96 [b]	4.45	6.30 ± 0.82 [c]	6.44	4.74 ± 0.46 [b]	4.90	4.68 ± 0.68 [b]	4.85	2.52 ± 0.68 [b]	2.52
5	20 (0)	7.8 (0)	5.29 ± 0.45 [b]	5.34	4.16 ± 0.25 [a,b]	4.20	5.62 ± 0.75 [a,b]	5.77	4.45 ± 0.84 [b]	4.45	4.44 ± 0.41 [b]	4.40	2.41 ± 0.41 [b]	2.45
6	20 (0)	24.9 (1)	5.56 ± 0.50 [b]	5.44	4.53 ± 0.71 [b]	4.49	6.36 ± 0.53 [c]	6.07	4.89 ± 0.77 [b]	4.65	4.74 ± 0.63 [b]	4.51	2.63 ± 0.57 [b]	2.60
7	40 (1)	2.3 (−1)	4.51 ± 0.71 [a]	4.50	3.93 ± 0.66 [a]	3.85	5.30 ± 0.91 [a]	5.53	3.48 ± 0.51 [a]	3.65	3.44 ± 0.79 [a]	3.46	1.78 ± 0.34 [a]	1.81
8	40 (1)	7.8 (0)	4.45 ± 0.33 [a]	4.41	3.45 ± 0.98 [a]	3.53	5.33 ± 0.35 [a]	4.86	3.52 ± 0.63 [a]	3.15	3.15 ± 0.52 [a]	2.99	1.81 ± 0.45 [a]	1.75
9	40 (1)	24.9 (1)	4.62 ± 0.81 [a]	4.66	3.62 ± 0.36 [a]	3.62	4.93 ± 0.49 [a]	5.17	3.03 ± 0.78 [a]	3.23	2.89 ± 0.64 [a]	3.04	1.90 ± 0.51 [a]	1.93
10	20 (0)	7.8 (0)	5.20 ± 0.52 [b]	5.34	4.23 ± 0.21 [b]	4.20	5.66 ± 0.62 [a,b]	5.77	4.32 ± 0.71 [b]	4.45	4.31 ± 0.41 [b]	4.40	2.47 ± 0.47 [b]	2.45
11	20 (0)	7.8 (0)	5.31 ± 0.43 [b]	5.34	4.19 ± 0.29 [a,b]	4.20	5.87 ± 0.77 [b]	5.77	4.50 ± 0.87 [b]	4.45	4.38 ± 0.46 [b]	4.40	2.43 ± 0.51 [b]	2.45

* $n = 72$; SD—standard deviation; [#] Color, odor, texture, flavor and overall acceptability are based on the 9 hedonic rating scale system, with anchoring point, 1—"disliked extremely" and 9—"liked extremely"; [##] Purchase intent is based on the 5 scoring scale with anchoring point, 1—"certainly would not buy", 5—"certainly would buy"; Exp.—experimental data; Pred.—predicted values; different letters within the same column indicate significant differences between the studied biscuits' sensory characteristics (one-way ANOVA and Duncan test, $p < 0.05$).

The consumers' results revealed that the mean sensory scores for color, odor, texture, flavor, and overall acceptability of the non-supplemented biscuits prepared by using different types of fats ranged between 5.81 and 6.10, 5.73–6.10, 5.52–6.88, 5.85–6.89, and 6.07–6.79, respectively, and corresponded to the classifications "liked slightly" and "liked moderately" on the hedonic scale where the maximum score is 9. In the context of the proposed products' purchase intent frequency, the biscuits without RPC received the highest scores (3.19–3.33) on a 5-point scale and qualification as "would buy".

On the contrary, with increasing RPC content in the dough, the fortified biscuits' color changed gradually from light (biscuits without RPC) to dark for biscuits with the highest amount of RPC flour (40 g). Therefore, these samples scored the lowest for color (4.45–4.62), odor (3.45–3.93), flavor (3.03–3.52), and overall acceptability (2.89–3.44). The biscuits with the highest RPC level were characterized by a typical intensity of bitter and grassy flavor due to the high concentration of phenolic compounds present in RPC.

Nevertheless, insignificant differences in the mean texture scores were found between control samples (without RPC) and biscuits after the replacement of WF with 20 g of RPC flour (Table 3, Duncan test). However, biscuits incorporated with the highest RPC amount (40 g) scored the lowest for texture (4.93–5.33). For this reason, biscuits formulated with 40 g of RPC flour had the lowest purchase intent scores (1.78–1.90), being largely qualified with "would not buy". The biscuits enriched with 20 g of RPC flour were predominantly qualified as "would probably buy", but there were significant differences with the qualification "would buy" for the control samples without RPC (Table 3, Duncan test).

The Duncan test indicated insignificant differences in scores for color, odor, flavor, overall acceptability, and purchase intent between biscuits obtained from different types of fats with the same content of RPC (Table 3). Only biscuits without RPC and with 20 g of RPC flour prepared by using margarine had significantly lower texture scores than those baked from dough containing rapeseed and coconut oils.

The obtained results suggest that the replacement of WF with varying RPC levels in the recipe affects the overall acceptability of the functional biscuits due to atypical attributes such as bitter and grassy flavor and strange odor. A new component such as chocolate could probably mask the negative attributes and increase the overall acceptability of biscuits fortified with RPC.

3.4. Optimization Process

3.4.1. Fitting the Models

Experimental results of the responses, including the DPPH values and sensory characteristics of biscuits fortified with RPC, were fitted to the CCD, and the least-squares technique was used for calculation of the regression coefficients of the individual linear, quadratic, and interaction terms (Table 4).

Table 4. Regression coefficients of quadratic polynomial models for the studied response variables.

Response Variables	Model Coefficients					
	Intercept	RPC	SAFA	RPC^2	SAFA^2	RPC × SAFA
DPPH$_{CE}$ [μmol TE/100 g]	2227.6 **	148.3 **	-383.3 **	1.7 *	12.7 **	2.2×10^{-2}
DPPH$_{UAE}$ [μmol TE/100 g]	1909.4 *	102.7 *	-325.7 *	9.5×10^{-1}	11.3 *	7.4×10^{-1}
Color	6.2 ***	-2.5×10^{-2} *	-4.8×10^{-2}	-4.1×10^{-4} *	1.4×10^{-3}	4.4×10^{-4}
Odor	5.9 ***	-8.7×10^{-2} ***	-6.2×10^{-2} *	9.7×10^{-4} **	2.8×10^{-3} **	-6.1×10^{-4} *
Texture	6.9 ***	1.7×10^{-2}	-1.8×10^{-1} *	-1.0×10^{-3} *	6.2×10^{-3} *	4.6×10^{-6}
Flavor	6.8 ***	-9.2×10^{-2} **	-1.2×10^{-1} *	4.9×10^{-4}	4.2×10^{-3} *	-3.5×10^{-4}
Overall acceptability	6.9 ***	-9.6×10^{-2} **	-1.2×10^{-1} *	4.5×10^{-4} *	3.9×10^{-3} *	-1.7×10^{-4}
Purchase intent	3.3 ***	-4.1×10^{-2} **	-2.4×10^{-2}	8.5×10^{-5}	9.4×10^{-4} *	7.5×10^{-5}

Significant at the * $p < 0.05$; ** $p < 0.01$; *** $p < 0.001$.

The second-order polynomial equations generated were used to predict the responses (Tables 2 and 3). The ANOVA results for the predicted response quadratic models are listed in Table 5. ANOVA test revealed that the quadratic polynomial models adequately represent responses of DPPH for extracts obtained by conventional extraction and UAE as well as sensory properties due to an insignificant lack-of-fit (F ranged between 2.89 and 18.83, $p > 0.05$) for each estimated response (Table 5).

Table 5. Analysis of variance (ANOVA) results for the studied responses of biscuits: $DPPH_{CE}$, $DPPH_{UAE}$, color, odor, texture, flavor, overall acceptability, and purchase intent.

Model Parameters	Degree of Freedom	Sum of Squares	Mean Square	F Value	Sum of Squares	Mean Square	F Value
			$DPPH_{CE}$			$DPPH_{UAE}$	
Regression	5	1.13×10^8	2.27×10^7	957.63 **	4.33×10^7	8.67×10^6	292.67 *
Residual	5	1.17×10^6	2.34×10^5		1.47×10^6	2.94×10^5	
Lack-of-fit	3	1.12×10^6	3.74×10^5	15.80	1.41×10^6	4.71×10^5	15.90
Pure error	2	4.73×10^4	2.37×10^4		5.92×10^4	2.96×10^4	
Total	10	1.15×10^8			4.48×10^7		
R^2			0.9902			0.9689	
Adjusted R^2			0.9804			0.9378	
			Color			Odor	
Regression	5	3.06	6.11×10^{-1}	178.04 *	7.90	1.58	1281.56 **
Residual	5	8.85×10^{-2}	1.77×10^{-2}		3.07×10^{-2}	6.15×10^{-3}	
Lack-of-fit	3	8.16×10^{-2}	2.72×10^{-2}	7.92	2.83×10^{-2}	9.42×10^{-3}	7.64
Pure error	2	6.87×10^{-3}	3.43×10^{-3}		2.47×10^{-3}	1.23×10^{-3}	
Total	10	3.14			7.93		
R^2			0.9737			0.9962	
Adjusted R^2			0.9477			0.9925	
			Texture			Flavor	
Regression	5	2.79	5.59×10^{-1}	154.89 *	14.17	2.83	328.31 *
Residual	5	7.16×10^{-1}	1.43×10^{-1}		5.05×10^{-1}	1.01×10^{-1}	
Lack-of-fit	3	6.89×10^{-1}	2.27×10^{-1}	12.56	4.88×10^{-1}	1.63×10^{-1}	18.83
Pure error	2	3.61×10^{-2}	1.80×10^{-2}		1.73×10^{-2}	8.63×10^{-3}	
Total	10	3.51			14.68		
R^2			0.7751			0.9664	
Adjusted R^2			0.5503			0.9328	
			Overall acceptability			Purchase intent	
Regression	5	15.59	3.12	736.33 *	3.04	6.08×10^{-1}	651.15 *
Residual	5	1.93×10^{-1}	3.85×10^{-2}		9.95×10^{-3}	1.99×10^{-3}	
Lack-of-fit	3	1.84×10^{-1}	6.14×10^{-2}	14.49	8.08×10^{-3}	2.69×10^{-3}	2.89
Pure error	2	8.47×10^{-3}	4.23×10^{-3}		1.87×10^{-3}	9.33×10^{-4}	
Total	10	15.78			3.05		
R^2			0.9882			0.9969	
Adjusted R^2			0.9763			0.9938	

* Significant at the $p < 0.05$ level; ** Significant at the $p < 0.001$ level; DPPH—2,2-diphenyl-1-picrylhydrazyl method; CE—methanolic extracts obtained by conventional extraction; UAE—methanolic extracts obtained by ultrasound-assisted extraction.

Moreover, high F-values ranged between 154.89 and 1281.56, and the probability values less than 0.05 and 0.001 indicated that the empirical models were significant (Table 5). For this reason, these proposed mathematical models are valid and convenient for predicting the antioxidant potential and sensory properties of functional biscuits prepared under any combination of RPC amount and fat type.

Additionally, determination coefficients (R^2) and adjusted R^2 were calculated to estimate the proposed models' goodness of fit (Table 5). The R^2 values (0.9664–0.9969) ensure a satisfactory fit of the proposed models to represent actual relationships between the responses ($DPPH_{CE}$, $DPPH_{UAE}$, color, odor, flavor, overall acceptability, and purchase intent) and the independent variables (RPC content and SAFA content). However, high values of adjusted R^2 = 0.9328–0.9938 indicated a close agreement between experimental and predicted results. On the contrary, the values of R^2 = 0.7751 and adjusted R^2 = 0.5503 for the response of biscuits' texture suggest that a high proportion of variability cannot be explained by the model, because R^2 should be at least 0.80 for a good fit.

All linear and quadratic parameters of the empirical models were highly significant (F = 19.73–4554.65, p = 0.00022–0.047) for the DPPH results of methanolic extracts obtained by conventional extraction and overall acceptability of the biscuits enriched with RPC, while the interaction between amounts of RPC and SAFA in functional biscuits produced a significant negative effect only on their odor (F = 66.85, p = 0.015). However, the quadratic parameter of the SAFA content and linear term of RPC of the models significantly influenced (F values ranged between 19.10 and 3221.88, p < 0.05) on flavor and purchase intent of the prepared biscuits. In addition, only two parameters (RPC and RPC^2) had significant effects on color of the novel biscuits (F = 19.64–846.11, p = 0.047–0.0012). Both linear (RPC, SAFA) and quadratic ($SAFA^2$) parameters of the models were highly significant (F = 27.44–1332.02 and p = 0.035–0.00075) for DPPH of methanolic extracts after UAE, whereas linear term of SAFA and interaction between independent variables (RPC × SAFA) caused insignificant effects (F = 0.00026–10.98, p = 0.080–0.99) on the texture of studied samples.

It is noteworthy that the variable with the largest positive effect on the antioxidant potential of the baked biscuits determined by DPPH assay was the linear term of RPC content (F = 1332.02–4554.65, p < 0.001). Nevertheless, this independent contributed variable the most negatively to all sensory characteristics of the fortified biscuits (F = 76.73–5902.97, p < 0.05).

3.4.2. Analysis of the Response Surfaces

The effects of the two independent variables (amounts of RPC and SAFA) on antioxidant potential of biscuits extracted by classical extraction and analyzed by the DPPH assay as well as their five sensory characteristics (color, odor, flavor, overall acceptability, and purchase intent) were illustrated using surface response and contour plots of the quadratic polynomial models (Figure 2).

As can be seen, the generated shapes of the response surfaces for AC results differ from those obtained for sensory scores. The DPPH of baked biscuits significantly increased with the increasing concentration of RPC in the dough (Figure 2a). The parabolic shape of the response surface for DPPH was caused by the positive values of the quadratic terms of RPC amount and content of SAFA in the fat used.

On the contrary, scores for color, odor, flavor, and overall acceptability and purchase intent of the biscuits analyzed were the highest when RPC was not added to WF and rapeseed oil or coconut oil were used for dough preparation (Figure 2b–f). The surface plots of odor, flavor, overall acceptability, and purchase intent indicate that the negative linear of RPC content and positive quadratic term of SAFA content were significant (Figure 2c–f). The elliptical contour of the odor plot confirms that there was an interaction between independent variables (Figure 2c), whereas the significant negative linear and quadratic effects of RPC amount resulted in a decrease of the score for biscuit color after the fortification with a higher RPC amount (Figure 2b).

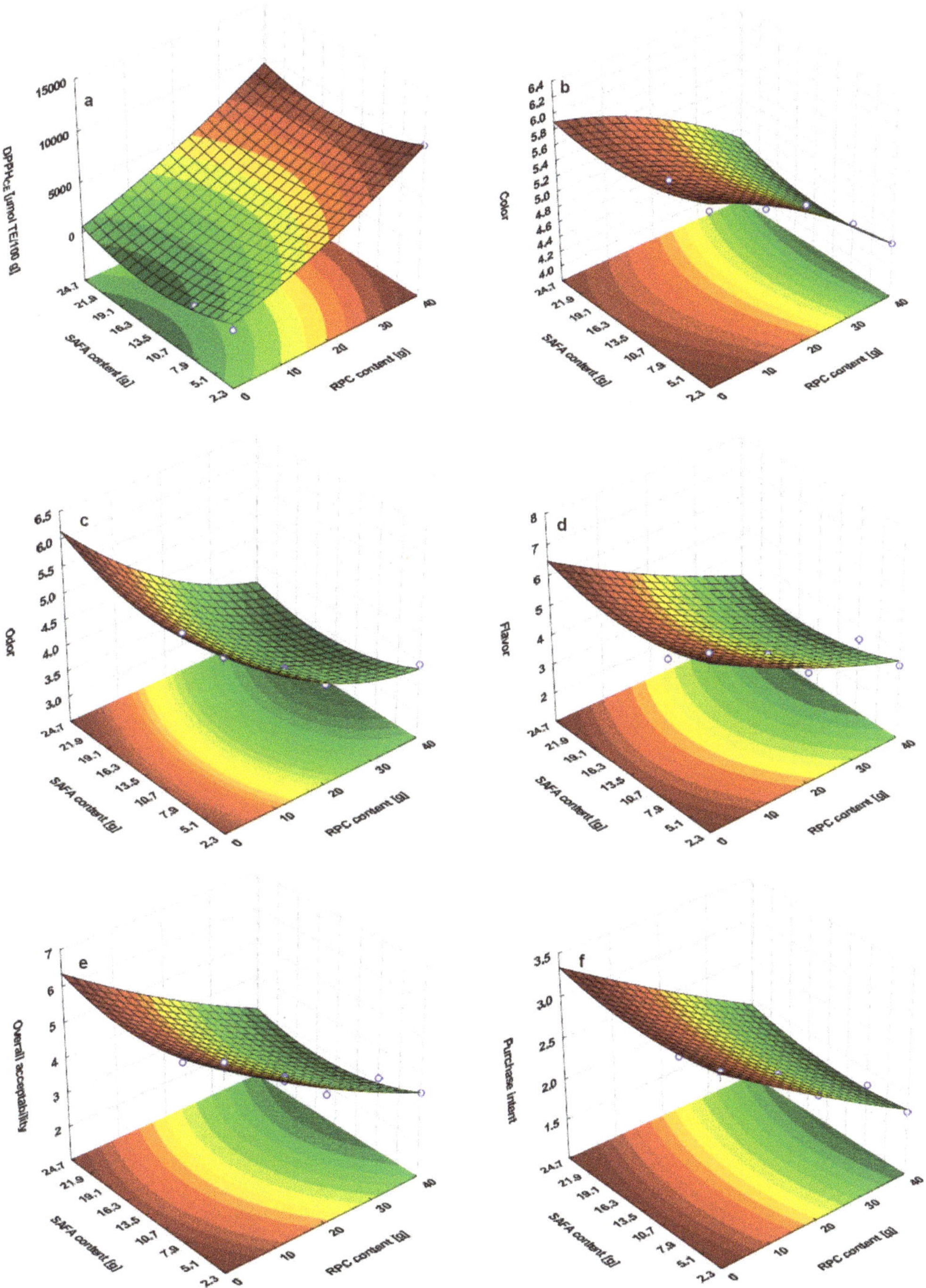

Figure 2. Response surfaces and contour plots for: (**a**) DPPH results of extracts obtained by conventional extraction (DPPH$_{CE}$), (**b**) color, (**c**) odor, (**d**) flavor, (**e**) overall acceptability, and (**f**) purchase intent of functional biscuits expressed as a function of the rapeseed press cake (RPC) content and saturated fatty acids (SAFA) content.

3.4.3. Verification of Predictive Models

Optimization of the prepared functional biscuits was performed to measure the optimum levels of independent variables (amounts of RPC and SAFA) to achieve the desired response goals. The antioxidant potential and sensory characteristics were specified as desired to be maximized. To evaluate the sufficiency of the proposed mathematical models, verification experiments were carried out at the predicted conditions derived from the RSM analysis. The predicted and experimental results are presented in Table 6.

Table 6. Predicted and experimental values of the studied responses for the optimum ingredients' amount.

Response Variables	Optimum Ingredients' Amount		Pred. Values	Exp. Values * $\pm$ SD	p Values
	RPC Content [g]	SAFA Content [g]			
$DPPH_{CE}$ [μmol TE/100 g]	40	2.3	10,050	10,151 $\pm$ 100	0.476
$DPPH_{UAE}$ [μmol TE/100 g]	40	2.3	6787	6875 $\pm$ 67	0.375
Color [#]	0	24.9	5.89	5.96 $\pm$ 0.11	0.632
Odor [#]	0.2	24.7	6.11	6.17 $\pm$ 0.13	0.736
Texture [#]	8.2	2.3	6.58	6.60 $\pm$ 0.14	0.913
Flavor [#]	0	24.9	6.46	6.52 $\pm$ 0.18	0.785
Overall acceptability [#]	0	24.9	6.34	6.39 $\pm$ 0.05	0.514
Purchase intent [##]	0.1	24.8	3.33	3.48 $\pm$ 0.29	0.693

* n = 3; SD—standard deviation; [#] 9-point quality scale; [##] 5-point quality scale.

Insignificant differences ($p > 0.05$) between the predicted and experimental response values confirmed that the proposed models were accurate and adequate for the optimization of amounts of two ingredients to baking functional biscuits with acceptable levels of RPC and sensory characteristics.

3.4.4. Physical Properties of Biscuits for Optimum Levels of Ingredients

The doughs were prepared by using RPC flour and SAFA at optimum levels for antioxidant potential (RPC = 40 g, SAFA = 2.3 g), and overall acceptability (RPC = 0 g, SAFA = 24.9 g) and physical properties of the baked biscuits were determined (Table 7).

Table 7. Physical properties of biscuits for the optimum ingredients' amount.

Optimum Levels	Diameter * $\pm$ SD [mm]	Thickness * $\pm$ SD [mm]	Spread Ratio * $\pm$ SD	Weight * $\pm$ SD [g]
RCP = 40 g SAFA = 2.3 g	67.00 $\pm$ 1.15 [a]	6.78 $\pm$ 0.26 [a]	9.90 $\pm$ 0.44 [a]	12.1644 $\pm$ 0.26 [a]
RCP = 0 g SAFA = 24.9 g	69.75 $\pm$ 1.50 [b]	7.75 $\pm$ 0.65 [b]	9.04 $\pm$ 0.65 [a]	15.1991 $\pm$ 0.65 [b]

* n = 6; SD—standard deviation; Different letters within the same column (a,b) indicate significant differences between physical parameters of biscuits.

The Duncan test indicated that RPC flour incorporation significantly decreased the diameter, thickness, and weight of the studied biscuits.

The decrease in diameter and thickness of cookies and biscuits with increasing amounts of added goji berry by-product (10–40%), okara powder (20–40%), prickly pear peel (10–30%), and sunflower seed flour (18–36%) was observed by other authors [10,13–15]. In contrast, the diameter value (52.36 mm) of control biscuit was lower than those (52.52–55.88 mm) measured for biscuits with 10–30% of white grape pomace, whereas the thickness of the studied samples decreased significantly (13.83–9.05 mm) after supplementation [8].

However, insignificant differences in the mean spread ratio were found between samples without RPC containing coconut oil and biscuits prepared with 40 g of RPC and rapeseed oil.

The insignificant increase in spread ratio and the significant contraction of diameter after addition of RPC flour can be attributed to the reduction of total gluten content as well as the increase in protein amount [10]. Probably, the weight of fortified biscuits was lower due to low oil absorption capacity of RPC flour [28]. For comparison, the spread ratio of biscuits before (48.9) and after (48.0–48.2) the addition of prickly pear peel did not differ significantly [14]. On the other hand, the changes in spread ratio of biscuits with white grape pomace (3.79–6.18) were significant at substitution levels between 10% and 30% [8].

4. Conclusions

The incorporation of RPC flour and the replacement of margarine by rapeseed oil into the formulation of WF-based biscuits resulted in a nutritionally enhanced product with a higher amount of antioxidants. Moreover, the UAE of ingredients used for dough preparation produced higher recoveries of total antioxidants in comparison with the conventional extraction.

A higher percentage of RPC in the formula increased the antioxidant potential of the proposed biscuits. However, these biscuits had lower sensory scores for color, odor, texture, flavor, overall acceptability, and purchase intent.

The estimated parameters, RPC content, and SAFA content affected the DPPH results, and sensory characteristics of the baked biscuits. However, the amount of RPC flour had a greater positive effect on the antioxidant potential of biscuits than the fat type used for dough preparation. Nevertheless, the added level of RPC was a more negatively effective independent variable on the sensory properties of the biscuits studied.

The good agreement between the predicted values and experimental results verified the validity of the proposed models and the optimal ingredient amounts for the baking of biscuits incorporating by-products of rapeseed oil industry.

Author Contributions: Conceptualization, A.S.-C., S.P. and M.M.; Data curation, A.S.-C., S.P. and M.M.; Formal analysis, S.P. and M.M.; Investigation, A.S.-C., S.P. and M.M.; Methodology, A.S.-C., S.P. and M.M.; Supervision, A.S.-C.; Visualization, A.S.-C., S.P. and M.M.; Writing—original draft, A.S.-C., S.P. and M.M.; Writing—review & editing, A.S.-C. All authors have read and agreed to the published version of the manuscript.

Funding: This research received no external funding.

Institutional Review Board Statement: Not applicable.

Informed Consent Statement: Not applicable.

Data Availability Statement: The data presented in this study are available on request from the corresponding author.

Acknowledgments: The authors would like to thank Angelika Gawrońska for analysis of antioxidant capacity.

Conflicts of Interest: The authors declare no conflict of interest.

References

1. Bolek, S. Olive stone powder: A potential source of fiber and antioxidant and its effect on the rheological characteristics of biscuit dough and quality. *Innov. Food Sci. Emerg. Technol.* **2020**, *64*, 102423. [CrossRef]
2. Nasir, G.; Chand, K.; Azad, Z.R.A.A.; Nazir, S. Optimization of finger millet and carrot pomace based fiber enriched biscuits using response surface methodology. *J. Food Sci. Technol.* **2020**, *57*, 4613–4626. [CrossRef]
3. Pasqualone, A.; Bianco, A.M.; Paradiso, V.M.; Summo, C.; Gambacorta, G.; Caponio, F. Physico-chemical, sensory and volatile profiles of biscuits enriched with grape marc extract. *Food Res. Int.* **2014**, *65*, 385–393. [CrossRef]
4. Abreu, J.; Quintino, I.; Pascoal, G.; Postingher, B.; Cadena, R.; Teodoro, A. Antioxidant capacity, phenolic compound content and sensory properties of cookies produced from organic grape peel (Vitis labrusca) flour. *Int. J. Food Sci. Technol.* **2019**, *54*, 1215–1224. [CrossRef]

5. de Oliveira Silva, F.; Miranda, T.G.; Justo, T.; da Silva Frasão, B.; Conte-Junior, C.A.; Monteiro, M.; Perrone, D. Soybean meal and fermented soybean meal as functional ingredients for the production of low-carb, high-protein, high-fiber and high isoflavones biscuits. *LWT-Food Sci. Technol.* **2018**, *90*, 224–231. [CrossRef]
6. Tańska, M.; Roszkowska, B.; Czaplicki, S.; Borowska, E.J.; Bojarska, J.; Dąbrowska, A. Effect of fruit pomace addition on shortbread cookies to improve their physical and nutritional values. *Plant Foods Hum. Nutr.* **2016**, *71*, 307–313. [CrossRef] [PubMed]
7. Tagliani, C.; Perez, C.; Curutchet, A.; Arcia, P.; Cozzano, S. Blueberry pomace, valorization of an industry by-product source of fibre with antioxidant capacity. *Food Sci. Technol.* **2019**, *39*, 644–651. [CrossRef]
8. Mildner-Szkudlarz, S.; Bajerska, J.; Zawirska-Wojtasiak, R.; Górecka, D. White grape pomace as a source of dietary fibre and polyphenols and its effect on physical and nutraceutical characteristics of wheat biscuits. *J. Sci. Food Agric.* **2013**, *93*, 389–395. [CrossRef] [PubMed]
9. Toledo, N.M.V.; Mondoni, J.; Harada-Padermo, S.S.; Vela-Paredes, R.S.; Berni, P.R.A.; Selani, M.M.; Canniatti-Brazaca, S.G. Characterization of apple, pineapple, and melon by-products and their application in cookie formulations as an alternative to enhance the antioxidant capacity. *J. Food Process. Preserv.* **2019**, *43*, e14100. [CrossRef]
10. Bora, P.; Ragaee, S.; Abdel-Aal, E.S.M. Effect of incorporation of goji berry by-product on biochemical, physical and sensory properties of selected bakery products. *LWT-Food Sci. Technol.* **2019**, *112*, 108225. [CrossRef]
11. de Barros, H.E.A.; Natarelli, C.V.L.; de Carvalho Tavares, I.M.; de Oliveira, A.L.M.; Araújo, A.B.S.; Pereira, J.; Carvalho, E.E.N.; de Barros Vilas Boas, E.V.; Franco, M. Nutritional clustering of cookies developed with cocoa shell, soy, and green banana flours using exploratory methods. *Food Bioprocess Technol.* **2020**, *13*, 1566–1578. [CrossRef]
12. Mas, A.L.; Brigante, F.I.; Salvucci, E.; Pigni, N.B.; Martinez, M.L.; Ribotta, P.; Wunderlin, D.A.; Baroni, M.V. Defatted chia flour as functional ingredient in sweet cookies. How do processing, simulated gastrointestinal digestion and colonic fermentation affect its antioxidant properties? *Food Chem.* **2020**, *316*, 126279.
13. Lee, D.P.S.; Gan, A.X.; Kim, J.E. Incorporation of biovalorised okara in biscuits: Improvements of nutritional, antioxidant, physical, and sensory properties. *LWT-Food Sci. Technol.* **2020**, *134*, 109902. [CrossRef]
14. Bouazizi, S.; Montevecchi, G.; Antonelli, A.; Hamdi, M. Effects of prickly pear (*Opuntia ficus-indica* L.) peel flour as an innovative ingredient in biscuits formulation. *LWT-Food Sci. Technol.* **2020**, *124*, 109155. [CrossRef]
15. Grasso, S.; Omoarukhe, E.; Wen, X.; Papoutsis, K.; Methven, L. The use of upcycled defatted sunflower seed flour as a functional ingredient in biscuits. *Foods* **2019**, *8*, 305. [CrossRef]
16. Troise, A.D.; Wilkin, J.D.; Fiore, A. Impact of rapeseed press-cake on Maillard reaction in a cookie model system. *Food Chem.* **2018**, *243*, 365–372. [CrossRef]
17. Barba, F.J.; Boussetta, N.; Vorobiev, E. Emerging technologies for the recovery of isothiocyanates, protein and phenolic compounds from rapeseed and rapeseed press-cake: Effect of high voltage electrical discharges. *Innov. Food Sci. Emerg. Technol.* **2015**, *31*, 67–72. [CrossRef]
18. Deng, Q.; Zinoviadou, K.G.; Galanakis, C.M.; Orlien, V.; Grimi, N.; Vorobiev, E.; Lebovka, N.; Barba, F.J. The effects of conventional and non-conventional processing on glucosinolates and its derived forms, isothiocyanates: Extraction, degradation, and applications. *Food Eng. Rev.* **2015**, *7*, 357–381. [CrossRef]
19. Szydłowska-Czerniak, A.; Tułodziecka, A.; Karlovits, G.; Szłyk, E. Optimisation of ultrasound-assisted extraction of natural antioxidants from mustard seed cultivars. *J. Sci. Food Agric.* **2015**, *95*, 1445–1453. [CrossRef]
20. Stone, H.; Sidel, J.L. Descriptive analysis. In *Sensory Evaluation Practices*; Stone, H., Sidel, J.L., Eds.; Academic Press: London, UK, 1985; pp. 202–226.
21. Macfie, H.J.; Bratchell, N.; Greenhoff, K.; Vallis, L.V. Designs to balance the effect of order of presentation and first-order carryover effects in hall tests. *J. Sens. Stud.* **1989**, *4*, 129–148. [CrossRef]
22. Apak, R.; Özyürek, M.; Güçlü, K.; Çapanoğlu, E. Antioxidant activity/capacity measurement. 2. Hydrogen atom transfer (HAT)-based, mixed-mode (electron transfer (ET)/HAT), and lipid peroxidation assays. *J. Agric. Food Chem.* **2016**, *64*, 1028–1045. [CrossRef] [PubMed]
23. Picó, Y. Ultrasound-assisted extraction for food and environmental samples. *Trac-Trends Anal. Chem.* **2013**, *43*, 84–99. [CrossRef]
24. Szydłowska-Czerniak, A.; Tułodziecka, A.; Szłyk, E. A silver nanoparticle-based method for determination of antioxidant capacity of rapeseed and its products. *Analyst* **2012**, *137*, 3750–3759. [CrossRef] [PubMed]
25. Casoni, D.; Simion, I.M.; Sârbu, C. A comprehensive classification of edible oils according to their radical scavenging spectral profile evaluated by advanced chemometrics. *Spectroc. Acta Part A Molec. Biomolec. Spectr.* **2019**, *213*, 204–209. [CrossRef] [PubMed]
26. Liu, R.; Lu, M.; Zhang, T.; Zhang, Z.; Jin, Q.; Chang, M.; Wang, X. Evaluation of the antioxidant properties of micronutrients in different vegetable oils. *Eur. J. Lipid Sci. Technol.* **2019**, *122*, 1900079. [CrossRef]
27. Gogate, P.R.; Prajapat, A.L. Depolymerization using sonochemical reactors: A critical review. *Ultrason. Sonochem.* **2015**, *27*, 480–494. [CrossRef]
28. Yadav, R.B.; Yadav, B.S.; Dhull, N. Effect of incorporation of plantain and chickpea flours on the quality characteristics of biscuits. *J. Food Sci. Technol.* **2012**, *49*, 207–213. [CrossRef] [PubMed]

MDPI

St. Alban-Anlage 66

4052 Basel

Switzerland

Tel. +41 61 683 77 34

Fax +41 61 302 89 18

www.mdpi.com

Applied Sciences Editorial Office

E-mail: applsci@mdpi.com

www.mdpi.com/journal/applsci

9 783036 537337